RYAN MEMORIAL LIBRARY
ST. CHARLES SEMINARY
OVERBROOK, PHILA., PA. 19151

SPACE, TIME AND GEOMETRY

# SYNTHESE LIBRARY

MONOGRAPHS ON EPISTEMOLOGY,

LOGIC, METHODOLOGY, PHILOSOPHY OF SCIENCE,

SOCIOLOGY OF SCIENCE AND OF KNOWLEDGE,

AND ON THE MATHEMATICAL METHODS OF

SOCIAL AND BEHAVIORAL SCIENCES

*Editors:*

DONALD DAVIDSON, *The Rockefeller University and Princeton University*

JAAKKO HINTIKKA, *Academy of Finland and Stanford University*

GABRIËL NUCHELMANS, *University of Leyden*

WESLEY C. SALMON, *Indiana University*

# SPACE, TIME AND GEOMETRY

*Edited by*

PATRICK SUPPES
*Stanford University*

D. REIDEL PUBLISHING COMPANY

DORDRECHT-HOLLAND / BOSTON-U.S.A.

*First printing:* December 1973

Library of Congress Catalog Card Number 73-86097

Cloth edition: ISBN 90 277 0386 8
Paperback edition: ISBN 90 277 0442 2

Published by D. Reidel Publishing Company,
P.O. Box 17, Dordrecht, Holland

Sold and distributed in the U.S.A., Canada and Mexico
by D. Reidel Publishing Company, Inc.
306 Dartmouth Street, Boston,
Mass. 02116, U.S.A.

All Rights Reserved
Copyright © 1973 by D. Reidel Publishing Company, Dordrecht-Holland
No part of this book may be reproduced in any form, by print, photoprint, microfilm,
or any other means, without written permission from the publisher

Printed in The Netherlands by D. Reidel, Dordrecht

TABLE OF CONTENTS

PREFACE                                                               VII

INTRODUCTION                                                            X

## PART I / CAUSALITY AND TIME

ZOLTAN DOMOTOR / Causal Models and Space-Time Geometries                3

GEORGE BERGER / Temporally Symmetric Causal Relations in
   Minkowski Space-Time                                                56

JOHN EARMAN / Notes on the Causal Theory of Time                       72

BAS C. VAN FRAASSEN / Earman on the Causal Theory of Time              85

ROBERT PALTER / Kant's Formulation of the Laws of Motion               94

ROBERT WEINGARD / On Travelling Backward in Time                      115

P. J. ZWART / The Flow of Time                                        131

## PART II / GEOMETRY OF SPACE AND TIME

JULES VUILLEMIN / Poincaré's Philosophy of Space                      159

CRAIG HARRISON / On the Structure of Space-Time                       178

CLARK GLYMOUR / Topology, Cosmology and Convention                    193

MICHAEL FRIEDMAN / Grünbaum on the Conventionality of
   Geometry                                                           217

ARTHUR FINE / Reflections on a Relational Theory of Space             234

ADOLF GRÜNBAUM / The Ontology of the Curvature of Empty
   Space in the Geometrodynamics of Clifford and Wheeler              268

MICHAEL FRIEDMAN / Relativity Principles, Absolute Objects and Symmetry Groups 296

ROBERT W. LATZER / Nondirected Light Signals and the Structure of Time 321

RICHARD H. HUDGIN / Coordinate-Free Relativity 366

PATRICK SUPPES / Some Open Problems in the Philosophy of Space and Time 383

ERNEST W. ADAMS / The Naive Conception of the Topology of the Surface of a Body 402

# PREFACE

The articles in this volume have been stimulated in two different ways. More than two years ago the editor of *Synthese*, Jaakko Hintikka, announced a special issue devoted to space and time, and articles were solicited. Part of the reason for that announcement was also the second source of papers. Several years ago I gave a seminar on special relativity at Stanford, and the papers by Domotor, Harrison, Hudgin, Latzer and myself partially arose out of discussion in that seminar. All of the papers except those of Grünbaum, Fine, the second paper of Friedman, and the paper of Adams appeared in a special double issue of *Synthese* (**24** (1972), Nos. 1–2). I am pleased to have been able to add the four additional papers mentioned in making the special issue a volume in the *Synthese Library*. Of these four additional articles, only the one by Fine has previously appeared in print (*Synthese* **22** (1971), 448–481); its relevance to the present volume is apparent.

In preparing the papers for publication and in carrying out the various editorial chores of such a task, I am very much indebted to Mrs. Lillian O'Toole for her extensive assistance.

INTRODUCTION

The philosophy of space and time has been of permanent importance in philosophy, and most of the major historical figures in philosophy, such as Aristotle, Descartes and Kant, have had a good deal to say about the nature of space and time. This classical tradition of analysis of spatial and temporal concepts has been relatively informal and accessible to the general philosopher. It is probably correct to say that the level of accessibility of the philosophy of mathematics and the philosophy of physics is about the same in the writings of important systematic philosophers from Aristotle to Kant.

Beginning in the latter part of the nineteenth century with the work of Frege and others, the philosophy of mathematics took a sharp turn away from this broad general tradition and became a more technical and more deeply developed and specialized subject. Today, neither philosophers nor mathematicians who do not have some special interest in foundations of mathematics attempt to contribute to the large and continually developing literature. Until fairly recently, the same line of deeper development was not characteristic of the philosophy of physics. Important philosophers of physics like Mach and Bridgman, writing over the same time span that runs from Frege to Gödel and Tarski, presented a body of ideas and methodological concepts that were clearly on a less satisfactory intellectual level than work in the philosophy of mathematics. Of course, there were notable exceptions to the relatively mediocre intellectual standard set by Mach and Bridgman. Analysis of some of that early work may be found in various articles in the present volume. For example, Domotor discusses the important work of Robb on the geometry of special relativity, and Vuillemin emphasizes aspects of Poincaré's philosophy of space that are not as well known as they should be. By mentioning only Robb and Poincaré, I do not mean to suggest that they are the only exceptions to the Mach-Bridgman standard. The literature on the philosophy of physics is enormous, and many good things can be found in that literature in the last half of the nineteenth century and the early decades of this century.

All the same, when one looks at the development of the foundations of geometry over this period it is surprising and disappointing that this development has not had the impact on the philosophy of space and time that the axiomatic theory of the natural numbers and axiomatic set theory have had on the philosophy of mathematics.

Fortunately, that time seems to have passed and in the articles contained in this volume alone, we can see a renaissance of interest in the philosophy of space and time among younger philosophers. What is most pleasing is their ability to deal with the geometry of space and time in a manner that does not avoid the technical issues and results, but that still places an emphasis on matters of philosophical interest. Except for Grünbaum, who is more or less the dean of American philosophers interested in space and time, Vuillemin, who represents a return of French philosophy to the intellectual traditions of Descartes and Poincaré, and myself, all authors of the articles in this volume are, I believe, under forty or forty-five, and, with the exception of Hudgin and Latzer, all are philosophers.

The seven articles of Part I deal with the many issues about causality and time that have much occupied recent literature of the philosophy of physics. In the first four articles of Domotor, Berger, Earman and van Fraassen, most of the issues of current interest are pushed back and forth in the implicit defense and criticism of the main views now current among philosophers about these matters. The article by Palter analyzes Kant's formulation of the laws of motion, and in doing so, brings attention to Kant's *Metaphysical Foundations of Natural Science*, which has been too much neglected in discussions of Kant's philosophy of space and time. The last two articles of Part I, by Weingard and Zwart, discuss some of the issues that have arisen about time reversibility and the flow of time.

The longer Part II contains eleven papers on the geometry of space and time. Vuillemin's article on Poincaré has already been mentioned. Harrison discusses problems about the structure of space-time that go back to Newton and end with contemporary discussions about the status of the relational theory of time in the theory of relativity. Glymour discusses conventionality in the philosophy of space and time from the standpoint of topology and, thereby, offers some new ways of analyzing this venerable topic. The first article by Friedman and the article by Fine take a critical look at some of the issues that Grünbaum has discussed extensively in previous publications. Grünbaum's own article sets forth his views on the ontology of the curvature of empty space, especially in the geometrodynamics of Clifford and Wheeler. Friedman's second article has much to say about principles of relativity and the ontological status of objects. The articles of Latzer and Hudgin present some axiomatic results of foundational interest. Those who have followed the axiomatic efforts in connection with the geometry of special relativity will find especially interesting Latzer's alternative to Robb's approach in his use of a binary primitive for nondirected light signals. In the next article I formulate some open problems in the philosophy of space and time and, in this formulation, stress the desirability of a still closer interaction between the foundations of geometry and the philosophy of space and time. In the final article Adams analyzes the geometry and topology of the surfaces of physical objects.

PATRICK SUPPES

# PART I

# CAUSALITY AND TIME

ZOLTAN DOMOTOR

# CAUSAL MODELS AND SPACE-TIME GEOMETRIES*

## I. INTRODUCTION

There are many points of view from which one finds it appropriate to look at the notion of causality. These include what I call the *logical* (linguistic), *probabilistic* (stochastic), *analytic* (physical, system-theoretic), and *geometric* (relativity-theoretic) viewpoints. In all of them, throughout the rather long and confusing intellectual history of the subject, the main intention has been to explicate and then to answer in the best possible way the following two basic questions:

$Q_1$      What curious entities, usually called *causes* and *effects*, are supposed to be in a causal relationship?

$Q_2$      What kind of relation, operator, or perhaps something else, called *causal relation* or *causal operator*, is associated with or attributed to the causal entities?

The desirable answer to question $Q_1$, structurally speaking, should result in an adequate *formal model*. The standard literature on causality clearly indicates that there is not too much agreement on whether the causal entities should be propositions, time or space-time dependent propositions, sets of propositions – as logicians and linguists propose; probabilistic events, random variables, experiments, stochastic processes – as the inductively oriented people keep suggesting; events, acts, signals, time-dependent functions – as physicists and system-theorists think; attributes of space-time points and space-time regions – as the specialists in relativity theory assume, or perhaps still something else. But then there is no obvious reason why all these ontologies should not be possible, especially in cases when the domain of possible applications radically varies. Their coexistence is frequently reinforced by certain so-called *representation* or *completeness theorems* which provide a constructive or formal translation of one ontological framework into another. Apparently, what one has here is nothing but species of 'ontological coordinate systems' with respect to which the description of causality is in some way relativized. Naturally, the main interest, then, lies in the *invariants* of these 'coordinate systems'. Once the representation theorems are known, it does really matter which ontology is being used.

Regardless of what the ontological primitives might be, we insist firmly on their precise set-theoretic description. By means of this formal request we are forced willy-nilly to bring all the 'hidden parameters' of our assumptions into the open air, so to speak, so that everyone can examine them. I know very well that many people reject such a special requirement because they want to hear about the 'real meaning' of causal entities, and not about a more or less nice *model*. But I know also that it is better to work out the idealized situation first – and then to retrench later when one can appreciate thoroughly what is gained and lost in the exhibited abstraction. It is clear that there is much more information present in the data concerning any problem than we actually want to deal with in our solution. Consequently, we must *decide* in each specific case how much structure we want to represent in our thinking, and how accessible we want to make each piece of information. The decision, a proper delineation and isolation of the relevant from the irrelevant, and the explicit identification of the dependencies between structural features concerning the problem are best accomplished in the language of models. At any rate, at the present moment I know of no serious or better way of handling difficult and tricky logical or foundational problems, such as causality, than by building a (set-theoretic) model. A by-product of this methodology is the credo that one cannot hope to have only one single universally good model of causality. But, rather, several models will eventually be necessary (just like in physics, where one deals with many different physical models), depending on the field of application and the purposes of the model builder.

Question $Q_2$ is just what the foundations of causality are all about. Its adequate answer should result in a *method* of *causal description* or *causal inference*.

My purpose in this paper is, first, to present as clearly as possible the particular points of view mentioned above and to examine their merits, demerits, and implications for the specific fields of empirical science; second, to study the algebraic behavior of the causal relation in Minkowski spaces; and third, to explore a representation theorem stating that every causal lattice is isomorphic to the lattice of all subspaces of a Minkowski space, where the causal ordering on the given lattice agrees with the causal ordering of the space. The significance of this investigation lies not only in the radical simplification and generalization of Robb's theory of space and time, but also in highlighting the intimate interplay between the geometry of physics and lattice algebra. In the course of displaying the results, several problems have emerged that seem rather tricky to solve.

It is hoped that this study will not only indicate how to organize the species of causal models, but also that it will influence in some manner the posing and solving of some of the intricate methodological and foundational problems of causality.

I shall not make an attempt to analyze the so-called *ideological questions* of causality such as Russell's or Mach's claim about the eliminability of the term 'cause' in physics, nor shall I investigate the difficult problems of causal explanation. For the purposes of this paper I feel perfectly comfortable with the fact that both natural and scientific languages convey a rich enough spectrum of causal notions, and that physics, social science, and philosophy of language are gradually converting it into a serious subject of specialized reasoning. A direct evidence for this is the vast amount of currently published work on causality. As a sample, see, for instance, the literature in theoretical physics and system theory: Alonso and Yndurazin (1967), Barucchi and Teppati (1967), Basri (1966), Cole (1968), Csonka (1970a, b), Falb and Freedman (1969), Finkelstein (1969), Fox *et al.* (1970), Gamba and Luzzatto (1964), Gheorghe and Mihul (1969), Goodman (1965), Hawking (1967), Kronheimer and Penrose (1967), Peres (1970), Saeks (1969), Slavnov (1969), Teppati (1968), Windeknecht (1967), and Zeeman (1964). Then recall the references in social science: Blalock (1961), Pearson (1971), Simon (1953, 1954), Strotz and Wold (1960), Witsenhausen (1971), and Wold (1956, 1958, 1960). And finally, notice the contributions in the philosophy of language and philosophy of science: Bunge (1968), Cole (1968), Czerwinski (1960), Davidson (1963, 1967), Fagot (1971), Good (1961, 1962), Kaplan (1965), Nagel (1965), Simon and Rescher (1966), and Suppes (1970).

The published results and problems show beyond a doubt that causality will continue to be a fundamental ingredient of both scientific and ordinary discourse.

The paper is divided into sections where Section II discusses the highlights of *basic models* of causality (logical, probabilistic, analytic, and geometric). The final part of Section II brings up the case of *derived* (compound, hybrid) *models*. The plan of writing for Section II was to set out the basic semantical frameworks of causality, hoping that within these, precise specialized causal structures could be worked out. In fact, this paper is in part a report on work in progress. Certain results concerning causality in automaton algebra and space-time will be published elsewhere.

Section III presents a concise theory of the metric, order, and affine structure of Minkowski spaces. It has a review flavor and the novelty is only in the way of presentation.

Section IV provides a lattice-theoretic study of the causal structure of Minkowski spaces. The results are presented mostly without proofs, but they can be completed in a straightforward fashion by means of knowledge gained in Section III. The paper concludes with a list of open problems.

Before we plunge into the specialized description of particular viewpoints, it seems worthwhile to give a flavor of at least some aspects of the aforementioned problems. If nothing else, we can certainly settle the necessary notation which will make things visible, clear, organized, and efficient.

Let us start by introducing certain entities, called *events* or *propositions*. For now it does not matter what these entities are. Their exact nature will be determined in subsequent sections. What is important is to agree that they can be collected into one *domain*, a set that we may call $\mathfrak{A}$. Much confusion is removed if one thinks of $\mathfrak{A}$ as nonempty and fixed in advance. This does not mean that one knows all the elements of $\mathfrak{A}$ in any constructive sense, for one may only know some rather general property specifying $\mathfrak{A}$. At any rate we want $\mathfrak{A}$ to be the domain of possible or *conceivable* events. Here 'possible' means possible with respect to some *a priori* conception. So much for the notation concerning question $Q_1$.

For purposes of question $Q_2$, the most natural thing to do is to take the causal relation to be a binary relation $\mapsto$ defined on $\mathfrak{A}$, so that if $A$ and $B$ are events, then $A \mapsto B$ means that $A$ causes $B$. This is a perfectly good atomic formula of a formal language of causality (which we hope to develop as we go along), where $A$ is called the *cause* and $B$ is called the *effect*. Another natural step is to take the causal operator $\mathscr{S}$ as a mapping $\mathscr{S}: \mathfrak{A} \longrightarrow \mathfrak{A}$ with special properties.

In order to save time and space, we shall bring in another piece of notation. It was Hume who emphasized that besides the *contiguity* and *constant conjunction*, the *succession* in time is one of the most important features of causality. In other words, the cause always precedes its effect in time (assuming that we do not intend to handle tachyons). Thus, we are going to introduce and consider another fixed set, call it $\mathcal{T}$, which will stand for the set of time instances. This set is not just a plain collection; it is structured by a before-after time-ordering relation $\prec$, where $s \prec t$ means that the time instant $s$ is before time instant $t$ ($s$ and $t$ are from $\mathcal{T}$). If equality of time instances is also admitted, then we put $s \preccurlyeq t$ instead of $s \prec t$.

If we care to make visible the fact that event $A$ depends on time, we can write $A(t)$ rather than just $A$, meaning that $A$ is a function of time or that $A$ is a time-dependent event. Still another way of putting it would be to say that $A$ is a mapping from the time set into the event set $A: \mathcal{T} \longrightarrow \mathfrak{A}$.

Now it may still not be very clear when an event does or does not occur. The thing is that $A(t)$ is merely a description of an event, depending on time $t$. The notion of *occurrence* is a new predicate, call it $\Vdash$, acting on events. (We use the yield sign because occurrence seems to resemble in many respects the notion of entailment.) So the fact that event $A$ occurs at time $t$ will be denoted by $\Vdash_t A$.

Assuming that question $Q_1$ has been somehow answered, question $Q_2$ will be answered provided that we exhibit the *desirable properties* of the causal connector $\mapsto$ (or causal operator $\mathcal{S}$) and at the same time its *semantics*, controlling the correctness of the properties.

Most people would agree to the properties displayed below:

(1)        $\neg A \mapsto A$   (Irreflexivity)
(2)        $A \mapsto B \Rightarrow \neg (B \mapsto A)$   (Asymmetry)
(3)        $A \mapsto B \ \& \ B \mapsto C \Rightarrow A \mapsto C$   (Transitivity)
(4)        $A(s) \mapsto B(t) \Rightarrow s \prec t$   (Time coherence).

On the other hand, one would probably not want properties like

(5)        $A \mapsto C \Rightarrow A * B \mapsto C$,

where $A * B$ denotes the *conjoint* of $A$ and $B$;

(6)        $A \mapsto B \ \& \ C \mapsto D \Rightarrow A * C \mapsto B * D$.

The reason is that in (5) event, for instance, $B$ could be something that would neutralize the effect of $A$, but all this seems debatable. The point is that we cannot hope to select correctly the properties of $\mapsto$ until we have seen its intended semantics. And even when the desired semantics is somehow known, still the necessary axioms may not be obvious. The only control device we have is a *representation* or *completeness theorem*, which translates the proposed axioms into a better known framework. People working in modal logic know very well how difficult it is to formalize much simpler notions, such as 'necessary', 'can' or 'probably', when the desire is to capture the use in natural languages. How much more difficult it must be to axiomatize causality.

The question is how to delineate intelligently the right kind of properties of the causal relation or 'tune-up' and 'mesh' the causal axioms without running into contradictions or working on something useless. The possible answers are highlighted in the next section.

## II. CAUSAL MODELS

### 1. *Logical Models*

The logical (linguistic) viewpoint advocates to represent the causal entities by sentences or formulas of a sufficiently well-defined language. The idea is that instead of considering events as things in themselves it is better to consider their *articulated descriptions*. Thus, the set $\mathfrak{A}$ is a collection of well-formed formulas of a specific language. The language is appropriately selected for the purpose of talking about empirical objects of some sort in terms of their time-dependent attributes and relationships. This is how Carnap and Reichenbach (among others, and according to Davidson, 1967, even Hume) thought question $Q_1$ should be answered. If one thinks linguistically, the question is not whether one should consider real flesh-and-blood events or just their descriptions, but rather, whether the events should be captured by formulas or by terms. Think of the sentences "The fact that there was a short circuit caused it to be the case that there was a fire" and "The short circuit caused the fire." In the first sentence we use formulas as a tool of description, while in the second sentence we use terms. The conjoint events are represented by conjunctions of formulas describing the particular events.

As we already know, question $Q_2$ is much harder. Burks (1951), for instance, Carnap, and long before him Kant, hoped to answer it in terms of a suitable 'empirical' or 'physical' necessity operator $\Box$. More specifically, Burks proposed (rather naively, I think) defining the causal relation by putting

$$A \mapsto B =_D \Box(A \Rightarrow B),$$

where $A, B \in \mathfrak{A}$.

Logicians realized rather quickly that this definition could not possibly answer our question, no matter what the meaning of the 'box' $\Box$ might be. For one thing, $\mathfrak{A}$ is just a set of propositional formulas without any use of a time set or other auxiliary index set. Second, the tautologies (such as $A \Rightarrow A$) will produce undesirable causal statements. And third, formulas such as $A \mapsto (A \mapsto B)$ if $A \mapsto B$, are also true, yet one can hardly imagine what the empirical significance of such causal statements should be. The semantics of the 'empirical' necessity operator is at least as mysterious as that of $\mapsto$. Despite these shortcomings, Burk's system has some significance in that it seems to be the first attempt to answer the second question by means of a systematic logical modeling.

Czerwinski (1960) considered a more realistic approach. He identified $\mathfrak{A}$ with the set of first-order formulas of a formalized multisorted language whose alphabet contains: (a) individual variables running over a set of objects (participating in causal interactions), (b) another system of individual variables for time instances, and (c) a collection of time-dependent predicates (such as to be at home, to be ill). All this can be made visible by putting $A(x, t)$ to mean that event $A$ is associated with object $x$ at time $t$. Actually Czerwinski uses time intervals which allow him to define the notion of occurrence $\Vdash$. Then the definition is as follows:

$$A \mapsto B =_D \forall x, s [\Vdash_s A(x) \Rightarrow \exists y, t(s \prec t \,\&\, \Vdash_t B(y))].$$

More or less informally we read it as "Event $A$ causes event $B$ precisely when for arbitrary object $x$ if the associated event $A$ occurs at time $s$, then there will exist a later time $t$ and perhaps another object $y$ such that the associated event $B$ will also occur."

Czerwinski claimed that by means of this definition he is able to reconstruct Mill's canons of inductive logic. One wonders how that is possible, when simple counterexamples can be raised. For instance, if $\Vdash_t B(y)$ means that person $y$ will die at time $t$, then $A \mapsto B$ becomes true as soon as we substitute the time of death of $y$, no matter what $A$ stands for. In other words, everything is a cause of death, because one day we are all going to die. Moreover, there is no constraint to tell when $t$ should follow $s$, so that the effect may occur arbitrarily long after the occurrence of the cause. A third source of difficulties arises when we begin to substitute certain analytic sentences for $A$ and $B$.

This should be sufficient for demonstrating how difficult it is to design a satisfactory logical model of causality.

D. Scott in his course on modal logic at Stanford in 1967 proposed a different course of action. The dictum was not to do anything with causality unless we had a proper semantics for it. He proposed to use a Kripke type of semantics $\mathbf{T} = \langle \mathcal{T}, \mathcal{R}, \{\mathcal{P}_i\}_{i \geq 1} \rangle$, where $\mathcal{T}$ denotes the set of so-called *possible worlds* (or index set, representing time, space-time, individual agents, and possibly many other things), $\mathcal{R}$ is a ternary relation, a kind of *alternativeness* (relevance) relation, used in Kripke models (think of the betweenness relation on the time set), and $\mathcal{P}_i \subseteq \mathcal{T}$ represents the collection of worlds in which the atomic formula $A_i$ of the language $\mathfrak{A}$ is true ($i = 1, 2, \ldots$).

Now if we write $\Vdash_t A$ to mean that $A$ is true (occurs) at $t$ in $\mathbf{T}$, then the definition of truth will be shown below:

(i)    $\Vdash_t A_i$    iff    $t \in \mathcal{P}_i$, where $A_i$ denotes the ith *atomic formula in* $\mathfrak{A}$;
(ii)    Not $\Vdash_t A$    iff    $A$ denotes a self-contradictory formula in $\mathfrak{A}$;
(iii)    $\Vdash_t A \Rightarrow B$    iff    $\Vdash_t A$ implies $\Vdash_t B$;
(iv)    $\Vdash_t A \mapsto B$    iff    $\forall s, v \in \mathcal{T} [\mathcal{R}(s, t, v)$ and $\Vdash_s A$ implies $\Vdash_v B]$.

Thus, the idea of semantics of $\mapsto$ is very much like that of necessity in the Kripke semantics, except that instead of binary alternative relations one considers ternary relations. According to this we are not very far from Burks. In fact, we are closer than perhaps we would like to be.

As far as the syntax is concerned, Scott proposed to start with the following two dual causal inference rules and the rules of standard logic (the latter are assumed to be known):

($\alpha$)    $\dfrac{B_1 \,\&\, \ldots \,\&\, B_n \Rightarrow C}{(D \mapsto B_1) \,\&\, \ldots \,\&\, (D \mapsto B_n) \Rightarrow (D \mapsto C)}$

($\beta$)    $\dfrac{C \Rightarrow B_1 \vee \ldots \vee B_n}{(B_1 \mapsto D) \,\&\, \ldots \,\&\, (B_n \mapsto D) \Rightarrow (C \mapsto D)}$.

The reader, dismayed by the austerity of the above, should work out carefully the particular cases. More concretely, from these two inference rules he can easily derive some special ones (e.g., if $B$ is true, then $A \mapsto B$ is true; if $\neg A$ is true, then $A \mapsto B$ is true) and by selecting an appropriate alternativeness relation $\mathscr{R}$, he can delineate even more specifically the behavior of $\mapsto$. For instance, the *reflexivity* of $\mathscr{R}$, i.e., $\mathscr{R}(t, t, t)$ for all $t \in \mathscr{T}$ induces the axiom $A \mapsto B \Rightarrow (A \Rightarrow B)$, saying that the causal implication is stronger than the material implication; the *functionality of* $\mathscr{R}$, that is, $\mathscr{R}(s, t, v) \Rightarrow s = v$ for all $s, t, v \in \mathscr{T}$, induces the reflexivity axiom $A \mapsto A$; the symmetry $\mathscr{R}(s, t, v) \Leftrightarrow \mathscr{R}(v, t, s)$ induces the contraposition property $(A \mapsto B) \Rightarrow (\neg B \mapsto \neg A)$, etc. The reader can readily check the validity of these consequences by substituting $\mathscr{R}$ into scheme (iv), displaying the semantic definition of $\mapsto$. He can even discover some further examples by taking $\mathscr{T}$ to be a time set and putting $\mathscr{R}(s, t, v) \Leftrightarrow (s \prec t \,\&\, t \prec v)$.

It is seen that the real burden is being put on the model builder. Before he can do anything, he has to select an appropriate relevance relation $\mathscr{R}$ and then he must figure out eventually the desired properties of $\mapsto$. Thinking of natural languages, he may find the selection of $\mathscr{R}$ at least as difficult as that of $\mapsto$. In the case of scientific languages, $\mathscr{R}$ has to be defined in terms of axioms of a scientific theory, otherwise the connection between cause and effect, in general, would not be lawlike.

Stalnaker and Thomason (1970) and also the Scandinavian modal logicians, L. Åqvist and B. Hansson, have noticed another alternative of a possible semantics for conditional sentences. Although they planned it for counterfactual conditionals, it can be adapted and adopted for causal statements as well. In an attempt to identify the so-called *basic conditional logic*, Chellas (1970) rewrote the Stalnaker-Thomason semantics into that of Scott's and replaced rule ($\beta$) by a special case of ($\beta$) (keeping ($\alpha$) unchanged): If $A \Leftrightarrow B$ is true, then $(A \mapsto C) \Leftrightarrow (B \mapsto C)$ is true.

The semantics in question is almost the same as before, except that $\mathscr{R}$ is this time a heterogeneous ternary relation: $\mathscr{R} \subseteq \mathscr{T} \times \mathscr{T} \times \mathbf{P}(\mathscr{T})$ ($\mathbf{P}(\mathscr{T})$ = the set of all subsets of $\mathscr{T}$). Putting $\|A\| = \{t \in \mathscr{T} \mid \Vdash_t A\}$, we can define the new semantics of causality as follows:

$$\Vdash_t A \mapsto B \quad \text{iff} \quad \forall s \in \mathscr{T}\,[\mathscr{R}(s, t, \|A\|) \text{ implies } \Vdash_s B].$$

For a nonspecialist the above formula must seem not only bewildering but also like a piece of useless abstraction.

The relational atomic formula $\mathscr{R}(s, t, \|A\|)$ is a kind of meaning postulate, to use Carnap's terminology, consisting of a conjunction of semantic rules, describing the constraints imposed on the possible connections between $A$ and $B$. Naturally, the main task, namely, the decision for the appropriate meaning postulates, remains open. Analogously as in the case of Scott's proposal, one should explore the 'interesting' examples of $\mathscr{R}$ first, and then enter into a serious discussion concerning the adequacy of the selected model.

Stalnaker and Thomason (1970) phrased their semantics somewhat differently:

$$\Vdash_t A \mapsto B \quad \text{iff} \quad t \circ \|A\| \subseteq \|B\|,$$

where the model is a triple $\langle \mathscr{T}, \circ, \{\mathscr{P}_i\}_{i \geq 1}\rangle$ in which $\circ$ is an external operation of the form $\circ : \mathscr{T} \times \mathbf{P}(\mathscr{T}) \longrightarrow \mathbf{P}(\mathscr{T})$. This operation is sometimes called a *choice function*. It was Hansson (1968) who noticed that the well-known fact of measurement theory, namely, that a binary relational structure $\langle \mathscr{T}, \mathscr{R}\rangle$ is representable under certain constraints in terms of a choice structure $\langle \mathscr{T}, \bullet\rangle$, is actually also useful in modal logic. The partially ordered set of possible worlds is replaced by a possible world system with a suitable choice function. Speaking more to the point, the before-after time relation may not be complicated enough to capture the underlying time structure of events. Therefore one replaces it by a choice function on time instances.

The reader, familiar with the details of Kripke semantics, will recognize the striking analogy of the previous two definitions with the semantic definition of necessity. Thus, Carnap and Kant were essentially correct after all in viewing causality as something logical and akin to necessity. It appears that the last three proposals for the semantics of causal relation are in the right direction as an attempt to answer question $Q_2$. The next important step is to work out the particular details of causal logics with specific data about the alternativeness relation and the choice function, with special emphasis on varieties of time and space-time sets.

## 2. *Probabilistic Models*

In the framework of a probabilistic approach the causal entities are represented by probabilistic events, which is to say, $\mathfrak{A}$ is a Boolean algebra ($\sigma$-algebra to be sure) or, equivalently, a field of subsets of a fixed set.

The idea of using events rather than sentences is an old one and is shared by a number of people, though with differing views concerning the specific structure of the algebra of events (Bunge, 1968; Mehlberg, 1969; Suppes, 1970).

Notice that things get mildly complicated when the time set is taken into account. Instead of $\mathfrak{A}$ we have to consider $\mathfrak{A}^{\mathcal{T}}$, the set of mappings $A: \mathcal{T} \longrightarrow \mathfrak{A}$. We can adapt an index-free notation, if we put $[A \cap B](t) = A(t) \cap B(t)$, $[A \cup B](t) = A(t) \cup B(t)$, and $[\bar{A}](t) = \bar{A}(t)$ for all $A, B \in \mathfrak{A}^{\mathcal{T}}$ and $t \in \mathcal{T}$.

Question $Q_2$ is answered by modeling the causal connection between causes and effects in terms of probabilistic laws. The collection of laws is captured by a specific probability measure $P$ on the algebra of events $\mathfrak{A}^{\mathcal{T}}$. In other words, we have at our disposal probabilistic statements of the sort $P[A(t)] = \alpha$ or $P[B(t)/A(s)] = \beta$ ($s \prec t$), saying, respectively, that the probability of the occurrence of event $A$ at time $t$ is $\alpha$ and the probability of occurrence of event $B$ at time $t$, given event $A$ at time $s$ is $\beta$ ($\alpha$ and $\beta$ are reals from the unit interval and $P[A(s)] \neq 0$).

Hume captured the role of time in causality by making a careful distinction between the so-called *token-events* and *type-events*. The difference is best seen in the examples: "Joe's eating spoiled fish caused him to be ill," and "Eating spoiled fish causes illness." The first sentence refers to token-events while the second refer to type-events. Today we know that type-events are just abstraction classes of token-events, induced by a suitable congruence relation on the set of token-events. Hume also had another complication in mind; he took ideas (images) and impressions as entities referring to objects and events. Then, of course, we have to distinguish between token-impressions and type-impressions. The relations among ideas, impressions, and events form a triangle well known in general semantics. But this is quite auxiliary to what follows.

It appears that Hume viewed the causal connection between token-events $a \mapsto b$ as an objective statistical relationship. If we define a *conjoint* of two type-events $A$ and $B$ as $A * B = \{a \in A \mid \exists b \in B(a \prec b)\}$, then Hume's definition of causality seems to amount to this:

$$a \mapsto b =_D a \prec b \ \& \ \exists \ A, B\{P[A*B/A] > 1 - \varepsilon \ \& \ a \in A \ \& \ b \in B\},$$

where $\prec$ is a time ordering on token-events and $\varepsilon$ is a small positive real number. The measure $P$ is to be interpreted statistically. The constant conjunction $A \mapsto B$, as Hume calls it, holds between all representatives of $A$ and $B$, in past and future, observed or unobserved. In other words, we define the causal relation on type-events simply by lifting it from token-events to the corresponding abstraction classes.

This interpretation of Hume's is always open to dispute. But one thing is certain: We have enough empirical problem situations in which events exhibit definite types of *probabilistic* (stochastic) *dependence*.

Suppes (1970), and somewhat earlier Good (1961, 1962), Mehlberg (1969), and as is well known, Reichenbach, realized the deep significance of probabilistic dependence for causality and experimented with several different definitions. Specifically, by making the dependence asymmetric by means of time ordering, Suppes (1970) proposed the straightforward notion of what he calls *prima facie cause*:

$$A(s) \mapsto B(t) =_D s \prec t \ \& \ P[B(t)/A(s)] > P[B(t)].$$

Using the axioms of standard probability theory and the above definition of causality, one can prove that the desirable conditions (1)–(4) in Section I are satisfied with the exception of transitivity.

Although this model sounds very sympathetic and considerably simplifies Hume's proposal, nevertheless it requires some modifications. For instance, an earlier or later event could eventually be found which would account for the cause just as well. Suppes realized that and therefore introduced at least two other definitions (the so-called *spurious causes*) which, unfortunately, are of rather complicated nature. They utilize the above definition of causality and put some additional constraints on the cause and effect involved – the purpose being to block the occurrence of the spotted undesirable features of the prima facie cause. There is an obvious need here for further experimentation with causal definitions.

Due to the special form of the definiens (algebraic combination of probabilistic inequalities), one can exhibit another set of definitions of causality, using comparative (qualitative) probabilities. Moreover, one can switch the ontology from probabilistic events to random variables. All this can be found in Suppes (1970).

My standpoint is that the scheme

$$(A \mapsto B)(t) =_D \forall s, v [\mathcal{R}(s, t, v) \Rightarrow P[B(v)/A(s)] > P[B(v)]]$$

in which $\mathscr{R}$ is a suitable ternary relation (in the simplest case just $s \prec t \prec v$), comprising additional constraints on the measure and related probabilistic theory, contains enough unexplored causal information and will eventually serve as an adequate basis for one class of probabilistic causal models, apparently of a rather general species. Note that in our proposal the time instance $t$ refers to the causal connection $A \mapsto B$. Without this arrangement we cannot switch very well from token-events to type-events. Obvious inclusion relationships comprised in $\mathscr{R}$ will make the causality transitive. The choice of $\mathscr{R}$ resembles in many directions the choice of the alternativeness relation in the context of logical models. In fact, the connection is remarkably intimate. Take $\mathscr{I}$ to be the real unit interval and consider a ternary relation $\Vdash \subseteq \mathfrak{A} \times \mathscr{I} \times \mathfrak{A}$ with the intended interpretation

$$A \Vdash_\alpha B \Leftrightarrow \alpha < P[B/A]$$
$$P[B/A] = \operatorname{Sup}\{\alpha \mid A \Vdash_\alpha B\},$$

where $A, B \in \mathfrak{A}$, and $\alpha \in \mathscr{I}$.

In this framework, the probabilistic inequalities can be translated into first-order logical formulas with atoms $A \Vdash_\alpha B$. For instance, the probabilistic inequality in our scheme takes the form

$$\forall_\alpha [\Omega \Vdash_\alpha B \Rightarrow A \Vdash_\alpha B].$$

After substituting this translation into the scheme, the reader will certainly notice the striking analogy to the modal semantics discussed before. The author of this paper has axiomatized $\Vdash$ on $\mathfrak{A} \times \mathscr{I} \times \mathfrak{A}$ in a somewhat different context and has shown its representation in terms of probabilities in the fashion exhibited above. For complicated causal models one has to introduce certain operations on $\mathscr{I}$. We shall not address ourselves to further details of this model.

Blalock (1964), Simon (1953), and Wold (1956) proposed treating causal relations on random variables. It turns out that systems of linear equations of random variables induce a rather interesting and intuitively appealing network of causal connections in the set of random variables occurring in the equations.

Consider, for example, the equations of random variables representing interactions among political attributes (Alker, 1966):

$$A_1 = a_{14} A_4 + b_{12} B_2 + b_{13} B_3 + c_1$$
$$A_2 = a_{21} A_1 + b_{21} B_1 + c_2$$

$$A_3 = a_{32}A_2 + b_{32}B_2 + c_3$$
$$A_4 = a_{42}A_2 + a_{43}A_3 + b_{41}B_1 + c_4.$$

If we denote by $A \mapsto B$ the causal connection from random variable $A$ to random variable $B$, then the above system of equations has the following causal representation.

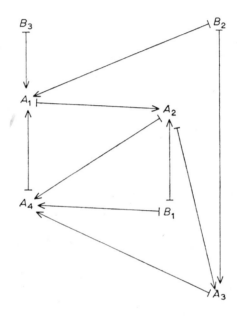

The method of construction is trivial: Variable $A_1$ is influenced by variables $A_4$, $B_2$, and $B_3$. Variable $A_2$ is influenced by variables $A_1$ and $B_1$, etc. Now depending on the signature of the coefficients $a_{ij}$ and $b_{ij}$, the causal influence may be positive or negative. As the representation shows, one may have even situations with causal loops.

Since Strauch (1971) made certain justified observations about inadequacies in the Simon-Wold model of causality, I shall not go into further details.

Watanabe (1970) proposed to take events as elements of the lattice of experiments over a fixed sample space. Then the causal relation between experiments is identified in terms of entropy and information measures.

In the same way that logical models are important for the study of natural languages (logic of conditional modes), the probabilistic models are fundamental in social science theorizing. The notion of correlation, probabilistic, and informational dependence do not capture adequately our intuitions about the causal interactions between systems of random variables. Much remains to be done in tracking down the proper schemes of probabilistic causal thinking.

## 3. *Analytic Models*

The analytic style of thinking on causality is quite different from the two previous approaches, as far as the formalism is concerned. The causal entities are represented by (physical) signals of some kind. (In the logical framework we would say that we consider terms rather than formulas as carriers of causality.) The standard technique is to identify the signals with vectors of a sufficiently well-structured space $\mathfrak{A}$ (mostly Hilbert space and Banach space). For simplicity, one can always think of $\mathfrak{A}$ as a function space $\mathbb{R}^{\mathcal{T}}$ ($\mathbb{R}$ denotes the set of reals), defined by the collection of mappings of the form $A: \mathcal{T} \to \mathbb{R}$, in which the vector addition, multiplication by a scalar from $\mathbb{R}$, and limit are defined coordinatewise:

$$[A + B](t) = A(t) + B(t), \quad [\alpha A](t) = \alpha A(t), \quad \text{and}$$
$$[\lim_n A_n](t) = \lim_n A_n(t), \quad \text{where} \quad A, B \in \mathbb{R}^{\mathcal{T}},$$
$$\alpha \in \mathbb{R}, \quad \text{and} \quad t \in \mathcal{T}$$

The canonical image behind this formalism is as follows. We are given a collection of physical systems that interact with each other in terms of energy and information exchange by means of signals. Every system has an *input*, which is reserved for receiving and decoding the incoming signals and an *output*, which is used for emitting and encoding the signals. A system also has a set of *internal states* which, together with the freshly arrived input signal determines the corresponding output signal. So much for clarifying the possible answer to question $Q_1$.

Formally, every system can be represented by an *operator* $\mathcal{S}: \mathfrak{A} \to \mathfrak{A}$, transforming the input signals into output signals. Concretely, $\mathcal{S}$ can be a polynomial of differential or integral operators, extracted from integro-differential equations describing the system in question.

We visualize this by thinking of a particular system $\mathcal{S}$ located in the ocean of signals where signals reaching the input of $\mathcal{S}$ are immediately transformed into other signals and emitted into the ocean. Now the problem is this. Thinking of systems, we can certainly imagine a large class of those which are clairvoyant (or have some other ESP) to the effect that they will *anticipate* the incoming signal and emit a response *before* the input signal occurs. But we would not like to call these systems physical or physically realizable.

It was Kalman (1965) who first introduced the notion of a *nonanticipatory system* for purposes of system theory. It is a system in which the output at any given time does not anticipate future values of the input. Physicists were familiar with this notion long before Kalman. In fact, Mach (and later Russell) held the view that causality is nothing more than a nonanticipatory type of constraint on physical systems. Much later Foures and Segal (1955) and Segal (1958) formalized this notion for the purposes of classical physics and quantum mechanics. Needless to say, this was possible due to von Neumann's operator formalism of physics and to the experience from relativistic physics.

In order to define the notion of a *causal* (nonanticipatory) system formally, we shall use an extra operator, called *truncation* (projection) *operator* $\mathscr{E}^t : \mathfrak{A} \to \mathfrak{A}$, defined as follows:

$$\mathscr{E}^t(A(s)) = \begin{cases} A(s) & \text{if } s \leqslant t \\ 0 & \text{otherwise}. \end{cases}$$

The meaning is clear: Operator $\mathscr{E}^t$ for $t \in \mathscr{T}$ cuts off the signals after time instance $t$.

We call $\mathscr{S} : \mathfrak{A} \to \mathfrak{A}$ a *causal opertor* if and only if

$$\mathscr{E}^t(A) = \mathscr{E}^t(B) \Rightarrow \mathscr{E}^t \circ S(A) = \mathscr{E}^t \circ S(B)$$

for all $A, B \in \mathfrak{A}$ and $t \in \mathscr{T}$ (the circle ∘ stands for the composition of operators).

The above definiton makes precise the phrases "No output can occur before the input" or "Past and present output values do not depend on future input values." In other words, whenever signals $A$ and $B$ coincide on any initial interval or time segment, then the corresponding responses are equal on the same time segment.

The notion of a causal operator has been generalized in many different ways (from Hilbert space to Banach space, from linear operators to nonlinear operators, from standard time sets to ordered groups, from system operators to automata, etc.). The reader interested in details is referred to Minerbo (1971), Porter and Zahn (1969), Saeks (1969), Sandberg (1965), Windeknecht (1967), and to the sources listed in the introduction.

The formal essence of the causality condition imposed on (physical) operator $\mathscr{S}$ may be summarized as a requirement that operator $\mathscr{S}$ commute in some sense with the family of truncation operators, whose role is closely related to that of $\mathscr{R}(s, t, v)$ and $s \prec t$, utilized in the previous species of causal models.

If $\mathfrak{A}$ is just a finite-dimensional space, then the causal operator may be taken as lower triangular matrices over an appropriate basis in $\mathfrak{A}$. It requires only a little familiarity with the theory of random variables to notice that $\mathfrak{A}$ can be, as an example, a linear space of random variables and $\mathscr{S}$ will be then a *causal stochastic operator*. The hybrid cases are discussed later.

Using the dual $\mathscr{E}_t = 1 - \mathscr{E}^t$ of the truncation operators $\mathscr{E}^t$, one can define the notion of an *anticausal operator* $\mathscr{S}: \mathfrak{A} \to \mathfrak{A}$:

$$\mathscr{E}_t(A) = \mathscr{E}_t(B) \Rightarrow \mathscr{E}_t \circ \mathscr{S}(A) = \mathscr{E}_t \circ \mathscr{S}(B),$$

where $A, B \in \mathfrak{A}$ and $t \in \mathscr{T}$.

An operator is called *memoryless* iff it is both causal and anticausal. Various theorems, such as "$\mathscr{S}, \mathscr{S}'$ are causal $\Rightarrow \mathscr{S} + \mathscr{S}', \mathscr{S}\mathscr{S}', \mathscr{S}$ are causal," or "$\mathscr{S}_n$ is causal $\Rightarrow \lim_n S_n$ is causal," are known for sufficiently general signals and operators. Moreover, it can be shown that every operator $\mathscr{S}$ can be decomposed into a sum of a causal $\mathscr{S}^+$, anticausal $\mathscr{S}^-$, and a memoryless operator $\mathscr{S}^0$:

$$\mathscr{S} = \mathscr{S}^+ + \mathscr{S}^- + \mathscr{S}^0.$$

In addition, the decomposition is unique up to an additive memoryless term $\mathscr{S}^0$.

Many systems encountered in practice contain an integration delay, and as a consequence, they cannot react instantaneously to inputs. For modeling systems of this type one introduces the notion of *strong causality*. Operator $\mathscr{S}$ is called *strongly causal* iff it is causal and for arbitrary $\varepsilon > 0$ and $t, s \in \mathscr{T}$ such that $s \leqslant t$ there exists a $v$ with

$$\mathscr{E}^s(A) = \mathscr{E}^s(B) \Rightarrow \|\mathscr{E}^{s+v}(\mathscr{S}(A) - \mathscr{S}(B))\| \leqslant \varepsilon \|\mathscr{E}^{s+v}(A - B)\|$$

for all $A, B \in \mathfrak{A}$ ($\| \ \|$ stands for the norm in the space $\mathfrak{A}$).

Now one may wonder how these models are related to the foundational questions of causality. As pointed out, Mach, Russell, and also Einstein held the view that causality is best modeled in terms of functional (operator) relationships. In other words, given a space of signals $\mathfrak{A}$ together with a binary relation $\mapsto$, we call this relation *causal* just in case there exists a physical system, described by a *causal* operator $\mathscr{S}$, such that

$$A \mapsto B \Leftrightarrow \mathscr{S}(A) = B \quad \text{for all } A, B \in \mathfrak{A}.$$

Using the properties of causal operators, one can show that properties (1)–(4) are satisfied.

Due to the tremendous impact of physics and system theory on applied mathematics, the analytic models of causality seem to be the most developed.

## 4. Geometric Models

In the category of geometric models of causality the events are represented by space-time points or by *attributes* of space-time points (space-time regions). What are the space-time points? It is certainly a good question and, in fact, so good that we should not even try to answer it. We shall take space-time points as primitives, call them point events, and collect them into a fixed set $\mathfrak{A}$. On this set we shall consider the causal relation $\mapsto$ with the following intended interpretation:

$A \mapsto B$  iff  The event $A$ is directly perceived strictly before event $B$.

If we say nothing more, then the causal relation is sort of a dense partially ordered relation with no first and last element.

Extending our notation to $A \mid B =_D \neg (A \mapsto B) \,\&\, \neg (B \mapsto A) \,\&\, A \neq B$, where $A, B \in \mathfrak{A}$ and $\mathfrak{A}$ = the set of directly perceived events, we may interpret $A \mid B$ as a formula stating that "Exactly one of the events $A$ and $B$ is directly perceived and the remaining one is perceived indirectly." Hence their time relationship cannot be determined.

The above *subjective* interpretation may be not only disturbing, but also rather remote from the issue of causality. Though it is not true, still, we shall give another interpretation, this time a physical one:

$A \mapsto B$  iff  Event $A$ can cause event $B$.

The modality "can cause" has many different meanings and what we want to do may still not be very clear. Therefore we shall give a *geometric* interpretation:

$A \mapsto B$ iff Space-time point $B$ lies inside or on the boundary of the future light cone with vertex $A$ of the four-dimensional Minkowski space.

Now we are ready to state the semantics of relativistic causality. We take $\mathfrak{A}$ to be an abstract set of events, $\mapsto$ is a binary relation on $\mathfrak{A}$ such that there is a mapping $\mathscr{D}: \mathfrak{A} \to \mathbb{R}_4$ from $\mathfrak{A}$ into the four-dimensional real vector space satisfying

$A \mapsto B$ iff $\mathscr{D}_0(A) < \mathscr{D}_0(B)$ & $\{[\mathscr{D}_1(A) - \mathscr{D}_1(B)]^2$
$+ [\mathscr{D}_2(A) - \mathscr{D}_2(B)]^2 + [\mathscr{D}_3(A) - \mathscr{D}_3(B)]^2$
$\leqslant [\mathscr{D}_0(A) - \mathscr{D}_0(B)]^2\}$,

where $\mathscr{D}(A) = \langle \mathscr{D}_0(A), \mathscr{D}_1(A), \mathscr{D}_2(A), \mathscr{D}_3(A)\rangle$ and $A, B \in \mathfrak{A}$. In addition, given two such mappings

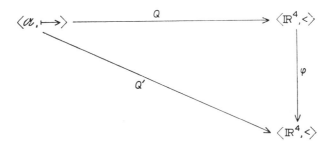

there should exist a transform $\varphi$ from the Lorentz group such that $\mathscr{D}' = \varphi \circ \mathscr{D}$.

The idea of viewing causality in this fashion and trying to base relativity theory on it has been advocated by Robb (1914, 1921). The point was to take the causal relation as a primitive in the set of point events and then axiomatize it in such a fashion that the axioms would imply the Minkowski structure of space-time points (point events) with a metric unique up to a Lorentz group transform precisely in the same way as stated above. Robb tried to accomplish this task by using 30 axioms (including the duals) and more than 200 theorems of a rather involved nature.

Though Robb never stated explicitly in any representation theorem that the axioms for causal relation are sufficient for finding the Minkowski metric with the property discussed above, nevertheless from his work it is quite obvious that such a theorem can be shown. Unfortunately, no physicist or mathematician responded effectively to Robb's challenge. Much later, when Einstein's theory of relativity was fully accepted, Robb's proposal fell entirely into the hands of philosophers. Reichenbach, Carnap, and Mehlberg used it as a model of their early positivistic theory of space and time.

In the last decade or so, a number of Robb's observations have been rediscovered or further developed in a more compact setting and modern terminology (Antoine and Gleit, 1971; Finkelstein, 1969; Gheorghe and Mihul, 1969; Kronheimer and Penrose, 1967; Noll, 1964; Suppes, 1959; Teppati, 1968; Zeeman, 1964).

The geometric interpretation we have given to $A \mapsto B$, namely, that $A$ can cause $B$ or that a signal emitted at point $A$ *can* reach point $B$, is in some sense causal, but not quite. Specifically, one wants a relation which will mean 'signal emitted at point $A$ *will* produce an effect on point $B$'. In order to model such a causal relation, the geometric structure of the space $\mathfrak{A}$ has to be enriched by a physical structure. More concretely, measures for mass distribution and force are needed. This would cut quite deeply into relativistic physics and bring the geometric point of view closer to the analytic standpoint. In the present paper I shall not address myself to this problem.

The significance of geometric causality lies in the fact that causal space-time geometry (i.e., Minkowski geometry in which the metric is replaced by a causal ordering) brings together the vast knowledge of ordered spaces with the species of metric geometries. As a result, one might hope to achieve a better understanding of relativistic space-time and a better evaluation of the significance of the hypothetical particles moving faster than light (tachyons). As is known, tachyons can play a positive role in physical explanation of certain microphenomena. On the other hand, their existence seems to violate the laws of geometric causality (Csonka, 1970a, b; Fox *et al.*, 1970; Israel, 1966; Minerbo, 1971). It seems that appropriate changes in the models of space-time would remove the inconsistency, analogously as the change of the algebra of events in quantum mechanics makes the quantum model consistent. The causal space-time may turn out to be a model in which the desired changes become explicitly visible. Many qualitative space-time geometries are more general than the Minkowski geometry, and therefore, require less specific information to start with.

The importance of the concept of space-time and its applications in relativity motivates us to reexamine its standard formalization. One of the alternatives of materializing this project is the study of causal geometries.

## 5. Hybrid (Derived) Models

In formal science it is typical to consider crossroads in the field of so-called *mother models*. We know about the possible mixtures of algebraic and topological concepts (topological groups, topological lattices, topological vector spaces, etc.) or about the order structures and algebras (partially ordered vector spaces, Archimedean ordered groups, etc.). The hybrid structures frequently possess interesting properties not present in the framework of 'pure' models. Thus it is natural to expect that in the domain of causal problems sooner or later we shall have to deal also with derived, compound, and hybrid models.

A rather trivial and simple hybridization occurs when questions $Q_1$ and $Q_2$ are answered in the languages of two different models. For example, we may adhere to the doctrine that events are essentially sentences, but refuse to handle the sentences by means of intensional logic. Instead, a probabilistic logic is used as a modeling device (Good, 1961, 1962).

Another more or less straightforward method is to mix geometric models with probabilistic ones. This is particularly useful when the random measurement error is to be taken into account (Ingraham, 1964). The probabilistic models can be mixed with the analytical ones as well. Take a linear space of random variables and study the features of stochastic causal operators on that space. In this situation the causal relations occur only in causal assertions of a probabilistic nature. Concretely, we have
$$P[A \mapsto B] = P[\mathscr{S}(A) = B],$$
where $A$ and $B$ are random variables, $\mathscr{S}$ is a stochastic causal operator, and $P$ is a suitable probability measure on a $\sigma$-field associated with the random variables.

Still another possibility is to transplant the idea of a nonanticipatory system into the framework of information structures (Witsenhausen 1971).

Since hybrid models have not received enough attention, there is no point in studying their merits and demerits.

It is just a conjecture that besides the so-called *basic models* of *causality* (of the sort discussed in the previous subjections), there soon will be a need for an unlimited number of *derived models*, manufactured analogously as one generates the group elements of a free group from a finite set of generators.

### III. THEORY OF SPACE-TIME GEOMETRIES

#### 1. *Metric Structure of Minkowski Spaces*

Much of the work in this section and in the following section consists in setting up a suitable configuration of geometric and algebraic notions, useful for the development of causal geometry. We review some of the most elementary facts concerning the metric, affine, order, and lattice structure of spaces and give an outline of the basic theory of Minkowski spaces. It is hoped that this will make it easier for the reader to pass back and forth between the *geometric* and *algebraic* concepts. In particular, the problem of representation will become more transparent.

Euclidean space stripped naked does not have coordinates or vector addition, or multiplication by scalars, as does, for instance, the $n$-dimensional real vector space $\mathbb{R}^n$. It has only points, lines, planes, linear subspaces, and hyperplanes. When thought of in this way, without any metric or vector space properties, it is referred to as a (real) *linear affine space*. The set of linear geometric objects, such as points, lines, planes, admits a lattice structure in which the partial ordering $\leq$ is just the inclusion between geometric objects, the meet $\cdot$ is identical with the set-theoretic intersection of the geometric objects, and the join $+$ is equal to the linear closure of the set-theoretic union of two objects. This so-called *affine lattice* was discovered independently by Menger and Birkhoff at the end of the 1920's and was further developed by Maeda (1951) and others.

One may wonder about the relevance of this kind of theory to our intuition about empirical space, which appears to be something quite different. One can perhaps admit that the 'real' space we live in (call it $\Omega$) consists of *points*. However, the lines, planes, etc., are not 'real', but rather special kinds of abstractions, built up from sets of point configurations in $\Omega$. They are identified as linear *invariants* (invariant subsets of $\Omega$) of certain transformations, acting from outside, as it were, on the set $\Omega$. More precisely, one has a supply of vectors, forming an algebraically structured set $\mathscr{V}$ which can be engaged into kinematic activities on $\Omega$, such as rotation and linear translation of the subsets of $\Omega$. The action of $\mathscr{V}$ is denoted by an external operation $+ : \Omega \times \mathscr{V} \longrightarrow \Omega$, assigning to a point $A$ and a vector $v$ another point $B$, namely, the result of translation of $A$ by $v$: $B = A + v$. In other words, the structure of the space $\Omega$ is given from outside by means of the vector space $\mathscr{V}$.

Now the question is whether the latter, perhaps a more realistic approach, is any better than the lattice-theoretic one. The answer is no. The discovery that every affine lattice of dimension greater than two is isomorphic to the lattice of all affine subspaces of a suitable space is certainly one of the best achievements of modern lattice theory and is comparable with Stone's Representation Theorem concerning Boolean algebras. One purpose of this paper is to discuss the extension of this representation from affine lattices to *causal lattices*. But before actually doing that we have to introduce some technical terminology concerning the spaces $\Omega$ and $\mathscr{V}$, and the lattices $\mathfrak{A}(\Omega)$ and $\mathfrak{A}(\mathscr{V})$.

DEFINITION 1.
*A pair $\langle \mathscr{V}, \cdot \rangle$ is said to be a finite-dimensional real quadratic* (inner product, bilinear) *space if and only if*

$V_1$     *$\mathscr{V}$ is a finite-dimensional real vector space and $\cdot : \mathscr{V} \times \mathscr{V} \longrightarrow \mathbb{R}$, defined by assignment $\langle v, w \rangle \longrightarrow v \cdot w$ for $v, w \in \mathscr{V}$, is a real-valued function, called the inner product* (bilinear form). *The elements of $\mathscr{V}$ are called vectors, and the elements of $\mathbb{R}$ are called scalars.*

$V_2$     *The following conditions are satisfied for all $v, w, w' \in \mathscr{V}$ and $\alpha \in \mathbb{R}$:*
  (i)   $v \cdot w = w \cdot v$ (*Symmetry*)
  (ii)   $v \cdot (w + w') = v \cdot w + v \cdot w'$ (*Additivity*)
  (iii)   $\alpha(v \cdot w) = \alpha v \cdot w$ (*Homogeneity*).

Using the above definition, one can develop rather quickly a useful algebra of vectors in quadratic spaces.

A pair $\langle \mathscr{V}', \cdot' \rangle$ is called a *quadratic subspace* of $\langle \mathscr{V}, \cdot \rangle$ when
(i) $\mathscr{V}'$ is a (linear) subspace of $\mathscr{V}$;
(ii) The inner product $\cdot'$ is equal to the restriction of $\cdot$ to $\mathscr{V}'$, i.e., $v \cdot' w = v \cdot w$ whenever $v, w \in \mathscr{V}'$.

Since in what follows we shall deal mostly with one fixed inner product $\cdot$ on $\mathscr{V}$ and its restrictions to various subspaces of $\mathscr{V}$, it is convenient (by abuse of language) to suppress its symbolic occurrence and write $\mathscr{V}$ instead of a more complicated $\langle \mathscr{V}, \cdot \rangle$. The fact that $\langle \mathscr{V}' \cdot' \rangle$ is a quadratic subspace of $\langle \mathscr{V}, \cdot \rangle$ will, then, be denoted by $\mathscr{V}' \leqslant \mathscr{V}$.

Analogously to what has been indicated above, the relation $\leqslant$ induces a lattice structure on the set of all linear or quadratic subspaces of $\mathscr{V}$. Call the linear lattice: $\mathfrak{A}(\mathscr{V})$, where the operations are the meet and join: $\mathscr{U} \cdot \mathscr{W} = \mathscr{U} \cap \mathscr{W}$, $\mathscr{U} + \mathscr{W} = \{u + w \mid u \in \mathscr{U} \,\&\, w \in \mathscr{W}\}$, where $\mathscr{U}, \mathscr{W} \subseteq \mathscr{V}$. The dimension (length) of this lattice is equal to $\dim \mathscr{V} - 1$. In fact, $\langle \mathfrak{A}(\mathscr{V}), \cdot, +, \leqslant \rangle$ is at least a complete complemented modular lattice; precisely, it is a *projective lattice* (Baer, 1952), whose structure will be further explored in the following sections.

The connection between the space $\mathscr{V}$ and the lattice $\mathfrak{A}(\mathscr{V})$ is very intimate. Specifically, the linear homomorphisms (isomorphisms) between two vector spaces $\mathscr{V}$ and $\mathscr{W}$ induce corresponding lattice homomorphisms (isomorphisms) between the lattices $\mathfrak{A}(\mathscr{V})$ and $\mathfrak{A}(\mathscr{W})$, as the following diagram indicates.

$$\begin{array}{ccc} \mathscr{V} & \xrightarrow{\varphi} & \mathscr{W} \\ \mathfrak{A} \downarrow & & \downarrow \mathfrak{A} \\ \mathfrak{A}(\mathscr{V}) & \xrightarrow{\mathfrak{A}(\varphi)} & \mathfrak{A}(\mathscr{W}) \end{array}$$

A more interesting, but much less trivial situation occurs, when the homomorphisms between two lattices $\mathfrak{A}_1$ and $\mathfrak{A}_2$ induce homomorphisms between the vector spaces whose lattices are isomorphic with $\mathfrak{A}_1$ and $\mathfrak{A}_2$, respectively.

Note that our lattice $\mathfrak{A}(\mathscr{V})$ captures only the vector space structure of $\mathscr{V}$ and the metric (quadratic) structure remains untouched. Our next task will be to geometrize and latticize the inner product. For that purpose we need some new notations.

Two vectors $v$ and $w$ are called *orthogonal* iff their inner product is equal to zero. In symbols:

$v \perp w =_D v \cdot w = 0.$

This notion becomes very efficient if we accept the conventions:

$$v \perp \mathscr{S} =_D \forall s[s \in \mathscr{S} \Rightarrow v \perp s];$$
$$\mathscr{S} \perp \mathscr{S}' =_D \forall s[s \in \mathscr{S} \Rightarrow s \perp \mathscr{S}'];$$
$$\mathscr{W}^\perp =_D \{v \in \mathscr{V} \mid v \perp \mathscr{W}\},$$

where $v \in \mathscr{V}$, $\mathscr{S}, \mathscr{S}' \subseteq \mathscr{V}$, and $\mathscr{W} \leqslant \mathscr{V}$. $\mathscr{W}^\perp$ is called the *orthogonal complement of $\mathscr{W}$ in $\mathscr{V}$*.

The following elementary theorem reveals the geometric and lattice-theoretic pattern of the notions exhibited above.

THEOREM 1.
*The following clauses are valid when $v, w, w' \in \mathscr{V}$, $\alpha \in \mathbb{R}$, and $\mathscr{U}, \mathscr{W} \leqslant \mathscr{V}$:*

(1)   $v \perp w \Rightarrow w \perp v$;
(2)   $v \perp w \Rightarrow \alpha v \perp w$;
(3)   $v \perp w \ \& \ v \perp w' \Rightarrow v \perp w + w'$;
(4)   $v \cdot v = w \cdot w \Rightarrow v + w \perp v - w$;
(5)   $\mathscr{W}^\perp \leqslant \mathscr{V}$;
(6)   $\mathscr{W} \leqslant \mathscr{W}^{\perp\perp}$;
(7)   $\mathscr{W} = \mathscr{W}^{\perp\perp}$ iff $\mathscr{V}^\perp \leqslant \mathscr{W}$;
(8)   $\mathscr{W} \leqslant \mathscr{U} \Rightarrow \mathscr{U}^\perp \leqslant \mathscr{W}^\perp$;
(9)   $\mathscr{U} \perp \mathscr{W}$ iff $\mathscr{U} \leqslant \mathscr{W}^\perp$ iff $\mathscr{W} \leqslant \mathscr{U}^\perp$;
(10)  $\dim \mathscr{W} + \dim \mathscr{W}^\perp = \dim \mathscr{V} + \dim(\mathscr{W} \cdot \mathscr{V}^\perp)$;
(11)  $\dim \mathscr{W}^{\perp\perp} - \dim \mathscr{W} = \dim \mathscr{V}^\perp - \dim(\mathscr{W} \cdot \mathscr{V}^\perp)$;
(12)  *Operation* $(\cdot)^{\perp\perp}: \mathscr{V} \to \mathscr{V}$ *satisfies the standard closure axioms*;
(13)  *The set of all closures $\mathscr{W}^{\perp\perp}$ forms a complete lattice with respect to $\leqslant$, set-theoretic intersection, and closure* $(\mathscr{W} \cup \mathscr{U})^{\perp\perp}$.

One must admit that the standard intuitions about orthogonality (perpendicularity) in Euclidean spaces may lead us astray in general quadratic spaces. At any rate, the set of all subspaces of the quadratic space $\mathscr{V}$ forms an orthocomplemented modular lattice $\langle \mathfrak{A}(\mathscr{V}), \perp \rangle$, provided that $\mathscr{V}^\perp = \{0\}$. This will be of considerable use in the theory of Minkowski space. The relation $\perp$, which is a generalization of the geometric notion of perpendicularity, is one of the most important notions in the description of quadratic spaces. Many subtle metrical aspects of spaces are captured precisely by means of this notion. However, there is still enough information in the inner product · which remains untouched by $\perp$. This is so because $\perp$ refers only to *equalities* between inner product values. *Inequalities* will lead to further geometric concepts, namely, *cones*.

By means of inner product inequalities the space $\mathscr{V}$ is partitioned (up to the zero vector 0) into three fundamental sets:

$$\mathscr{V}^+ = \{v \mid v \cdot v > 0\} \cup \{0\};$$
$$\mathscr{V}^- = \{v \mid v \cdot v < 0\} \cup \{0\};$$
$$\mathscr{V}^0 = \{v \mid v \cdot v = 0\},$$

where the (physically oriented) terminology is as follows: $\mathscr{V}^+$, $\mathscr{V}^-$, and $\mathscr{V}^0$, respectively, are called the *space cone*, *time cone*, and *light cone* of the space $\mathscr{V}$, and thin elements are called *spacelike*, *timelike*, and *lightlike vectors*. Analogously, a subset $\mathscr{S} \subseteq \mathscr{V}$ is called *spacelike*, *timelike*, and *lightlike*, depending on whether $\mathscr{S} \subseteq \mathscr{V}^+$, $\mathscr{S} \subseteq \mathscr{V}^-$, or $\mathscr{S} \subseteq \mathscr{V}^0$.

Although the last notion is not very lattice-theoretic nevertheless it produces an efficient *classification* of quadratic spaces, and it will lead to an ordering relation on $\mathfrak{A}(\mathscr{V})$, as we shall see later. For that purpose, let us put

$$\operatorname{Rad}(\mathscr{W}) =_D \{v \in \mathscr{W} \mid v \perp \mathscr{W}\} = \mathscr{W} \cdot \mathscr{W}^\perp;$$
$$\operatorname{sing} \mathscr{W} =_D \dim \operatorname{Rad}(\mathscr{W});$$
$$\operatorname{ind} \mathscr{V} =_D \operatorname{Max}\{\dim \mathscr{W} \mid \mathscr{W} \leqslant \mathscr{V} \ \& \ \mathscr{W} \subseteq \mathscr{V}^-\},$$

where $\mathscr{W} \leqslant \mathscr{V}$, and $\operatorname{Rad}(\mathscr{W})$ is called the *radical* of $\mathscr{W}$, $\operatorname{sing} \mathscr{W}$ is called the *singularity* of $\mathscr{W}$, and $\operatorname{ind} \mathscr{W}$ is called the (negative) *index* of $\mathscr{W}$. The ordered triple

$$\operatorname{sim} \mathscr{V} = \langle \dim \mathscr{V}, \operatorname{ind} \mathscr{V}, \operatorname{sing} \mathscr{V} \rangle$$

characterizes uniquely (up to isomorphism) every quadratic space, and is called the *similarity type* of $\mathscr{V}$.

The class of similarity types $\langle n, m, k \rangle$ leads to the following fundamental classification of finite-dimensional quadratic spaces (see table on the following page).

As it is seen, all the above classified spaces are associated with the *orthogonal geometry* based mostly on properties of $\perp$. We are totally ignoring the so-called *simplectic*, *unitary*, and many other geometries. It is important to realize that what has been said so far is not special to

| Space | Similarity Type | Remark |
|---|---|---|
| Euclidean | $\langle n, 0, 0 \rangle$ | $n > 0$ |
| Pseudo-Euclidean | $\langle n, m, 0 \rangle$ | $n \geq m > 0$ |
| Semi-Euclidean | $\langle n, 0, k \rangle$ | $n \geq k > 0$ |
| Negative Euclidean | $\langle n, n, 0 \rangle$ | $n > 0$ |
| Trivial (Null) | $\langle n, 0, n \rangle$ | completely isotropic |
| Kotelnikov | $\langle n, 0, 1 \rangle$ | $n > 0$ |
| Minkowski | $\langle n, 1, 0 \rangle$ | $n > 1$ |
| Hyperbolic | $\langle n, \tfrac{1}{2}n, 0 \rangle$ | $n$ is even |

Minkowski spaces, but to a much broader class of spaces. Many authors treat the above notions as if they were the speciality of relativistic geometry.

Minkowski spaces with $n = 4$ represent the underlying geometry of relativity physics, while Kotelnikov (1927) spaces with $n = 4$ give the correct four-dimensional geometry for classical physics.

A quadratic space $\mathscr{V}$ is called *regular* (nonsingular, nondegenerate) iff $\text{Rad}(\mathscr{V}) = \{0\}$ and it is called *isotropic* iff it contains at least one nonzero vector $v$ such that $v \perp v$; otherwise it is called *anisotropic*. The following theorem is a direct consequence of the definitions introduced above:

THEOREM 2.
*The following clauses are valid for all* $\mathscr{W} \leq \mathscr{V}$:

(1) $\mathscr{W} \subseteq \mathscr{V}^0 \Rightarrow \mathscr{W} \leq \mathscr{W}^\perp$;

(2) $\text{Rad}(\mathscr{V}) = \mathscr{V}^\perp \subseteq \mathscr{V}^0$;

(3) $\text{Rad}(\mathscr{W}) = \mathscr{W} \cdot \mathscr{W}^\perp$, i.e., $\mathscr{W}$ is regular iff $\mathscr{W} \cdot \mathscr{W}^\perp = \{0\}$;

(4) $\langle \text{Rad}(\mathscr{V}), \cdot \rangle$ *is a trivial quadratic space* ($v \cdot w = 0$ *for all* $v, w \in \text{Rad}(\mathscr{V})$).

For a compact description of the structure of quadratic spaces we find it useful to introduce the following operations:

Let $\mathscr{V}_1, \ldots, \mathscr{V}_n \leq \mathscr{V}$. Then

$$\mathscr{V}_1 + \ldots + \mathscr{V}_n = \{v_1 + \ldots + v_n \mid v_1 \in \mathscr{V}_1 \ \& \ \ldots \ \& \ v_n \in \mathscr{V}_n\};$$
$$\mathscr{V}_1 \oplus \ldots \oplus \mathscr{V}_n = \mathscr{V}_1 + \ldots + \mathscr{V}_n, \ \text{if} \ \mathscr{V}_i \cdot \mathscr{V}_j = \{0\}$$
$$\& \ 1 \leq i < j \leq n;$$
$$\mathscr{V}_1 \top \ldots \top \mathscr{V}_n = \mathscr{V}_1 \oplus \ldots \oplus \mathscr{V}_n, \ \text{if} \ \mathscr{V}_i \perp \mathscr{V}_j$$
$$\& \ 1 \leq i < j \leq n;$$
$$\mathbb{R}v = \{\alpha v \mid \alpha \in \mathbb{R}\},$$

is called, respectively, the *algebraic sum*, the (internal) *direct sum*, the *orthogonal sum*, and the *scalar product*, of the spaces $\mathscr{V}_1, ..., \mathscr{V}_n$. The operations $+$, $\oplus$, and $\top$ are commutative, associative, and $\{0\}$ acts as their zero element. Obviously, these are partial operations in $\mathfrak{A}(\mathscr{V})$.

A finite subset $\{v_1, ..., v_n\} \subseteq \mathscr{V}$ is called a *base* of $\mathscr{V}$ iff $\mathscr{V} = \mathbb{R}v_1 \oplus ... \oplus \mathbb{R}v_n$, and it is called an *orthogonal base of* $\mathscr{V}$ iff $\mathscr{V} = \mathbb{R}v_1 \top ... \top \mathbb{R}v_n$. Every finite-dimensional quadratic space $\mathscr{V} \neq \{0\}$ has at least one orthogonal base.

The above conventions lead to the following important theorem:

**THEOREM 3.**

(1) *Orthogonal splitting*

> Suppose $\mathscr{V} = \mathscr{V}_1 \oplus ... \oplus \mathscr{V}_n$, where $\mathscr{V}$ is a real vector space and $\langle \mathscr{V}_i, \cdot_i \rangle$ is a quadratic space. Then there is a unique (induced) inner product $\cdot$ on $\mathscr{V}$ such that $\mathscr{V} = \mathscr{V}_1 \top ... \top \mathscr{V}_n$ and $\cdot_i$ is equal to the restriction of $\cdot$ on $\mathscr{V}_i$ $(1 \leq i \leq n)$. Thus, $\text{Rad}(\mathscr{V}_1 \top ... \top \mathscr{V}_n) = \text{Rad}(\mathscr{V}_1) \top ... \top \text{Rad}(\mathscr{V}_n)$.

(2) *Sylvester's Theorem* (Inertia Law)

> $\mathscr{W} \subseteq \mathscr{V}^-$ & $\dim \mathscr{W} = \text{ind } \mathscr{V} \Rightarrow \mathscr{V} = \mathscr{W} \top \mathscr{W}^- $ & $\mathscr{W} \subseteq \mathscr{V}^+$.

(3) *Regularity*

(i) *If there is $\mathscr{W} \leq \mathscr{V}$ such that $\mathscr{W}^\perp$ is regular, then $\mathscr{V}$ is regular;*
(ii) *If $\mathscr{U}, \mathscr{W} \leq \mathscr{V}$ are regular subspaces, then $\mathscr{V} = \mathscr{W} \top \mathscr{W}^-$; and $\mathscr{V} = \mathscr{W} \top \mathscr{U} \Rightarrow \mathscr{U} = \mathscr{W}^\perp$;*
(iii) *If $\mathscr{V}$ is regular and $\mathscr{U}, \mathscr{W} \leq \mathscr{V}$, then $\dim \mathscr{V} = \dim \mathscr{W} + \dim \mathscr{W}^\perp$ and $\mathscr{W}^{\perp\perp} = \mathscr{W}$ with $\text{Rad}(\mathscr{W}) = \text{Rad}(\mathscr{W}^\perp) = \mathscr{W} \cdot \mathscr{W}^\perp$. Moreover, $\mathscr{U} \leq \mathscr{W}$ iff $\mathscr{W}^\perp \leq \mathscr{U}^\perp$, $(\mathscr{U} + \mathscr{W})^\perp = \mathscr{U}^\perp \cdot \mathscr{W}^\perp$, and $(\mathscr{U} \cdot \mathscr{W})^\perp = \mathscr{U}^\perp + \mathscr{W}^\perp$.*

(4) *Isotropy*

> The simplest and most important example of a regular isotropic space is the hyperbolic plane $\mathscr{H}$, defined by
> $\mathscr{H} = \mathbb{R}v \oplus \mathbb{R}w$ with $v \perp v$, $w \perp w$, and not $v \perp w$.
> Generally, a two-dimensional quadratic space is hyperbolic precisely when it is regular and isotropic. Every regular isotropic

*metric space $\mathscr{V}$ is split by a hyperbolic plane:*
$\mathscr{V} = (\mathbb{R}v \oplus \mathbb{R}w) \perp \mathscr{W}$ *for some* $\mathscr{W} \leqslant \mathscr{V}$.

This theorem will be useful for clarifying the nature of subspaces of the Minkowski space.

Now let us turn briefly to the external properties of quadratic spaces.

Let $\langle \mathscr{V}, \cdot \rangle$ and $\langle \mathscr{V}', \cdot' \rangle$ be two quadratic spaces. Then by an *isometry* from $\mathscr{V}$ onto $\mathscr{V}'$ we mean a linear one-to-one and onto mapping $\varphi: \mathscr{V} \to \mathscr{V}'$ such that

$$v \cdot w = \varphi(v) \cdot' \varphi(w)$$

for all $v, w \in \mathscr{V}$, and we write $\mathscr{V} \simeq \mathscr{V}'$. Note that $\mathscr{U} \simeq \mathscr{W} \Rightarrow \mathscr{U}^\perp \simeq \mathscr{W}^\perp$, whenever $\mathscr{U}, \mathscr{W} \leqslant \mathscr{V}$.

The set of all isometries on $\mathscr{V}$,

$$\mathbf{O}(\mathscr{V}) = \{\varphi: \mathscr{V} \to \mathscr{V} \mid \varphi \text{ is an isometry}\},$$

is called the *orthogonal group* of isometries of $\mathscr{V}$. It is indeed a group with respect to functional composition and, in fact, a subgroup of the group of all invertible linear transforms on $\mathscr{V}$. It is trivial to check that

$$v \perp w \quad \text{iff} \quad \varphi(v) \perp \varphi(w)$$

for any $\varphi \in \mathbf{O}(\mathscr{V})$.

The group $\mathbf{O}(\mathscr{V})$ is decomposable into two disjoint sets:

$$\mathbf{O}(\mathscr{V}) = \mathbf{O}^+(\mathscr{V}) \cup \mathbf{O}^-(\mathscr{V}),$$

where $\mathbf{O}^+(\mathscr{V})$ is called the group of *rotations* (it is indeed a group and, in fact, a normal subgroup of $\mathbf{O}(\mathscr{V})$), and $\mathbf{O}^-(\mathscr{V})$ is called the set of *reflections* (it is not a subgroup of $\mathbf{O}(\mathscr{V})$). Actually, $\mathbf{O}^+(\mathscr{V})$ contains precisely those transforms on $\mathscr{V}$ whose determinant is $+1$, and $\mathbf{O}^-(\mathscr{V})$ contains the transforms with determinants equal to $-1$. The description can be presented in a rather nice matrix language, but we shall ignore all further technicalities. It should be remembered, however, that if $\mathscr{V}$ is a Minkowski space, then $\mathbf{O}^+(\mathscr{V})$ is called the *Lorentz group* of rotations of $\mathscr{V}$.

From now on we shall narrow down our interest to Minkowski spaces, keeping in mind that many (including causal) notions remain meaningful in a much broader class of spaces.

## Definition 2.

A pair $\langle \mathscr{V}, \cdot \rangle$ is called a *finite-dimensional real Minkowski space* if and only if

$M_1$      $\langle \mathscr{V}, \cdot \rangle$ is a finite-dimensional real quadratic space;
$M_2$      $\mathscr{V}$ is regular $(\text{Rad}(\mathscr{V}) = \{0\})$; that is,
         $\forall v [v \in \mathscr{V} \Rightarrow v \cdot w = 0] \Rightarrow w = 0$;
$M_3$      Ind $\mathscr{V} = 1$, that is,
         the maximal dimension of any subspace of $\mathscr{V}$ included in the time cone $\mathscr{V}^-$ is 1.

From now on, $\langle \mathscr{V}, \cdot \rangle$ will always denote a finite-dimensional real Minkowski space, unless stated explicitly otherwise. It is a routine matter to check that the following clauses are valid for all nonzero vectors and subspaces of $\mathscr{V}$:

## Theorem 4. (Elementary Theorem of Minkowski Spaces)

(1)    $(v \cdot v)(w \cdot w) \leqslant (v \cdot w)^2$, where $v, w \in \mathscr{V}^-$;
(2)    $v, w \in \mathscr{V}^- \Rightarrow \neg\, v \perp w$;
(3)    $v \in \mathscr{V}^- \,\&\, w \in \mathscr{V}^0 \Rightarrow \neg\, v \perp w$;
(4)    $v \in \mathscr{V}^- \,\&\, v \perp w \Rightarrow w \in \mathscr{V}^+$;
(5)    $v \perp w$ iff $\mathbb{R}v = \mathbb{R}w$, where $v, w \in \mathscr{V}^0$;
(6)    $v \in \mathscr{V}^0 \Rightarrow \dim(\mathbb{R}v)^\perp = \dim \mathscr{V} - 1$, $(\mathbb{R}v)^\perp \cdot \mathscr{V}^- = \{0\}$, and $\dim(\text{Rad}(\mathbb{R}v)^\perp) = \dim \text{Rad}(\mathbb{R}v) = 1$;
(7)    $v \cdot v = w \cdot w \,\&\, v + w \in \mathscr{V}^0 \Rightarrow v + w \perp v \,\&\, v + w \perp w$;
(8)    $\text{Max}\{\dim \mathscr{W} \mid \mathscr{W} \leqslant \mathscr{V} \,\&\, \mathscr{W} \subseteq \mathscr{V}^0\} = 1$;
(9)    $\mathscr{V}$ has a splitting $\mathscr{V} = \mathscr{H} \top \mathscr{W}$, where $\mathscr{H}$ is a hyperbolic plane and $\mathscr{W} \subseteq \mathscr{V}^+$ is unique up to isometry;
(10)    $\mathscr{V}$ has another splitting $\mathscr{V} = (\mathbb{R}w) \top (\mathbb{R}w)^\perp$, where $0 \neq w \in \mathscr{V}^-$ and $(\mathbb{R}w)^\perp \subseteq \mathscr{V}^+$;
(11)    $\langle \mathscr{V}, \cdot \rangle$ is a Minkowski space iff $\exists\, v [v \cdot v < 0]$ and every two-dimensional subspace of $\mathscr{V}$ contains a vector $w$ with $w \cdot w > 0$.

Let $\langle \mathscr{V}, \cdot \rangle$ be a quadratic (not necessarily Minkowski) space. We keep calling $\mathscr{V}$ a 'quadratic' space, yet no motivation has been reserved for that term.

The obvious identity

$$\|v+w\|^2 = \|v\|^2 + \|w\|^2 + 2v\cdot w,$$

where $\|v\|^2 = v\cdot v$ and $v, w \in \mathscr{V}$, quickly removes the mystery. The real-valued function $\|\ \|^2 : \mathscr{V} \to \mathbb{R}$ is called the quadratic form associated with the inner product $\cdot$. As it is seen, there is a one-to-one correspondence between the inner product $\cdot$ and $\|\ \|^2$. Consequently, the pair $\langle \mathscr{V}, \|\ \|^2 \rangle$ is just another equivalent description of $\langle \mathscr{V}, \cdot \rangle$. Since the inner product is a bilinear function, we can call $\mathscr{V}$ a *bilinear space*.

If we put $\|v\| = \sqrt{v\cdot v}$ for $v \in \mathscr{V}$, we cannot call it a *norm* in the ordinary sense, because in general it would take not only negative values, but even complex values. Even if $\mathscr{V}$ is a Minkowski space, the standard triangular inequality $\|v-w\| \leq \|v-u\| + \|u-w\|$ fails to hold. Nevertheless we shall use both the quadratic form $\|\ \|^2$ and the norm $\|\ \|$ (the latter with care), because they will help us to associate certain Euclidean images with the Minkowski ones. Naturally, $\|\ \|$ will be used only when its values are real and nonnegative.

It is well known that any quadratic space $\mathscr{V}$ is decomposable into three subspaces

$$\mathscr{V} = \mathscr{V}_- \top \mathscr{V}_+ \top \mathscr{V}_0,$$

where $\mathscr{V}_0 = \text{Rad } \mathscr{V} (\|v\|^2 = 0$ for $v \in \mathscr{V}_0)$, $\mathscr{V}_+ = $ a subspace of maximal dimension such that the restricted inner product is positive definite ($\|v\|^2 > 0$ for $v \in \mathscr{V}_+ - \{0\}$), and $\mathscr{V}_- = $ a subspace of maximal dimension such that the restricted inner product is negative definite ($\|v\|^2 < 0$ for $v \in \mathscr{V}_- - \{0\}$). The decomposition is unique up to isometry.

If $\mathscr{V}$ is a Minkowski space, then $\text{Rad } \mathscr{V} = \{0\}$ and we have $\mathscr{V} = \mathscr{V}_- \top \mathscr{V}_+$, where $\dim \mathscr{V}_- = 1$, $\mathscr{V}_- \subseteq \mathscr{V}^-$, $\dim \mathscr{V}_+ = \dim \mathscr{V} - 1$, $\mathscr{V}_+ \subseteq \mathscr{V}^+$, and $\mathscr{V}_- \cdot \mathscr{V}_+ = \{0\}$. Consequently, for every vector $v$ we can write

$$v = v^- + v^+,$$

where $v^- \in \mathscr{V}^-$, $v^+ \in \mathscr{V}^+$, and $v^- \perp v^+$.

The restricted inner products induce restricted quadratic forms $\|\ \|^2_-$ and $\|\ \|^2_+$ such that

$$\|v\|^2 = \|v^-\|_- + \|v^+\|_+.$$

The quadratic form $\|\ \|^2_+$ leads to a negative Euclidena norm $-\|\ \|_-$, whereas $\|\ \|_+$ leads to a standard Euclidean norm $\|\ \|_+$. As pointed out, these norms are unique up to isometry in the class of norms on one-dimensional subspaces in $\mathscr{V}^-$ and (dim $\mathscr{V}$ − 1)-dimensional subspaces in $\mathscr{V}^+$. This is a direct consequence of the definition of the direct sum of quadratic spaces:

$$\langle \mathscr{V}_1, \cdot_1 \rangle \oplus \cdots \oplus \langle \mathscr{V}_n, \cdot_n \rangle = \langle \mathscr{V}_1 \oplus \cdots \oplus \mathscr{V}_n, \cdot \rangle,$$

where $(w_1 + \cdots + w_n) \cdot (v_1 + \cdots + v_n) = w_1 \cdot_1 v_1 + \cdots + w_n \cdot_n v_n$ and $\|w_1 + \cdots + w_n\|^2 = \|w_1\|_1^2 + \cdots + \|w_n\|_n^2$ for $v_i, w_i \in V_i$ and $\|w_i\|_i^2 = w_i \cdot_i w_i$ ($1 \leq i \leq n$).

Consider a Minkowski space $\mathscr{V}$ and introduce a coordinate system in it by means of the orthogonal basis $\{x_1, \ldots, x_n\}$:

$$\mathscr{V} = \mathbb{R}x_1 \top \ldots \top \mathbb{R}x_n.$$

The vectors $x_1, \ldots, x_n$ can be normalized so that we may as well put $\|x_i\|_+^2 = 1$ for $1 < i \leq n$ and $-\|x_1\|_-^2 = 1$. Then for any vector $v \in \mathscr{V}$ we have

$$v = \alpha_1 x_1 + \cdots + \alpha_n x_n;$$

hence,

$$v^2 = \|\alpha_1 x_1\|^2 + \cdots + \|\alpha_n x_n\|^2,$$

i.e.,

$$v^2 = -\alpha_1^2 + \alpha_2^2 + \cdots + \alpha_n^2.$$

Consequently,

$$\|v\| = \sqrt{-\alpha_1^2 + \alpha_2^2 + \cdots + \alpha_n^2},$$

which is a more familiar looking form of the *Minkowski norm*. We prefer the coordinate-free description because it is easier to see what is coordinate invariant and what is not. Furthermore, the notation is much simpler and more efficient.

Now it is clear that we can write

$$\begin{aligned}
v \in \mathscr{V}^0 &\quad \text{iff} \quad \|v\|^2 = 0 \quad \text{iff} \quad \|v^+\|_+ = \|v^-\|_-; \\
v \in \mathscr{V}^+ &\quad \text{iff} \quad \|v\|^2 > 0 \quad \text{iff} \quad \|v^+\|_+ > \|v^-\|_-; \\
v \in \mathscr{V}^- &\quad \text{iff} \quad \|v\|^2 < 0 \quad \text{iff} \quad \|v^+\|_+ < \|v^-\|_-.
\end{aligned}$$

The similarity types of the proper subspaces $\mathscr{W}$ of the $n$-dimensional Minkowski space $\mathscr{V}$ are: $\langle m, 0, 0 \rangle$, $\langle m, 1, 0 \rangle$, and $\langle m, 0, 1 \rangle$, where $1 \leq m \leq n - 1$. Using these, we can decompose the lattice $\mathfrak{A}(\mathscr{V})$ into three subsets:

$$\mathfrak{A}^+(\mathscr{V}) = \{\mathscr{W} \mid \text{sim } \mathscr{W} = \langle m, 0, 0\rangle \ \&\ 1 \leqslant m \leqslant n-1\};$$
$$\mathfrak{A}^-(\mathscr{V}) = \{\mathscr{W} \mid \text{sim } \mathscr{W} = \langle m, 1, 0\rangle \ \&\ 1 \leqslant m \leqslant n\ \ \ \};$$
$$\mathfrak{A}^0(\mathscr{V}) = \{\mathscr{W} \mid \text{sim } \mathscr{W} = \langle m, 0, 1\rangle \ \&\ 1 \leqslant m \leqslant n-1\}.$$

We may call $\mathfrak{A}^+(\mathscr{V})$, $\mathfrak{A}^-(\mathscr{V})$, and $\mathfrak{A}^0(\mathscr{V})$, respectively, the class of *spacelike*, *timelike*, and *lightlike* subspaces of $\mathscr{V}$. This classification is a generalization of the distinction that is usually made among the lines which are *outside*, *inside*, and *on* the light cone, determined by the norm $\|\ \|$. Later on we shall show that $\mathfrak{A}^+$ is a kind of ideal, and $\mathfrak{A}^-$ is a filter, in the lattice $\mathfrak{A}$. But before that we shall have to explore the order structure of $\mathscr{V}$.

## 2. Order Structure of Minkowski Spaces

A finite-dimensional real vector space $\mathscr{V}$ equipped with an ordering relation $\rightarrow$, compatible with its linear structure, is called an *ordered vector space*. More specifically, given a vector space $\mathscr{V}$ and a binary relation $\rightarrow$ on $\mathscr{V}$, then the couple $\langle \mathscr{V}, \rightarrow \rangle$ is called a *partially ordered vector space* iff

(i)    $\rightarrow$ is reflexive, antisymmetric, and transitive;
(ii)   $v \rightarrow w \Rightarrow v + u \rightarrow w + u$;
(iii)  $v \rightarrow w \Rightarrow \alpha v \rightarrow \alpha w$,

where $v, w, u \in \mathscr{V}$ and $\alpha \geqslant 0$.

An equivalent definition can be given in terms of a *convex cone*. Concretely, by a *positive cone* in $\mathscr{V}$ we mean a nonempty subset $\mathscr{C} \subseteq \mathscr{V}$ such that the following geometric properties are satisfied for all $v, w \in \mathscr{V}$:

(a)  $v, w \in \mathscr{C} \Rightarrow v + w \in \mathscr{C}$;
(b)  $\alpha \geqslant 0\ \&\ v \in \mathscr{C} \Rightarrow \alpha v \in \mathscr{C}$;
(c)  $v, -v \in \mathscr{C} \Rightarrow v = 0$.

The link between the ordering relation $\rightarrow$ and the cone $\mathscr{C}$ is given by

$$v \rightarrow w \quad \text{iff} \quad w - v \in \mathscr{C},$$

for all $v, w \in \mathscr{V}$.

Hence, the notion of a partially ordered vector space can be given equivalently both in the geometric language of cones $\langle \mathscr{V}, \mathscr{C}\rangle$ and the order-theoretic language of relations $\langle \mathscr{V}, \rightarrow \rangle$. In the following sections we shall use freely either of these characterizations of ordered spaces. Naturally, any further specification of the relation $\to$ leads to additional constraints on $\mathscr{C}$, and vice versa.

Let $\mathscr{V}$ be a Minkowski space and consider the cone $\mathscr{C} = \{v \in \mathscr{V} \mid v \notin \mathscr{V}^+\}$ with a binary relation

$$v \equiv w =_D \exists t \in \mathscr{V}^- - \{0\} [(v \cdot t)(w \cdot t) > 0],$$

where $v, w \in \mathscr{C}$.

Using Theorem 4, we can show that $\equiv$ is an equivalence relation on $\mathscr{C}$. This relation classifies the vectors of $\mathscr{C}$ into two disjoint classes $\mathscr{C}^+$ and $\mathscr{C}^-$. It turns out that these classes are *convex cones* to which we may adjoin the zero vector *ex post*. Thus,

$$\mathscr{C} = \mathscr{C}^+ \cup \mathscr{C}^- \quad \text{and} \quad \mathscr{C}^+ \cap \mathscr{C}^- = \{0\}.$$

The quotient $\mathscr{C}/\equiv\ = \{\mathscr{C}^+, \mathscr{C}^-\}$ represents the set of all *possible time directions*. If one of the elements of $\mathscr{C}/\equiv$, say $\mathscr{C}^+$, is singled out, then the Minkowski space $\mathscr{V}$ is said to be (positively) *directed in time* and we write $\langle \mathscr{V}, \mathscr{C}^+ \rangle$. We may call $\mathscr{C}^+$ the *future cone* and, dually, $\mathscr{C}^-$ the *past cone of* $\mathscr{V}$. It is important to realize that $\mathscr{C}^+$ has been selected by *convention* and *not* on the basis of some intrinsic properties of the Minkowski space. The question of whether there is a *privileged* choice in $\mathscr{C}/\equiv$ and if there is, what its mechanism is, is known as the so-called *time-arrow problem*. It seems clear that this problem cannot be decided solely on the grounds of the geometry of $\mathscr{V}$. An answer requires additional structure, some 'physics', usefully enriching the geometry of $\mathscr{V}$. The pair $\langle \mathscr{V}, \mathscr{C}^+ \rangle$ will often be referred to as a *causal Minkowski space*.

The following elementary theorem explores the basic properties of the cones $\mathscr{C}^+$ and $\mathscr{C}^-$:

THEOREM 5.
*The following clauses are valid for all $v, v', w \in \mathscr{V}$ and $\alpha \geq 0$:*

(1) $\quad v, w \in \mathscr{C}^+ \Rightarrow v + w \in \mathscr{C}^+\ \&\ \alpha v \in \mathscr{C}^+$;

(2) $\quad v \in \mathscr{C}^+\ \&\ -v \in \mathscr{C}^+ \Rightarrow v = 0$;

(3) $\quad v \in \mathscr{C}^+$ iff $-v \in \mathscr{C}^-$;

(4) $\quad \forall v \exists w [v - w \in \mathscr{C}^+\ \&\ w \in \mathscr{C}^+]$;

(5) $\quad \forall v, v' \exists w [v - w \in \mathscr{C}^+\ \&\ v' - w \in \mathscr{C}^+]$;

(6) $\quad \forall v, v' \exists w [w - v \in \mathscr{C}^+ \ \& \ w - v' \in \mathscr{C}^+]$;
(7) $\quad v, w \in \mathscr{V}^- \Rightarrow v + w \in \mathscr{V}^-,$ if $v \equiv w,$ and $v + w \in \mathscr{V}^+$ otherwise.

Thus, $\mathscr{C}^+$ is a convex cone in the sense of analytic geometry and the theory of partially ordered vector spaces: $\mathscr{C}^+ + \mathscr{C}^+ = \mathscr{C}^+$, $\mathbb{R}^+ \mathscr{C}^+ \subseteq \mathscr{C}^+$, $\mathscr{C}^+ \cap \mathscr{C}^- = \{0\}$, $\mathscr{C}^+ - \mathscr{C}^+ = \mathscr{V}$, and $\mathscr{C}^+ = -\mathscr{C}^-$. We can therefore switch to the ordering

$$v \to w =_D w - v \in \mathscr{C}^+,$$

where $v, w \in \mathscr{V}$.

The relation $\to$ is called the *causal relation* in the Minkowski space $\mathscr{V}$. From the properties of the cone $\mathscr{C}^+$ it follows that $\to$ is reflexive, antisymmetric, transitive, and (upward and downward) directed. One can even switch to the *strict causal relation* $\mapsto$ by putting

$$v \mapsto w =_D (v \to w) \ \& \ \neg (w \to v).$$

Since the pair $\langle \mathscr{V}, \mapsto \rangle$ is just another equivalent description of $\langle \mathscr{V}, \mathscr{C}^+ \rangle$, we can refer to it again as a *causal Minkowski space*.

Now one wonders what the relationship is between the cone $\mathscr{C}^+$ and the Minkowski metric $\| \ \|$. The answer is easy. Select a fixed vector, say $e$ in $\mathscr{C}^+$ such that $e \notin \mathscr{V}^0$, $\|e\|^2 = -\|e_-^2\|_- + \|e_+^2\|_+ = -\|e_-^2\|_- = 1$. Then obviously,

$$v \in \mathscr{C}^+ \quad \text{iff} \quad \|v^+\|_+ \leq \varepsilon_v \quad \text{iff} \quad \|v\|^2 \leq 0 \ \& \ v \cdot e \leq 0$$

and

$$v \in \mathscr{C} \quad \text{iff} \quad \|v\|^2 \leq 0,$$

where $v = \varepsilon_v e + \alpha_2 x_2 + \cdots + \alpha_n x_n$.

That is,

$$v \in \mathscr{C}^+ \quad \text{iff} \quad \sqrt{\alpha_2^2 + \cdots + \alpha_n^2} \leq \varepsilon_v.$$

The set

$$\dot{\mathscr{C}}^+ = \{ v \in \mathscr{V} \mid \|v^+\|_+ = \varepsilon_v \},$$

and

$$\dot{\mathscr{C}} = \{ v \in \mathscr{V} \mid \|v\|^2 = 0 \},$$

is called, respectively, the *boundary of the cone* $\mathscr{C}^+$ and $\mathscr{C}$.

In the language of relations we use

$$v \dot\to w =_D w - v \in \dot{\mathscr{C}}^+,$$

and call it the *signal relation*. If needed, we can use

$$v \mapsto w =_D v \dot\to w \ \& \ v \neq w.$$

Another relation, called *incompatibility relation* $|$, is of some interest:

$$v \mid w =_D w - v, v - w \in \mathscr{V}^+ - \{0\}.$$

Its more familiar definition is as follows:

$$v \mid w \quad \text{iff} \quad \neg(v \to w) \ \& \ \neg(w \to v).$$

All this can be said in the language of norms just as well. Hence, we have

$$v \to w \quad \text{iff} \quad \|v^+ - w^+\|_+ \leq \varepsilon_{v-w};$$
$$v \dot\to w \quad \text{iff} \quad \|v^+ - w^+\|_+ = \varepsilon_{v-w};$$
$$v \mid w \quad \text{iff} \quad \|v^+ - w^+\|_+ > \varepsilon_{v-w}.$$

The attentive reader will notice that the last definition of $\to$ is an abstract version of the definition of geometric causality presented in Section II. 4.

## 3. *Affine Structure of Minkowski Spaces*

The tendency has always been to define geometric spaces in terms of simple, strong, and intuitive axioms. In other words, one requires a rather small number of universal (first-order if possible) axioms, leading quickly to nontrivial theorems and expressing easily testable properties of the empirical space we live in. Besides that, one is led to prefer such axiomatizations of space which utilize standardized concepts of mathematics.

For example, analytically a Euclidean space is defined as a pair $\langle \mathscr{V}, \cdot \rangle$, where $\mathscr{V}$ is a vector space and $\cdot$ is a regular positive definite inner product on $\mathscr{V}$ ($\mathscr{V}^- \cup \mathscr{V}^0 = \{0\}$). This definition meets the above requirements.

Often we are not interested in vectors and their algebra, but rather in the behavior of linear geometric objects such as lines, planes, hyperplanes. Then the vector space plays only an auxiliary role. This kind of problem situation in the context of Minkowski spaces will be the topic of this section.

DEFINITION 3.
*A structure $\langle \Omega, \mathscr{V}, + \rangle$ is said to be a finite-dimensional space-time or simply a Robb space if and only if*

$R_1$    $\Omega$ *is a nonempty set, called the space and its elements are called points;*

$\mathscr{V}$ *is a finite-dimensional real Minkowski space and its elements are called translations and,*

$+ : \Omega \times \mathscr{V} \to \Omega$, *whose effect is defined by $\langle A, w \rangle \to A + w$, is an external operation, called act.*

$R_2$    *The following conditions are satisfied for all $v, w \in \mathscr{V}$ and $A, B \in \Omega$:*
  (i)   $A + 0 = A$ (Zero act)
  (ii)   $A + (v + w) = (A + v) + w$ (Additivity)
  (iii)   $A + v = A \Rightarrow v = 0$ (No fixed points)
  (iv)   $\exists w (A + w = B)$ (Transitivity).

Conditions (iii) and (iv) define uniquely a vector $w$ in terms of points $A$ and $B$. This vector is denoted by $B - A$.

One can see almost immediately that the following clauses are valid for all points in $\Omega$ and all translations in $\mathscr{V}$:

THEOREM 6

(1)    $A - B = 0$   iff   $A = B$;
(2)    $(A - B) + v = (A + v) - B$;
(3)    $(A + w) - (B + v) = (A - B) + (w - v)$;
(4)    $(A - B) + (B - C) = A - C$;
(5)    $B + (A - B) = A$.

It is convenient to put

$$A + \mathscr{S} = \{A + v \mid v \in \mathscr{S}\},$$

where $\mathscr{S} \subseteq \mathscr{V}$.

A subset $A + \mathscr{W} \subseteq \Omega$, where $\mathscr{W} \leqslant \mathscr{V}$ and $A \in \Omega$, is called an *affine subspace* (subspace-time) of $\langle \Omega, \mathscr{V}, + \rangle$, determined by $\mathscr{V}$ and $A$. The linear subspace $\mathscr{W}$ is called the *direction* of the affine subspace $A + \mathscr{W}$ and is denoted by $\mathrm{Dir}(A + \mathscr{W}) = \mathscr{W}$.

We put $\dim(A + \mathscr{W}) = \dim \mathscr{W}$. As a matter of fact, we should symbolize the subspaces by $\langle \Gamma, \mathscr{W}, + \rangle$ instead of $\Gamma$, where $\Gamma = A + \mathscr{W}$. However, if there is no danger of confusion, we shall use the simplified notation. Remember, $\Gamma$ uniquely determines $\mathscr{W}$. In addition, we shall often talk about the Minkowski space $\Omega$ instead of the finite-dimensional real space-time $\langle \Omega, \mathscr{V}, + \rangle$, keeping in mind that $\mathscr{V}$ and $+$ are fixed. The fact that $\Gamma$ is an affine subspace of $\Omega$ is expressed by $\Gamma \leqslant \Omega$. For two subspaces $\Gamma$ and $\Delta$ in $\Omega$, $\Gamma < \Delta$ means $\Gamma \leqslant \Delta \,\&\, \Gamma \neq \Delta$.

Often it is quite convenient to put $\Omega = \mathscr{V}$. Then an affine subspace of $\Omega$ is essentially identical with a *coset* of $\mathscr{V}$. In fact, every affine space is affine isomorphic to a space with the above identification: $\langle \mathscr{V}, \mathscr{V}, + \rangle$.

A subspace $\Gamma \leqslant \Omega$, where $\Gamma = A + \mathscr{W}$, $\mathscr{W} \leqslant \mathscr{V}$, and $A \in \Omega$, is called, respectively, a *point*, *line*, *plane*, and a *hyperplane*, if $\dim \mathscr{W} = 0$, $\dim \mathscr{W} = 1$, $\dim \mathscr{W} = 2$ or $\dim \mathscr{W} = \dim \mathscr{V} - 1$. It is convenient to consider the empty set $\emptyset$ also a subspace of $\Omega$ with $\dim \emptyset = -1$ and $\mathrm{Dir}(\emptyset) = \emptyset$.

The following theorem is a consequence of the aforementioned conventions:

THEOREM 7
*Let* $\Gamma = A + \mathscr{W}$ *and* $\Gamma' = B + \mathscr{W}'$ *with* $A, B \in \Omega$ *and* $\mathscr{W}, \mathscr{W}' \leqslant \mathscr{V}$. *Then the following conditions are valid*:

(1) *If we define* $(A + \mathscr{W}) \cdot (B + \mathscr{W}') =_D C + \mathscr{W} \cdot \mathscr{W}'$ *with* $C \in \Gamma \cdot \Gamma'$, *then* $\Gamma, \Gamma_1 \leqslant \Omega \;\Rightarrow\; \Gamma \cdot \Gamma' \leqslant \Omega$;
(2) *If we define* $(A + \mathscr{W}) + (B + \mathscr{W}') =_D A + \mathbb{R}(B - A) + (\mathscr{W} + \mathscr{W}')$ *with* $A \in \Gamma$ *and* $B \in \Gamma'$, *then* $\Gamma, \Gamma' \leqslant \Omega \;\Rightarrow\; \Gamma + \Gamma' \leqslant \Omega$;
(3) $\dim(\Gamma + \Gamma') = \dim \Gamma + \dim \Gamma' - \dim \Gamma \cdot \Gamma'$, *when* $\Gamma \cdot \Gamma' \neq \emptyset$;
(4) $\dim(\Gamma + \Gamma') = \dim \Gamma + \dim \Gamma' + 1$, *when* $\Gamma \cdot \Gamma' = \emptyset$;
(5) $A + \mathscr{W} = B + \mathscr{W}'$ *iff* $\mathscr{W} = \mathscr{W}' \,\&\, B \in A + \mathscr{W}$.

The last two clauses show the usefulness of the empty set being an extra additional affine subspace. The first and second clauses indicate that the set of all affine subspaces of $\Omega$ is a lattice with respect to the subspace relation $\leqslant$, and operations $\cdot$ and $+$. Moreover, $\emptyset$ and $\Omega$ serve as zero and unit element, respectively. This lattice will be denoted by $\mathfrak{A}(\Omega)$. It is a complete, atomic lattice of dimension (length) equal to dim $\Omega$. Its additional properties will be explored later. For the time being we can call it the *affine lattice* of $\Omega$. It is important to remember that it captures only the affine properties of $\Omega$ and omits all the metric structure of $\Omega$, which has not been explored yet anyway.

We shall transplant almost mechanically nearly all the metric properties of $\mathscr{V}$ into $\Omega$. Thus, we put ind $\Omega$ = ind $\mathscr{V}$, sing $\Omega$ = sing $\mathscr{V}$, etc. Moreover, we define

$$\mathfrak{A}^+(\Omega) =_D \{\Gamma \mid \text{Dir}(\Gamma) \in \mathfrak{A}^+(\mathscr{V})\};$$
$$\mathfrak{A}^-(\Omega) =_D \{\Gamma \mid \text{Dir}(\Gamma) \in \mathfrak{A}^-(\mathscr{V})\};$$
$$\mathfrak{A}^0(\Omega) =_D \{\Gamma \mid \text{Dir}(\Gamma) \in \mathfrak{A}^0(\mathscr{V})\},$$

and use the same terminology we used for the quadratic spaces.

Let $\langle \mathscr{V}, \mathscr{C}^+ \rangle$ be a causal Minkowski space. Then $A + \mathscr{C}^+ \subseteq \Omega$ for $A \in \Omega$ is called the *affine* (future) *cone* of $\Omega$. Next we put

$$A \to B =_D B - A \in \mathscr{C}^+,$$

and call it the *causal relation* on $\Omega$. Obviously,

$$A + \mathscr{C}^+ = \{B \mid A \to B\};$$
$$A \to B \quad \text{iff} \quad B + \mathscr{C}^+ \subseteq A + \mathscr{C}^+;$$
$$A = B \quad \text{iff} \quad A + \mathscr{C}^+ = B + \mathscr{C}^+;$$
$$A \to B \quad \text{iff} \quad (B-A) \cdot (B-A) \leqslant 0 \ \& \ (B-A) \cdot e \leqslant 0,$$

where $e \in \mathscr{C}^+$, $e \notin \mathscr{V}^0$, and $(B-A) \cdot e \leqslant 0$ has the same meaning as $B - A \equiv e$.

That is to say, precisely one affine cone corresponds to every point in $\Omega$, and moreover, $(A + \mathscr{C}^+) \cap (B + \mathscr{C}^+) \neq \emptyset$ for all $A, B \in \Omega$.

All the remaining causal notions, such as $\mapsto$, $\dot{\to}$, $\overset{\cdot}{\mapsto}$, and $|$, are easy to translate, and we shall use them without warning.

The couple $\langle \Omega, \mapsto \rangle$, where $\Omega$ is a space-time, and $\mapsto$ is a causal relation on $\Omega$, is called the *causal space-time*. It was Robb (1914) who studied carefully both the affine and metric properties of the structure $\langle \Omega, \mapsto \rangle$, with dim $\Omega = 4$, and indicated the existence of a corresponding Minkowski space $\langle \mathscr{V}, \| \ \|^2 \rangle$, induced by the axioms for the causal space-time.

For purposes of representation it is convenient to enrich the lattice $\mathfrak{A}(\Omega)$ by several geometric relations (parallelism $\|$, orthogonality $\perp$, and causality $\mapsto$) so as to distribute uniformly the axiomatic load on a larger system of primitives. How this can be done will be the subject of the next section. The suitably enriched lattice $\mathfrak{A}(\Omega)$ will be called the *causal lattice*.

## IV. CAUSAL GEOMETRIES

The purpose of this section is a lattice-theoretic study of the causal structure of Minkowski spaces and space-times.

As indicated in the previous sections, a causal space-time $\langle \Omega, \mapsto \rangle$, as well as the causal Minkowski space $\langle \mathscr{V}, \mapsto \rangle$, determines a causal geometry (causal lattice), namely, that of all subspaces of $\Omega(\mathscr{V})$, partially ordered by the causal relation, and equipped with further relations, such as orthogonality and parallelism.

The question arises under what conditions this activity can be reversed. That is, under what circumstances can one reconstruct the causal space-time (causal Minkowski space) from the causal lattice of its subspaces? This appears to be a fundamental problem in geometry and it can be answered on several different levels, involving varying degrees of difficulty.

(1) On a simpler level, where the causal lattice is given as the lattice of subspaces of some definite Minkowski space and where one wants to prove that the totality of subspaces of a Minkowski space forms a causal lattice.

(2) On a deeper level, where the causal lattice is given abstractly and nothing is supposed to be known about the underlying Minkowski space. Here the identification (construction) of a compatible causal Minkowski space is the principal objective. The degree of freedom involved is drastically reduced by the constraint demanding that the Minkowski norm be unique up to the group of Lorentz transforms.

(3) And finally, on a very deep level, where even the causal lattice is unknown. One starts with an abstract partially ordered set, intended as the set of points of space-time, and the aim is to reconstruct the compatible causal lattice and then the corresponding Minkowski space.

The first two (perhaps easier) alternatives of the problem, to the author's best knowledge, have not been studied, although the last and certainly the most difficult problem situation, was attacked by Robb at the beginning of this century. This seems almost strange because in Robb's theory the ordering $\mapsto$ on $\Omega$ carries all the axiomatic burden. One has to encode implicitly into $\mapsto$ not only the (affine) lattice of subspaces of $\Omega$, together with their dimensions, but also the corresponding auxiliary real quadratic space. And this means that one has to produce from $\mapsto$ and $\Omega$ a copy of a model of reals and a Minkowski metric. Eventually, one should arrive at a convex cone in the Minkowski space, matching the original partial ordering $\mapsto$.

Robb had special reasons to proceed this way. He thought that a rather simple ontology, consisting of a set of (perceptual) events $\Omega$ and a (subjective) before-after ordering $\mapsto$, would be entirely sufficient to capture the relativistic nature of space and time. It is certainly a remarkable idea considering (as it was shown much later) that one cannot axiomatize Euclidean geometry in terms of a *binary* relation alone.

The difficulty with these 'thin' ontologies lies in the fact that most of the axioms governing their properties are bound to be technical or highly 'nonempirical'. As a result, the conceptual encoding is inefficient and little information can be extracted from it for purposes of applications and deeper understanding. Consequently, in the course of modern geometry people experimented with more differentiated and, in some sense, more efficient ontologies, and by means of these they managed not only to track down the behavior of a tremendously large class of known geometric objects, but also they discovered many new ones. Think, as an example, of the discovery of continuous and finite geometries, and the applications of the latter. The geometric practice shows that it is good to keep in modeling of geometric entities a reasonable balance between the number of primitives and the complexity of structural axioms, encoding the properties and relationships of the objects involved.

In this section we shall follow the latter point of view and will concentrate on the behavior of the causal relation $\mapsto$ in lattices rather than in plain sets. Moreover, we shall bring into action also some other geometric relations, such as parallelism and orthogonality.

Following Robb (1914), let us start with an abstract nonempty set $\Omega$ whose elements will be called (point) events, together with a binary relation $\mapsto$ on $\Omega$, called the *causal relation*. For purposes of interpretation one can imagine that $\Omega$ is a set of geometric space-time points at which signals (information) can be both received and emitted. Then the atomic formula $A \mapsto B$ means that a signal emitted at place $A$ can reach (and be received at) place $B$. Most of the intended interpretations originate in the kinematics of relativistic physics (Nevanlinna, 1968). Our task will be to explore the geometric structure of $\Omega$ and the order-theoretic structure of $\mapsto$.

The intended application of $\mapsto$ suggests that it should be an irreflexive, asymmetric, transitive, and dense ordering, without first and last elements.

If no other constraints are added, the structure $\langle \Omega, \mapsto \rangle$ will certainly be a model for many things, including those which will be most remote from our conception of Minkowski spaces. Concretely, the axiom for complete simple ordering of certain subsets of $\Omega$, the Archimedean axiom, and upper and lower dimensionality axioms will have to be added, if one wishes to delineate the possible models to the desired ones.

Although by utilizing heavily the incomparability relation $A \mid B$, I managed to simplify Robb's axioms for $\langle \Omega, \mapsto \rangle$, due to the problems pointed out above, I propose to take a different course of action. Namely, unlike Robb, I shall assume that the *n*-dimensional affine lattice $\langle \mathfrak{A}_n(\Omega),$ $\emptyset, \Omega, \cdot, +, \leqslant \rangle$ of subspaces of $\Omega$, discussed in the previous sections, is known. On this lattice we shall consider parallelisms $\parallel$ and $\Vdash$, orthogonality $\perp$, and the causal relations $\mapsto$, $\leftarrowtail$, and $\mid$, defined as follows:

DEFINITION 4
*Let $\Gamma$ and $\Delta$ be elements of $\mathfrak{A}_n(\Omega)$, i.e., subspaces of $\Omega$. Then we define*

$\Gamma \parallel \Delta$ iff $\mathrm{Dir}(\Gamma) = \mathrm{Dir}(\Delta)$;
$\Gamma \Vdash \Delta$ iff $\mathrm{Dir}(\Gamma) \leqslant \mathrm{Dir}(\Delta)$;
$\Gamma \perp \Delta$ iff $\mathrm{Dir}(\Gamma) \perp \mathrm{Dir}(\Delta)$;
$\Gamma \mapsto \Delta$ iff $\forall A \in \Gamma \exists B \in \Delta (A \mapsto B)$;
$\Gamma \leftarrowtail \Delta$ iff $\forall A \in \Gamma \exists B \in \Delta (B \mapsto A)$;
$\Gamma \mid \Delta$ iff $\forall A \in \Gamma \forall B \in \Delta (A \neq B \Rightarrow A \mid B)$,

*where $\mathrm{Dir}(\Gamma)$ stands for the direction of $\Gamma$.*

The meaning of the first three relations is well known. The intended meaning of $\Gamma \mapsto \Delta$ is this: Every event in $\Gamma$ can cause an event in $\Delta$. Or, from every point of $\Gamma$ one can send a signal into some point of $\Delta$. Dually, the intended interpretation of $\Gamma \leftarrowtail \Delta$ is as follows: Every event in $\Gamma$ can be caused by some event from $\Delta$. Or, every point of $\Gamma$ can receive a signal from at least one point of $\Delta$. In a different context, $\Gamma \mapsto \Delta$ can be interpreted as a comparison of *complexities* of $\Gamma$ and $\Delta$. Finally, $\Gamma \mid \Delta$ just means that no event in $\Gamma$ can cause any event in $\Delta$ and vice versa, or that no signal can be emitted from $\Gamma$ and received in $\Delta$ and vice versa.

The significance of considering causality on linear geometric objects rather than on point events lies in the fact that once the causality becomes clear in lattices, it can be hoped that the results will be relevant not only to the lattices of convex regions in space-time (which forms a more realistic device than linear lattices), but also to lattices occurring in other fields such as logic, probability, and operator theory.

The causal notions are related to the order-theoretic description of $\langle \Omega, \mathscr{V}, + \rangle$ in the following way:

$$\Gamma \mapsto \Delta \text{ iff } \forall A[A \in \Gamma \Rightarrow (A + \mathscr{C}^+) \cap \Delta \neq \emptyset];$$
$$\Gamma \mapsto \Delta \text{ iff } \forall A[A \in \Gamma \Rightarrow (A + \mathscr{C}^-) \cap \Delta \neq \emptyset];$$
$$\Gamma \mid \Delta \text{ iff } \forall A[A \in \Gamma \Rightarrow (A + \mathscr{C}) \cap \Delta = \emptyset].$$

Naturally, $A + \mathscr{C}^- = \{B \mid B \to A\}$, $A + \mathscr{C}^+ = \{B \mid A \to B\}$, and $A + \mathscr{C} = \{B \mid A \to B \text{ or } B \to A\} = \{X \mid X - A) \cdot (X - B) \leq 0\}$.

Next we shall exhibit the basic properties of the causal lattice $\mathfrak{A}(\Omega)$.

THEOREM 8

Let $\mathfrak{A}(\Omega) = \mathfrak{A}$ be the causal lattice of a Robb space $\Omega$. Then the following conditions are valid for all variables running over $\mathfrak{A}$:

(1)      $\mathfrak{A}^+$ is a causal ideal in $\mathfrak{A}$:
 (i) $\emptyset \in \mathfrak{A}^+$;
 (ii) $\Gamma \in \mathfrak{A}^+ \& \Gamma' \leq \Gamma \Rightarrow \Gamma' \in \mathfrak{A}^+$;
 (iii) $\Gamma, \Gamma' \in \mathfrak{A}^+ \& \Gamma \mid \Gamma' \Rightarrow \Gamma + \Gamma' \in \mathfrak{A}^+$.

(2)      $\mathfrak{A}^-$ is a causal filter in $\Omega$;
 (i) $\Omega \in \mathfrak{A}^-$;
 (ii) $\Gamma \in \mathfrak{A}^- \& \Gamma \leq \Gamma' \Rightarrow \Gamma' \in \mathfrak{A}^-$;
 (iii) $\Gamma, \Gamma' \in \mathfrak{A}^- \& A \mapsto B \text{ or } B \mapsto A$ for some $A, B \in \Gamma \cdot \Gamma'$
 $\Rightarrow \Gamma \cdot \Gamma' \in \mathfrak{A}^-$, where $A \mapsto B =_D A \mapsto B \& \neg (B \mapsto A)$.

(3)      $\Gamma \in \mathfrak{A}^+ \& \Gamma \Vdash \Gamma' \Rightarrow \Gamma' \in \mathfrak{A}^+$;

(4)      $\Gamma \in \mathfrak{A}^- \& \Gamma' \Vdash \Gamma \Rightarrow \Gamma' \in \mathfrak{A}^-$;

(5)      $\Gamma \in \mathfrak{A}^+ \& \Delta \in \mathfrak{A} \Rightarrow \Gamma \cdot \Delta \in \mathfrak{A}^+$;

(6)      $\Gamma \in \mathfrak{A}^- \& \Delta \in \mathfrak{A} \Rightarrow \Gamma + \Delta \in \mathfrak{A}^-$;

(7)      $\Gamma \in \mathfrak{A}^0 \& \Gamma \leq \Gamma' \Rightarrow \Gamma' \in \mathfrak{A}^0 \cup \mathfrak{A}^-$;

(8)      $\Gamma \in \mathfrak{A}^0 \& \Gamma' \leq \Gamma \Rightarrow \Gamma' \in \mathfrak{A}^0 \cup \mathfrak{A}^+$;

(9)      $\Gamma, \Gamma' \in \mathfrak{A}^0 \Rightarrow \Gamma + \Gamma' \in \mathfrak{A}^0 \cup \mathfrak{A}^- \& \Gamma \cdot \Gamma' \in \mathfrak{A}^0 \cup \mathfrak{A}^+$;

(10)     Let us put $\Gamma^\perp = A + (\text{Dir}(\Gamma))^\perp$, where $A \in \Gamma$. Then
 (i) $\Gamma \in \mathfrak{A}_1^+ \Rightarrow \Gamma^\perp \in \mathfrak{A}_{n-1}^-$;
 (ii) $\Gamma \in \mathfrak{A}_1^- \Rightarrow \Gamma^\perp \in \mathfrak{A}_{n-1}^+$;
 (iii) $\Gamma \in \mathfrak{A}_1^0 \Rightarrow \Gamma^\perp \in \mathfrak{A}_{n-1}^0$.

The index $i$ in $\mathfrak{A}_i$ denotes the dimension of the spaces considered in $\mathfrak{A}(1 \leq i \leq n)$.

This theorem (whose proof is easy but tedious) is not only a generalization of some observations in Robb (1914), but also a lattice-theoretic systematization of the subspaces of $n$-dimensional space-times.

The ideal $\mathfrak{A}^+$ of spacelike subspaces of $\Omega$ contains all those objects which behave in Euclidean fashion. Any subspace belonging to $\mathfrak{A}^+$ lies outside any cone whose vertex is incident with the subspace.

The filter $\mathfrak{A}^-$ of timelike subspaces of $\Omega$ contains all those geometric objects which intersect any cone whose vertex is in the object in question.

The set $\mathfrak{A}^0$ of lightlike subspaces of $\Omega$ contains all the possible tangent subspaces to cones whose vertex is on the subspace under consideration.

The next theorem deals with orthogonality.

### Theorem 9

Let $\mathfrak{A}$ be the causal lattice of a Robb space $\Omega$. Then the following conditions are true when all variables run over $\mathfrak{A}$:

(1) $\emptyset \perp \Gamma, \neg \Omega \perp \Gamma$;

(2) $\Gamma \perp \Gamma \Rightarrow \Gamma \in \mathfrak{A}_1^0$, if $\dim \Gamma > 0$;

(3) $\Gamma \perp \Delta \Rightarrow \Delta \perp \Gamma$;

(4) $\Gamma \perp \Gamma'$ iff $\Gamma \parallel \Gamma'$, when $\Gamma, \Gamma' \in \mathfrak{A}_1^0$;

(5) $\Gamma \perp \Delta \ \& \ \Gamma' \perp \Delta \Rightarrow \Gamma + \Gamma' \perp \Delta$, if $\Gamma \cdot \Gamma' \neq \emptyset$;

(6) $\Gamma \perp \Gamma' \ \& \ \Gamma' \perp \Delta \ \& \ \Gamma + \Gamma' \perp \Delta \Rightarrow \Gamma \perp \Gamma' + \Delta$, if $\Gamma \cdot \Gamma' \neq \emptyset$ and $\Gamma' \cdot \Delta \neq \emptyset$;

(7) Let $\Gamma \in \mathfrak{A}_p^+, \Delta \in \mathfrak{A}_q^+, \Gamma \cdot \Delta = \emptyset$, and $\Gamma \perp \Delta$ with $1 \leq p, q \leq n-1$. Then $\Gamma + \Delta$ is in, respectively, $\mathfrak{A}_r^+, \mathfrak{A}_r^-$, and $\mathfrak{A}_r^0$ $(r = p + q + 1)$, depending on whether $A - B \in \mathscr{V}^+$, $A - B \in \mathscr{V}^-$, or $A - B \in \mathscr{V}^0$ for some $A \in \Gamma \ \& \ B \in \Delta$.

The proof is a matter of routine. Let us prove, as an example, clause (5).

*Assumptions*: $\Gamma = A + \mathscr{W}, \Gamma' = A + \mathscr{W}', \Delta = B + \mathscr{U}, \Gamma + \Gamma' = A + \mathscr{W} + \mathscr{W}'$, where $A \in \Gamma \cdot \Gamma'$ and $B \in \Delta$. Furthermore, $\mathscr{W} \perp \mathscr{U}$, and $\mathscr{W}' \perp \mathscr{U}$.

*Conclusion*: $(\mathscr{W} + \mathscr{W}') \perp \mathscr{U}$.

The parallelism has the usual properties, that is, $\parallel$ is an equivalence relation, and $\Vdash$ is a partial ordering with the usual additional properties, such as $(\Delta \Vdash \Gamma \ \& \ \Gamma' \leq \Gamma) \Rightarrow \Delta \Vdash \Gamma'$, where $\dim \Gamma' > 0$, or $(A \in \Delta \ \& \ \Delta \Vdash \Gamma) \Rightarrow (A + \Gamma) \cdot \Delta \parallel \Gamma$. The reader should notice the analogy with logic and the purpose of symbolization.

The notion of incomparability is new in this context and the following theorem gives its basic properties.

THEOREM 10

*In a causal lattice $\mathfrak{A}$ we have:*

(1) $\quad \emptyset \,|\, \Delta \quad$ and $\quad \neg\, \Omega \,|\, \Gamma \;(\Gamma \neq \emptyset)$;
(2) $\quad \Gamma \,|\, \Gamma \quad$ iff $\quad \Gamma \in \mathfrak{A}^+$;
(3) $\quad \Gamma \,|\, \Delta \Rightarrow \Delta \,|\, \Gamma$;
(4) $\quad \Gamma \,|\, \Delta \,\&\, \Gamma' \leqslant \Gamma \Rightarrow \Gamma' \,|\, \Delta$;
(5) $\quad \Gamma \,|\, \Delta \,\&\, \Gamma' \,|\, \Delta \Rightarrow \Gamma + \Gamma' \,|\, \Delta, \quad$ if $\quad \Delta \cdot \Gamma \neq \emptyset \,\&\, \Delta, \Gamma \in \mathfrak{A}^+$;
(6) $\quad \Gamma' \leqslant \Gamma \Leftrightarrow \forall \Delta\, [\Delta \,|\, \Gamma \Rightarrow \Delta \,|\, \Gamma']$;
(7) $\quad \Gamma \in \mathfrak{A}_k^+ \,\&\, A \,|\, \Gamma \Rightarrow A + \Gamma \in \mathfrak{A}_{k+1}^+ \quad (1 \leqslant k < n-1)$;
(8) $\quad \Gamma \in \mathfrak{A}_k^0 \,\&\, A \,|\, \Gamma \Rightarrow A + \Gamma \in \mathfrak{A}_{k+1}^0 \quad (1 \leqslant k < n-1)$;
(9) $\quad \Gamma, \Gamma' \in \mathfrak{A}_1^0 \,\&\, \Gamma, \Gamma' \leqslant \Delta \,\&\, \Delta \in \mathfrak{A}_k^0 \Rightarrow \Gamma \,|\, \Gamma', \quad$ if $\quad \Gamma \neq \Gamma'$ and $\quad 1 < k < n$;
(10) $\quad \Gamma \in \mathfrak{A}_k^0 \Rightarrow A \,|\, \Gamma \quad$ for some $A \quad (1 \leqslant k < n-1)$;
(11) $\quad \Gamma \in \mathfrak{A}_k^+ \Rightarrow A \,|\, \Gamma \quad$ for some $A \quad (1 \leqslant k < n-1)$;
(12) $\quad$ Let $\Gamma, \Gamma' \in \mathfrak{A}_0^1$. Then
$\quad\quad$ (i) $\neg\, \Gamma \,|\, \Gamma$;
$\quad\quad$ (ii) $\Gamma \,|\, \Gamma' \,\&\, \Gamma' \,|\, \Gamma'' \Rightarrow \Gamma \,|\, \Gamma''$, if $\Gamma \neq \Gamma''$;
(13) $\quad A, B \mapsto C, D \Rightarrow (A \,|\, B \quad$ iff $\quad C \,|\, D), \cdot \,$ if $\; C \neq D \,\&\, A \neq B$.

As one can see, the incompatibility relation shares some of the properties of the orthogonality relation. That is why one uses it in axiomatizing the metric aspects of Minkowski spaces. Specifically, the desired dimension of the Minkowski space is captured partly by condition (8). Of course, if one does not assume the lattice structure *a priori*, then condition (8) becomes a second-order axiom, containing sequences of incomparable points. Robb used clause (8) with $k - 2$ as one of his axioms, after he defined the notion of a lightlike plane.

THEOREM 11

*In a causal lattice $\mathfrak{A}$ the following clauses are valid:*

(1) $\quad \emptyset \mapsto \Gamma \quad$ and $\quad \Delta \mapsto \Omega$;
(2) $\quad \emptyset \leftarrowtail \Gamma \quad$ and $\quad \Delta \leftarrowtail \Omega$;
(3) $\quad \Gamma \mapsto \Gamma \Rightarrow \Gamma \notin \mathfrak{A}^+$;
(4) $\quad \Gamma \mapsto \Delta \,\&\, \Delta \mapsto \Lambda \Rightarrow \Gamma \mapsto \Lambda$;

(5) $\Gamma \mapsto \varDelta \ \& \ \varDelta \mapsto \Gamma \Rightarrow \Gamma \,||\, \varDelta$;
(6) $A \mapsto B \ \text{ or } \ B \mapsto A \Rightarrow A + B \notin \mathfrak{A}^+$;
(7) $A \stackrel{.}{\mapsto} B \ \text{ or } \ B \stackrel{.}{\mapsto} A \Rightarrow A + B \in \mathfrak{A}^0$;
(8) $\Gamma \in \mathfrak{A}_k^0 \ \& \ (A \mapsto \Gamma \ \text{ or } \ A \leftarrowtail \Gamma) \Rightarrow A + \Gamma \in \mathfrak{A}_{k+1}^-$
  $(1 < k \leqslant n-1)$;
(9) $\Gamma \in \mathfrak{A}_{n-1}^0 \Rightarrow A \mapsto \Gamma \ \text{ or } \ A \leftarrowtail \Gamma$;
(10) Let $\Gamma, \Gamma' \leqslant \varDelta \ \& \ \Gamma \,||\, \Gamma'$, where $\Gamma, \Gamma' \in \mathfrak{A}^+, \varDelta \in \mathfrak{A}^-$, and $A \in \Gamma$. Then
  (i) $A \mapsto \Gamma' \Rightarrow \Gamma \mapsto \Gamma'$;
  (ii) $A \leftarrowtail \Gamma' \Rightarrow \Gamma \leftarrowtail \Gamma'$;
(11) Let $\Gamma, \Gamma' \, \Gamma'' \in \mathfrak{A}_1^0$. Then
  (i) $\Gamma \mapsto \Gamma' \Rightarrow \Gamma = \Gamma' \ \text{ or } \ \Gamma \cdot \Gamma = \emptyset$;
  (ii) $\Gamma \mapsto \Gamma'$ iff $\Gamma' \leftarrowtail \Gamma$;
  (iii) $\Gamma \mapsto \Gamma$;
  (iv) $A \mapsto \Gamma \ \& \ A \leftarrowtail \Gamma \Rightarrow A \in \Gamma$;
  (v) $\Gamma \,|\, \Gamma'' \ \& \ \Gamma' \mapsto \Gamma'' \Rightarrow \Gamma' \mapsto \Gamma$;
  (vi) $\Gamma \,|\, \Gamma'' \ \& \ \Gamma'' \mapsto \Gamma' \Rightarrow \Gamma \mapsto \Gamma'$;
  (vii) $\Gamma \,||\, \Gamma'$ iff $(\Gamma \mapsto \Gamma' \ \text{ or } \ \Gamma' \mapsto \Gamma \ \text{ or } \ \Gamma \,|\, \Gamma')$.

The above theorems will give one a fair feeling for the menagerie of interesting properties of the elements in a causal lattice.

Next we shall discuss rather briefly the problem of representation. As pointed out, there are essentially three kinds of representation theorems which play a fundamental role in the theory of geometric causality. First, a relatively straightforward one stating that the lattice of all subspaces of every finite-dimensional real space-time is a causal lattice. It should be pointed out that just like in classical geometry, where we have affine and projective lattices (the former being the set of all cosets of the direction of the space in question and the latter being just the class of all subspaces of the direction) whose interconnection is reflected in Vilcox lattices, similarly, in Minkowski geometry we can consider the affine and projective version of causal lattices. Due to the metric structure of the space, the lattices are orthocomplemented. The lattices of Minkowski subspaces are, in addition, partially ordered. Any deeper and more specific explanation would force us into further technicalities on lattices.

The second and considerably more difficult representation theorem is displayed below:

## THEOREM 12

Let $\langle \mathfrak{A}, \perp, \mapsto \rangle$ be an abstract causal lattice of a finite dimension, greater than two. Then there exists a finite-dimensional vector space $\mathscr{V}$ over the field $\mathbb{R}^*$ (isomorphic with the field of reals) and a quadratic function $\| \ \|^2 : \mathscr{V} \to \mathbb{R}$ such that

(i) $\langle \mathscr{V}, \| \ \|^2 \rangle$ is a Minkowski space whose dimension is equal to the dimension of $\mathfrak{A}$;

(ii) The causal lattice $\mathfrak{A}(\mathscr{V})$ is isomorphic to the lattice $\mathfrak{A}$. Consequently, for any two elements $\Gamma, \Delta \in \mathfrak{A}$, the following conditions are satisfied:

$$\Gamma \perp \Delta \quad \text{iff} \quad \|\text{Dir } \varphi(\Gamma) + \text{Dir } \varphi(\Delta)\|^2 = \text{Dir } \varphi(\Gamma)\|^2 + \|\text{Dir } \varphi(\Delta)\|^2;$$

$$\Gamma \mapsto \Delta \quad \text{iff} \quad \|\varphi(\Delta) - \varphi(\Gamma)\|^2 \leq 0 \ \& \ \|\varphi(\Delta) - \varphi(\Gamma) + \overset{\circ}{\mathscr{C}}{}^+\|^2$$
$$\leq \|\varphi(\Delta) - \varphi(\Gamma)\|^2 + \|\overset{\circ}{\mathscr{C}}{}^+\|^2,$$

where $\varphi : \mathfrak{A} \to \mathfrak{A}(\mathscr{V})$ denotes the isomorphism, and $\overset{\circ}{\mathscr{C}}{}^+$ stands for the interior of $\mathscr{C}^+$ ($\overset{\circ}{\mathscr{C}}{}^+ = \mathscr{C}^+ - \overset{\circ}{\mathscr{C}}{}^+$).

The proof of this theorem is based on Baer's representation theorem saying that for any irreducible modular matroid lattice $\mathfrak{A}$ of dimension greater than three (for affine lattices it is enough to have a dimension greater than two) there exists a linear vector space $\mathscr{V}$ over a field $\mathbb{R}^*$ such that $\mathfrak{A}$ is isomorphic with $\mathfrak{A}(\mathscr{V})$.

The partial ordering $\mapsto$ in $\mathfrak{A}$ induces a partial ordering in $\mathfrak{A}(\mathscr{V})$, and therefore in $\mathscr{V}$. Once we have a partially ordered vector space, we can utilize certain theorems giving the necessary and sufficient conditions for the existence of linear functions $F : \mathscr{V} \to \mathbb{R}$ compatible with the ordering of $\mathscr{V}$ (see, for instance, Domotor, 1969, page 27, Theorem 2). But we need a quadratic function and not a linear one! This is obtained by showing the existence of a parametrized family of linear functions $\{F_v \mid v \in \mathscr{V}\}$. Next we force $F_v$ to behave additively with respect to the parameters. By means of this trick we obtain a bilinear function $F^* : \mathscr{V} \times \mathscr{V} \to \mathbb{R}$ from which the desired quadratic function is obtained in an obvious way.

The full description of the proof requires an explicit reference to causal and other lattices, and a rather specialized knowledge of partially ordered vector spaces. A rather short proof, but with quite strong assumptions, can be found in Gheorghe and Mihul (1969).

The representation serves the purpose of a *semantic test*, verifying the adequacy of the axiomatic way of presentation of causality.

We shall omit the discussion of the theorem which states that Robb's axioms of space-time imply the existence of a causal lattice. Instead, we shall address ourselves to the question of *uniqueness*. Concretely, we wish to know how unique the Minkowski metric $\| \ \|^2$ is in Theorem 12.

It was Zeeman (1964) who showed that if we have given a causal Minkowski space $\langle \mathscr{V}, \mapsto \rangle$ together with the group of all *causal automorphisms*

$$\mathbf{G} = \{f : V \xrightarrow{1-1} \mathscr{V} \mid \forall v, w [v \mapsto w \text{ iff } f(v) \mapsto f(w)]\},$$

then the group **G** is generated by translations, dilations, and orthochronous Lorentz transforms (preserving the future cone) of $\langle \mathscr{V}, \| \ \|^2 \rangle$. In other words, the Lorentz group can be deduced from purely order-theoretic (qualitative) assumptions, expressed in the language of causal relation, without any recourse to the Minkowski metric. Thus, the fundamental relativistic invariants actually follow from the simple principles of geometric causality. This observation leads to important applications in the research concerning the alleged particles traveling faster than light and known under the name *tachyons*. These particles seem to violate the logic of causality as described by the causal lattice $\mathfrak{A}(\mathscr{V})$. A deeper insight into the mechanism of causal lattices may suggest how to explain the causal coexistence of tachyons with classical particles.

Besides the causal group **G**, acting on $\langle \mathscr{V}, \mapsto \rangle$ and the Lorentz group $\mathbf{O}^+(\mathscr{V})$, acting on $\langle \mathscr{V}, \| \ \|^2 \rangle$, there is a third group **L**, acting on the causal lattice $\langle \mathfrak{A}, \mapsto \rangle$:

$$\mathbf{L} = \{g : \mathfrak{A} \xrightarrow{1-1} \mathfrak{A} \mid \forall_{\Gamma, \Delta} (\Gamma \mapsto \Delta \text{ iff } g(\Gamma) \mapsto g(\Delta))\}.$$

If we denote the representing mappings by $\varphi : \mathfrak{A} \to \mathfrak{A}(\mathscr{V})$ and $\psi : \langle \mathscr{V}, \mapsto \rangle \to \langle \mathscr{V}, \| \ \|^2 \rangle$, then the representation and uniqueness problem appears as:

$$\begin{array}{ccc} \langle \mathfrak{A}, \mapsto \rangle & \xrightarrow{g} & \langle \mathfrak{A}, \mapsto \rangle \\ \varphi \downarrow & & \varphi \downarrow \\ \langle \mathscr{V}, \mapsto \rangle & \xrightarrow{f} & \langle \mathscr{V}, \mapsto \rangle \\ \psi \downarrow & & \psi \downarrow \\ \langle \mathscr{V}, \| \ \|^2 \rangle & \xrightarrow{h} & \langle \mathscr{V}, \| \ \|^2 \rangle \end{array}$$

where $h \in \mathbf{O}^+, f \in \mathbf{G}, g \in \mathbf{L}$.

At present the causal lattice group **L** and its relationship to the other two groups is probably unknown.

Needless to say, any contribution in this direction would help us to organize the geometry of relativistic physics. The ability to identify the geometric attributes of objects and to distinguish them from others is accomplished precisely via the group which leaves them invariant. Actually, two groups are always involved. The one under which the geometric theorems (formulas) remain valid and another, a *normal subgroup* of the first one, under which the components (terms) are preserved.

## V. CONCLUDING REMARKS AND OPEN PROBLEMS

It is now a proper time and place to stand back and see what is common, if anything, to the causal models discussed in this paper.

In the first place, as a methodological policy we insisted that each model be tested semantically by exhibiting a representation whose purpose was to discard the 'bad' causal models and to verify the degree of adequacy of the 'good' ones.

Next, we have observed some common structural features reappearing in several different models. In particular, question $Q_1$ appears to be answerable in a very broad context in the language of lattices. For instance, in the case of logical models we know that the collection of formulas forms essentially a Boolean algebra. The probabilistic models use Boolean algebras directly. The signal space in the context of analytic models is a vector space of some sort and it is known, for example, that the set of closed subspaces of a Hilbert space forms an irreducible orthomodular lattice. Hence, the causal operator, acting on the signal space, can be translated appropriately into the world of lattices. Last, the geometric models of causality lead, again, to certain lattice structures.

In a nutshell, the *logic of causality* is always one of the partially ordered lattices $\langle \mathfrak{A}, \mapsto \rangle$, where the specific properties are delineated by the context in which the model is intended to be used.

In many directions we have been unable to accomplish more than a sketchy review. Many problems remained open. Some additional ones are displayed below:

(1) Find the so-called minimal model of causal logic and show the completeness theorem for it.

(2) Given an abstract alphabet which is partially ordered by a binary relation, give necessary and/or sufficient conditions for the existence of a deterministic automaton such that any two expressions will stand in (causal) relation precisely when for some state of the automaton (or a sequence of states), if the first expression is on the input of the automaton, the second one will be on the output of the automaton.

(3) Given the set of future light cones in a Minkowski space, define the meet of two cones as the greatest cone contained in both, and the join of two cones as the smallest cone containing both. The set of cones by these operations is converted into a lattice. By adding the causal relation, give a proper axiomatization of this lattice that leads to a representation by Minkowski spaces.

The causal lattice as presented in this paper is very general. It captures only the geometry of relativistic physics. Additional physical semantics is needed, using masses and forces, on the basis of which causality would become a relation directly dependent on a physical theory.

The author hopes to contribute to some of these problems in forthcoming papers.

### ACKNOWLEDGEMENTS

The author is deeply indepted to Patrick Suppes for his stimulation, advice, and great interest in this work. He also wishes to thank Duncan Luce and Dana Scott for valuable discussions concerning the methods of axiomatizing geometry.

*University of Pennsylvania*

### NOTE

\* The research presented here was supported by the NSF Grant GS-2936.

### BIBLIOGRAPHY

Alker, Jr., H. R., 'Causal Inference and Political Analysis', in J. L. Bernd (ed.), *Conference on Mathematical Applications in Political Science*, Vol. **2**, Arnold Foundation, Southern Methodist University, Dallas, Texas, 1966, pp. 7–43.

Alonso, J. L. and Yndurazin, F. J., 'On the Continuity of Causal Automorphisms of Space-Time', *Communications in Mathematical Physics* **4** (1967), 349–351.

Antoine, J.-P. and Gleit, A., 'Space-Time Structure and Measurement Theory', *International Journal of Theoretical Physics* **4** (1971), 197–216.

Baer, R., *Linear Algebra and Projective Geometry*, Academic Press, New York, 1952.
Barucchi, G. and Teppati, G., 'The Causality Group', *Nuovo Cimento* **52A** (1967), 50–61.
Basri, A. S., *A Deductive Theory of Space and Time*, North Holland, Amsterdam, 1966.
Blalock, Jr., H. M., *Causal Inferences in Nonexperimental Research*, University of North Carolina Press, Chapel Hill, 1964.
Blalock, Jr., H. M., *Theory Construction*, Prentice-Hall, Englewood Cliffs, New Jersey, 1969.
Bunge, M., 'Conjunction, Succession, Determination, and Causation', *International Journal of Theoretical Physics* **1** (1968), 299–315.
Burks, A. W., 'The Logic of Causal Propositions', *Mind* **60** (1951), 363–382.
Chellas, B. F., 'Basic Conditional Logic', Dittoed manuscript, University of Pennsylvania, Philadelphia, 1970.
Cole, M., 'On Causal Dynamics Without Metrization', *International Journal of Theoretical Physics* **1** (1968), 115–151.
Csonka, P. L., 'Causality and Faster than Light Particles', *Nuclear Physics* **B21** (1970), 436–444. (a)
Csonka, P. L., 'Implications of Full Causality for Neutrino and Other Particle Production Rates' *The Physical Review* **D2** (1970), 1923–1925. (b)
Czerwinski, Z., 'On the Notion of Causality and Mill's Canons', *Studia Logica* **9** (1960), 37–62.
Davidson, D., 'Actions, Reasons, and Causes', *The Journal of Philosophy* **60** (1963), 685–700.
Davidson, D., 'Causal Relations', *The Journal of Philosophy* **64** (1967), 691–703.
Domotor, Z., 'Probabilistic Relational Structures and Their Applications', Technical Report No. 144, May 14, 1969, Stanford University, Institute for Mathematical Studies in the Social Sciences.
Fagot, A., 'Causal Versus Teleological Explanation of Behavior', Unpublished Ph. D. dissertation, Stanford University, Stanford, 1971.
Falb, P. L. and Freedman, M. I., 'A Generalized Transform Theory for Causal Operators', *SIAM Journal on Control* **7** (1969), 452–471.
Finkelstein, D., 'Space-Time Code', *The Physical Review* **184** (1969), 1261–1270.
Foures, Y. and Segal, I. E., 'Causality and Analyticity', *Transactions of the American Mathematical Society* **78** (1955), 385–405.
Fox, R., Kuper, C. G., and Lipson, S. G., 'Faster-Than-Light Group Velocities and Causality Violation', *Proceedings of the Royal Society, London* **A316** (1970), 515–524.
Gamba, A. and Luzzatto, G., 'Causality and Conformal Invariance', *Nuovo Cimento* **33** (1964), 1732–1733.
Gheorghe, C. and Mihul, E., 'Causal Groups of Space-Time', *Communications in Mathematical Physics* **14** (1969), 165–170.
Good, I. J., 'A Causal Calculus, I', *The British Journal for the Philosophy of Science* **11** (1961), 305–318.
Good, I. J., 'A Causal Calculus, II', *The British Journal for the Philosophy of Science* **12** (1962), 43–51.
Goodman, R., 'A Group-Theoretic Approach to Causal Systems', in J. Fox (ed), *Proceedings of the Symposium on System Theory*, Vol. **15**, Polytechnic Press, Brooklyn, New York, 1965, pp. 169–174.
Hansson, B., 'Choice Structures and Preference Relations', *Synthese* **18** (1968), 443–458.
Hawking, S. W., 'The Occurrence of Singularities in Cosmology, III. Causality and

Singularities', *Proceedings of the Royal Society, London* **A300** (1967) 187–201.
Ingraham, R. L., 'Stochastic Space-Time', *Nuovo Cimento* **34** (1964), 182–197.
Israel, W., 'Causal Anomalies in the Extended Schwarzschild Manifold', *Physics Letters* **21** (1966), 47–49.
Kalman, R. E., 'Algebraic Theory of Linear Systems', in M. E. Van Valkenburg (ed.), *Proceedings Third Annual Allerton Conference on Circuit and System Theory*, Dept. of Electrical Engineering, Urbana, Ill., 1965, pp. 563–577.
Kaplan, A., 'Non-Causal Explanation', in D. Lerner (ed.), *Cause and Effect*, Free Press, New York, 1965, pp. 145–155.
Kotelnikov, A. P., 'Principles of Relativity and Lobatchevskian Geometry', *In Memoriam Lobatchevski*, Vol. **2**, University Press, Kazan, Soviet Union, 1927, pp. 37–66.
Kronheimer, E. H. and Penrose, R., 'On the Structure of Causal Space', *Proceedings of the Cambridge Philosophical Society* **63** (1967), 481–501.
Maeda, F., 'Lattice-Theoretic Characterization of Abstract Geometries', *Journal of Science of Hiroshima University, Ser A* **15** (1951), 87–96.
Mehlberg, H., 'The Problem of Causality in an Indeterministic Science, *International Journal of Theoretical Physics* **2** (1969), 351–372.
Minerbo, G., 'Causality and Analyticity in Formal Scattering Theory', *The Physical Review* **D3** (1971), 928–932.
Nagel, E., 'Types of Causal Explanation in Science', in D. Lerner (ed.), *Cause and Effect*, Free Press, New York, 1965, pp. 11–32.
Nevanlinna, R., *Space-Time and Relativity*, Addison-Wesley, London, 1968.
Noll, W., 'Euclidean Geometry and Minkowskian Chronometry', *The American Mathematical Monthly* **71** (1964), 129–144.
Pearson, II, A. T., 'Causal Statements and Correlation', Ph. D. Thesis, Cornell University, 1971.
Peres, A., 'Bogolubov Causality in S-Matrix Theory', *Nuclear Physics* **323** (1970), 125–154.
Porter, W. A. and Zahn, C. L., 'Basic Concepts in System Theory', Technical Report No. 44, 1969, The University of Michigan, Department of Electrical Engineering.
Robb, A. A., *The Absolute Relations of Space and Time*, University Press, Cambridge, 1921.
Robb, A. A., *A Theory of Time and Space*, University Press, Cambridge, 1914, Second ed., 1936.
Saeks, R., 'Causality in Hilbert Space', Memorandum EE-6814C, 1969, University of Notre Dame, Electrical Engineering Department.
Sandberg, I. W., 'Conditions for the Causality of Nonlinear Operators Defined on a Function Space', *Quarterly of Applied Mathematics* **23** (1965), 87–91.
Segal, I. E., 'Direct Formulation of Causality Requirements on the S-operator', *The Physical Review* **109** (1958), 2191–2198.
Simon, H. A., 'Causal Ordering and Identifiability, in W. C. Hood and T. C. Koopmans (eds.), *Studies in Econometric Method*, Wiley, New York, 1953, pp. 49–74.
Simon, H. A., 'Spurious Correlation: A Causal Interpretation', *Journal of American Statistical Association* **49** (1954), 467–492.
Simon, H. A. and Rescher, N., 'Cause and Counterfactual', *Philosophy of Science* **33** (1966), 323–340.
Slavnov, D. A., 'Causality and Locality Conditions', *Soviet Journal of Nuclear Physics* **9** (1969), 511–514.

Stalnaker, R. C. and Thomason, R. H., 'A Semantic Analysis of Conditional Logic', *Theoria* **36** (1970), 23–42.

Strauch, R. D., 'The Fallacy of Causal Analysis', Technical Report No. P-4618, April 1971, The Rand Corporation, Santa Monica, California.

Strotz, R. H. and Wold, H., 'Recursive vs. Nonrecursive Systems: An Attempt of Synthesis', *Econometrica* **28** (1960), 417–427.

Suppes, P., 'Axioms for Relativistic Kinematics With or Without Parity', in L. Henkin, P. Suppes and A. Tarski (eds.), *The Axiomatic Method, with Special Reference to Geometry and Physics*, North Holland, Amsterdam, 1959, pp. 291–307.

Suppes, P., *A Probabilistic Theory of Causality* (*Acta Phil. Fennica* **24**), North-Holland Publishing Company, Amsterdam, 1970.

Teppati, G., 'An Algebraic Analogue of Zeeman's Theorem', *Nuovo Cimento* **54A** (1968), 800–804.

Watanabe, S., *Knowing and Guessing*, Wiley, New York, 1970.

Windeknecht, T. G., 'Mathematical Systems Theory – Causality', *Mathematical Systems Theory* **1** (1967), 279–288.

Witsenhausen, H. S., 'On Information Structures, Feedback and Causality', *SIAM Journal on Control* **9** (1971), 149–160.

Wold, H., 'Causal Inference from Observational Data. A Review of Ends and Means', *Journal of the Royal Statistical Society* **A119** (1956), 28–61.

Wold, H., 'A Case Study of Interdependent Versus Causal Chain Systems', *Review of the International Statistical Institute* **26** (1958), 5–25.

Wold, H., 'A Generalization of Causal Chain Systems', *Econometrica* **28** (1960), 443–463.

Zeeman, E. C., 'Causality Implies the Lorentz Group', *Journal of Mathematical Physics* **5** (1964), 490–493.

GEORGE BERGER

# TEMPORALLY SYMMETRIC CAUSAL RELATIONS IN MINKOWSKI SPACE-TIME

### I. INTRODUCTION

This paper is a contribution to the causal theory of special-relativistic (Minkowski) space-time.[1] My aim is to give an axiomatic characterization of temporally symmetric causal relations in Minkowski space-time, to show precisely the relation of causal to metrical structure, and to explicate the thesis that space-time 'has' a causal and topological structure which 'underlies' its metrical structure.[2] The paper is largely mathematical in content, but many comments concerning the system appear as notes. Acquaintance with basic topological notions and with the Minkowski formalism of special relativity is presupposed.[3] I shall first state the main result to be established and will then provide the philosophical motivation for the construction.

Let the set $S$ be the four-dimensional differentiable manifold (4-manifold) of *space-time points*,[4] and let $K$ be a binary symmetric and reflexive relation of *causal connectibility* defined on $S$ by axioms to be given in Section II. That is, we call points $x, y \in S$ $K$-related ($K(x, y)$) if and only if an event occurring at $x$ *can* either influence or be influenced by an event occurring at $y$. $K$ is explained as being a *modal* and *temporally symmetric* causal relation: for the claim that $K(x, y)$ conveys no information about the point ($x$ or $y$) at which the later (or earlier) event must occur for $x$ and $y$ to be *actually* causally connected.[5]

Let $G$ be the group (under functional composition) of all maps $g$ of $S$ onto itself which leave invariant the dimensionality of subsets of $S$ (provided that the subset in question has a well-defined dimension number in the sense of dimension theory (Hurewicz and Wallman, 1941; Menger, 1943)), and which are such that for all $x, y \in S$, $g \in G$, $K(x, y)$ if and only if $K(g(x), g(y))$. I will define, in Section II, two properties[6] of nonempty subsets of $S$, which are to be systematic counterparts of two properties of world lines familiar from relativity theory. I want one to characterize a subset of $S$ if and only if that subset (presystematically) comprises all and only the points of a possible path of causal propagation in $S$, i.e., if it may be a *trajectory* in $S$ for a material particle or a light ray. I want the other, a subproperty of the first, to characterize a subset of $S$ if and only if that subset is a path along which a light ray may pass (a *null trajectory*).

The proof (in Sections III and IV) that the defined properties *do* characterize the subsets just mentioned will also be a proof that the system is materially adequate to describe the spatiotemporal and causal structure of possible processes in Minkowski space-time. This proof has two parts, the first of which has two subparts:

(A) I will show that

(1) all elements of $G$ map trajectories into trajectories, and in particular map null trajectories and only null trajectories into null trajectories,

(2) $G$ specifies a privileged family $\mathscr{F}$ of homeomorphisms of $S$ onto $R_4$, the set of all ordered quadruples of real numbers endowed with a Euclidean metric, such that

(a)  each map in $\mathscr{F}$ is the systematic counterpart of an 'inertial reference frame';

(b)  any two members of $\mathscr{F}$ are related, for all members of $S$, by an affine transformation modulo a transformation-dependent scale factor. These transformations form a group, each element of which is the systematic counterpart of an element of the inhomogeneous nonorthochronous Lorentz group of coordinate transformations (the *Poincaré group* of Anderson, 1967, pp. 141–151; see also Roman, 1960, Ch. 1) modulo the scale factor.[7]

(B) I then argue that the scale factors are physically irrelevant to the provision of a description of the causal structure of space-time.

In the language of special relativity, A2 (a)–(b) imply that any two members of $\mathscr{F}$ are related by a transformation that leaves the *relativistic interval* (to be defined below) between any two points of $S$ invariant modulo scale changes. As Mehlberg (1966) and Suppes (1968) have argued, *strict* interval invariance under the Poincaré group should be guaranteed by any account of Minkowskian space-time structure. Thus if the argument for the irrelevancy of the scale factor is correct, then my three primitive notions, $S$, $K$, and $G$, can serve to describe the causal structure of possible processes in Minkowski space-time. The *metrical* structure of space-time will be described by the relativistic interval.

*Terminological note:* The claim that three primitive notions, $S$, $K$, and

$G$, provide a *description* of the causal structure of space-time shall mean the following. I will specify a mathematical structure (the *topological-causal structure*) incorporating $S$, $K$, and $G$, such that an important class of statements about causality in space-time can be interpreted into the language of this structure. This structure will need only topological and causal concepts for its description.

The philosophical import of this result lies in its relation to recent attempts by Grünbaum (1963, Ch. 7; 1967, pp. 56–64), Mehlberg (1966) and van Fraassen (1970; Ch. 5) to formulate versions of the causal theory of time (more generally of space-time) using temporally symmetric causal relations. Expanding upon Grünbaum's work, van Fraassen shows that a ternary relation of *temporal betweenness* (or a quaternary relation of *temporal separation*) can be defined on the set of *events* in the history of any physical system on the basis of axioms concerning causal connectibility. A relation of *genidentity*, which applies to an unordered pair of events just in case the two events are 'in' the world line of a single *thing* (Lewin, 1923), is also used as a primitive notion. More or less explicit appeal to a thing-event ontology is thereby invoked.

It seems that recourse to a (counterintuitive) symmetric causal relation was suggested by Grünbaum's successful critique (1963, pp. 181–187) of Reichenbach's attempt (1958, pp. 135–149) to define an asymmetric relation *later than* on the set of events participating in certain causal processes. Another motivation (especially for Grünbaum, I think) was the belief that some or all spatiotemporal and/or metrical relations exhibited by space-time (or by the set of events) 'result from', or are 'definable in terms of', causal (and perhaps also topological) properties and relations. A precise rendition of a thesis of this kind will be given in Section III. A third motivation (for Mehlberg at least) was a desire to support and to make precise the thesis (Mehlberg, 1961, 1966) that Minkowski space-time exhibits no 'intrinsic' temporal asymmetries, so that (it is argued) temporal asymmetries must result from the physical laws and/or boundary conditions governing processes undergone by certain kinds of physical systems (Grünbaum, 1963, Ch. 8; Mehlberg, 1961). It is felt that an adequate formulation of this thesis should employ temporally symmetric causal relations.[8]

My version of the causal theory follows Mehlberg's (1966, pp. 449–453) in spirit. If I interpret him correctly, he rejects use of a relation of genidentity on the grounds that the concepts *temporally persisting material thing* and *rigid body* are generally inapplicable in the subatomic realm. I follow him in this rejection. Like him, I show that axioms concerning $K$ (called by Mehlberg "collision connectibility") can help to characterize special relativistic space-time.[9] I differ from Mehlberg by taking the 4-manifold $S$ as primitive instead of the set of events. This choice accords with standard physical practice (Anderson, 1967, pp. 407; Trautman, 1965, pp. 101-103), although in the informal explanation of my primitive notions I spoke of events occurring *at* members of $S$. Such locutions play no *formal* role in the system (nor does the relation *later than* used in that explanation), since I intend only to elucidate the spatiotemporal-causal structure of possible processes *in* $S$. I therefore impose a relation $K$ directly on $S$, and do not speak of the events involved.

I hope that the system to be presented will provide a perspicuous exhibition of the interrelations between the (temporally symmetric) causal, topological, and metrical structure of space-time. Such an exhibition might be of value in the discussion of problems in the philosophy of special relativity. But to preclude misunderstanding of my intentions, note that I will use topological, metrical, and causal notions as primitives of my system. Thus I do not claim to provide a *relational* theory of space-time based upon causal notions. That is, I do not show that spatiotemporal (metrical and topological) notions can be *defined* in terms of causal notions, nor that a representation theorem for what I will call a topological-causal structure can be given. I count members of $S$ and sets of such as individuals, while one version of the relational conception of space-time requires that there be no spatiotemporal individuals: just things, events, and causal relations among the latter.[10] Also, I have made little attempt to reduce my axiom set to a demonstrably independent set.

I adopted this liberal policy for the following reason. The only other fully adequate system of temporally symmetric causal relations, Mehlberg's, employs analytic-geometrical notions in its axioms. Thus it is not quite correct to say that this system concerns *only* the relation $K$. Moreover, that system considers alternative coordinate systems (for the set of events) and transformations between them. If Mehlberg's system is to be of physical interest, then he must assume the existence of such coordinate systems. This comes close to postulating that the set of events is a 4-manifold. Some such postulate may be unavoidable, although the work of Suppes (1959) and Zeeman (1964, 1967) shows that weaker ones may suffice. We must at least, it seems, assume that $S$ is Hausdorff, and that transformation equations hold between coordinate functions. I thus felt free to make explicit the topological assumptions that I need, and which seem to be implicitly used by Mehlberg. As will be seen below, this strategy led to an elegant axiom which asserts the existence of null trajectories.

## II. DEFINITIONS AND AXIOMS

I shall use the following logical and set-theoretical notation:

$\rightarrow$ (material conditional), $\leftrightarrow$ (material equivalence), $x$, $y$, etc. (individual variables), $A$, $B$, etc. (set variables), $-$ (complementation), 0 (functional composition), $\emptyset$ (null set).

Other symbols are standard. I use 'iff' for 'if and only if'. Definition numbers are prefixed by 'D', theorem numbers by 'T'. The style of axiomatization follows closely the recommendations of Suppes (1957, Ch. 12). By A1 of D4 below, the standard topological concepts of *open set*, *closed set*, *ball*, *boundary* (I let $BDY(A)$ be the boundary of the set $A \subset S$), *interior*, and *dimension n* ($0 \leqslant n \leqslant 4$) are immediately applicable in $S$. If $f$ is a function from a set $X$ into $R_n$, I let, for each $x \in X$, $f(x) = \langle f_1(x), ..., f_n(x) \rangle$. $f_i$ is called the $i$th *component function of f*. Finally, the $i$th $n$-ary *projection function* (for any $0 < i \leqslant n$) is that function $\varphi_i^n$ with domain the set of all ordered $n$-tuples, such that $\varphi_i^n(\langle x_i, ..., x_n \rangle) = x_i$.

After these preliminaries I present the system by means of D1–D4. I intend the symbols used to be interpreted as in Section I.

D1. Let $x \in S$. Then the *closed causal cone* of $x$ in $S$, $C(x)$, is the set of all $z \in S$ such that $K(z, x)$. That is,

$$z \in C(x) \leftrightarrow K(z, x).^{11}$$

D2. A one-dimensional linear subset $A$ of $S$ is a *trajectory* in $S(T(A))$, iff
(a) for all distinct $x, y \in A$, $x \in C(y)$
(b) if $z \in S - A$ then there are distinct $u, v \in A$ such that

$$z \in C(u) \text{ and } z \notin C(v).$$

Trajectories are maximal one-dimensional classes of mutually $K$-related points, since by D1 and D2 (b) no point not in a trajectory is $K$-related to all points in that trajectory, while any point in the trajectory does have that property. Under the intended interpretation, D2 (b) makes trajectories 'infinitely long'. This will permit, in Section IV, parametrization of trajectories by parameters ranging over all the reals. D2 was suggested by van Fraassen (1970, pp. 184–185).

D3. Let $f = \langle f_1, f_2, f_3, f_4 \rangle$ be a homeomorphism of $S$ onto $R_4$, and let $c$ be a positive constant (the 'speed of light'). Then for all $x, y \in S$

$$Q_{f,c}(x, y) = \left[ \sum_{i=1}^{3} (f_i(x) - f_i(y))^2 - c^2 (f_4(x) - f_4(y))^2 \right]^{1/2}$$

is the *relativistic interval* for $x$ and $y$, in *coordinate system $f$*, for $c$. $Q_{f,c}$ may be called the *interval function* for $f$.

D4 contains the axioms of the system.

D4. An ordered quintuple $\chi = \langle S, K, G, \mathscr{F}, c \rangle$ is a *Minkowski Structure* (MS) iff there is a set $S$, a symmetric and reflexive binary relation $K$ on $S$, a group $G$ of maps of $S$ onto itself, a nonempty family $\mathscr{F}$ of homeomorphisms of $S$ onto $R_4$, and a positive constant $c$, such that the following axioms are satisfied:

A1. $S$ is a topological 4-manifold of class $C^2$, with family $\mathscr{F}$ of global coordinate functions. That is, the functions in $\mathscr{F}$ are related by twice-differentiable transformation equations.

A2. For any $x, y \in S$, $g \in G$, $A \subset S$
   (a) $x \in C(y) \leftrightarrow g(x) \in C(g(y))$
   (b) $A$ is $n$-dimensional ($1 \leqslant n \leqslant 4$) iff its image under $g$, $g[A]$, is $n$-dimensional.

A3. (a) $f, f' \in \mathscr{F}$ iff there is a unique $g \in G$ such that $f' = f \circ g$
   (b) for all $x, y \in S, f \in \mathscr{F}$
      $Q_{f,c}(x, y)^2 \leqslant 0 \leftrightarrow x \in C(y)$
   (c) for all $x \in S$, $BDY(C(x)) \neq \emptyset$.

A4. For all distinct $x, y \in S$, $x \in C(y)$ iff there is a trajectory $A$ such that $x, y \in A$.

A5. For all distinct $x, y \in S$, if $y \in BDY(C(x))$ there is just one trajectory (called a *null trajectory*) $A \subset BDY(C(x))$ such that
   (a) $x, y \in A$
   (b) for all $z, w \in A$, $z \in BDY(C(w))$.

The interpretation of most of the axioms seems clear. $G$ is a group under functional composition, and will be called the group of *causal automor-*

*phisms* of S (cf. Zeeman, 1964). The functions in $\mathscr{F}$, specified by A3 (a), (b) are called *global coordinate functions* since their domains are all of S. We may say (by (A3(a))) that an element $g \in G$ *transforms* an $f \in \mathscr{F}$ into $f \circ g \in \mathscr{F}$, and that this operation is a *transformation*. A3 (b) stipulates, under our interpretation and D1, that two points in S are K-related iff they are separated by a timelike or null interval, so that $C(x)$ is the full Minkowski light cone of the point x. c is the speed of light. The system's novelty lies in A5. Null trajectories are paths in space-time along which a light ray may pass, so the predicate *is a null trajectory* is a subproperty of $T$.[12] The figure below illustrates the two conditions of A5.

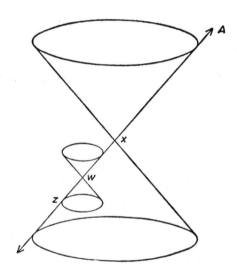

### III. TOPOLOGICAL AND CAUSAL THEOREMS

In this Section I present six theorems whose proofs require no assumptions concerning particular coordinate functions or values of the relativistic interval function. Also, we need not here assume that S is a 4-manifold. Four dimensionality of S will suffice for T1–T6. Hence T1–T6 are all of a *nonmetrical* character. In stating T1–T6 I omit the introductory hypothesis "Let $\chi = \langle S, K, G, \mathscr{F}, c \rangle$ be an MS," which I assume to hold for them.

T1. For any $x, y \in S$
  (a) $x \in C(x)$
  (b) $x \in C(y) \leftrightarrow y \in C(x)$.

These are immediate consequences of D1, the reflexivity, and the symmetry of $K$.

T2. Let $x \in S$. Then

$$BDY(C(x)) \subseteq C(x),$$

i.e., $C(x)$ is closed.

PROOF. Let $y \in BDY(C(x))$. By A5 there is a null trajectory $A$ such that $x, y \in A$. Since $T(A)$, D2 gives $y \in C(x)$.

T3. For all $x \in S$, $A \subset S$, $g \in G$

$$x \in A \leftrightarrow g(x) \in g[A].$$

PROOF: (a) Let $x \in A$. Since $g[A]$ is the image of $A$ under $g$, $g(x) \in g[A]$,
(b) Since $G$ is a group under $o$, $g^{-1} \in G$. Thus by (a)

$$g(x) \in g[A] \to (g^{-1} o g)(x) \in (g^{-1} o g)[A]$$
$$\to x \in A.$$

T4. Let $g \in G$. Then for all $x \in S$

$$C(g(x)) = g[C(x)],$$

i.e., causal automorphisms map closed causal cones into closed causal cones.

PROOF: Suppose that for some $y \in S$ we have $y \in C(g(x))$ but $y \notin g[C(x)]$. Then T3 and the second assumption give $g^{-1}(y) \notin C(x)$. From this and A2(a) follows $y \notin C(g(x))$, yielding a contradiction.

Now suppose that $y \in g[C(x)]$ but $y \notin C(g(x))$. By T3 and our first assumption, $g^{-1}(y) \in C(x)$. By A2(a) we get $y \in C(g(x))$, again yielding a contradiction.

T5. For any $x \in S$, $g \in G$

$$BDY(C(g(x))) = g[BDY(C(x))],$$

i.e., causal automorphisms map boundaries of closed causal cones into boundaries of closed causal cones.

PROOF: Let $y \in BDY(C(g(x)))$ but suppose $y \notin g[BDY(C(x))]$. Then by T3, $g^{-1}(y) \notin BDY(C(x))$, and by previous theorems and our first assumption $g^{-1}(y) \in C(x)$. Thus $g^{-1}(y)$ is in the interior of $C(x)$, and there is a ball $B$ about $g^{-1}(y)$, all points of which are in the interior of $C(x)$. So by T3 and A2(a) all points of $g[B]$ are in $C(g(x))$. But this contradicts our first assumption: for since $g[B]$ is a ball about $y \in BDY \times (C(g(x)))$ it must contain a point *not* in $C(g(x))$.

Now let $y \in g[BDY(C(x))]$ but suppose that $y \notin BDY(C(g(x)))$. Then by T3 $g^{-1}(y) \in BDY(C(x))$, yielding $g^{-1}(y) \in C(x)$ by T2. By A2(a), $y \in C(g(x))$. By this and the second assumption, $y$ must be in the interior of $C(g(x))$. Thus we know that there is a ball $B$ containing $y$, all points of which are in the interior of $C(g(x))$. By T4, $y \in B \subset g[C(x)]$. T3 then gives $g^{-1}(y) \in g^{-1}[B] \subset C(x)$. This contradicts our result $g^{-1}(y) \in BDY(C(x))$, since $g^{-1}[B]$ is a ball about $g^{-1}(y)$, and all points of $g^{-1}[B]$ are in $C(x)$.

T6. For all $x \in S$, $g \in G$, if $A$ is a null trajectory containing $x$, then $g[A]$ is a null trajectory containing $g(x)$; i.e., causal automorphisms map null trajectories into null trajectories.

PROOF: Let $A \subset S$ be a null trajectory containing $x$, and let $g \in G$. By T3, $g(x) \in g[A]$, so suppose that $g[A]$ is *not* null. By D2 and A2(a), (b), elements of $G$ map trajectories into trajectories, so $g[A]$ is a trajectory. By A5 there must be two points, $z, w \in g[A]$ such that $z \notin BDY(C(w))$. But since $T(g[A])$ we have $z \in C(w)$ by D2. $z$ must then be in the interior of $C(w)$. Thus by T4 and T5, $g^{-1}(z)$ is in the interior of $C(g^{-1}(w))$. Therefore $g^{-1}(z) \notin BDY(C(g^{-1}(w)))$. This contradicts A5, since $g^{-1}(z)$, $g^{-1}(w) \in A$, which is a null trajectory.

T1–T6 show that the causal automorphisms for an MS permute that structure's trajectories among themselves in such a manner that null trajectories are mapped into null trajectories. It follows from T6 that no nonnull trajectory gets mapped into a null trajectory.

It is now clear that my system gives a theory of causal relations in space-time. More precisely, I have given a *topological and causal* description of Minkowski space-time, and of the spatiotemporal structure of possible processes. Let $\chi = \langle S, K, G, \mathcal{F}, c \rangle$ be an MS and let A1' be the statement '$S$ is a four-dimensional Hausdorff space'. Then $\chi' = \langle \varphi_1^5(\chi), \varphi_2^5(\chi), \varphi_3^5(\chi) \rangle$ is a relational structure which, if characterized by A1', A2(a), (b), A3(c), A4, and A5, is a realization for T1–T6. The proofs of T1–T6 use no metrical notions. The adequacy of $\chi'$ (called the *topological causal structure* (TCS) for the MS $\chi$) as a description of temporally symmetric causal relations in space-time will be shown below by the completion of the demonstration and argument outlined in Section I.

The relation of a TCS to its MS makes precise *a* sense in which space-time 'has' a topological-causal structure which 'underlies' or is 'prior' to its metrical structure. Formally, it is a necessary condition for a quintuple to be an MS that *there exists* a triple that is a TCS corresponding to it in the manner just indicated.[13] Introduction of the usual Minkowski metric serves to *redescribe* (by A3 (a), (b)), in the language of analytical geometry, the *nonmetrical* properties of space-time and of its null and nonnull trajectories.[14,15]

## IV. METRICAL THEOREMS

T7. Let $\chi = \langle S, K, G, \mathscr{F}, c \rangle$ be an MS, let $A \subset S$ be a null trajectory, $f \in \mathscr{F}$, $g \in G$. Then if $x, y \in A$

$$Q_{f,c}(g(x), g(y))^2 = 0.$$

PROOF: Suppose $Q_{f,c}(g(x), g(y))^2 \neq 0$ for some $x, y \in A$. Two cases arise:

(a) $Q_{f,c}(g(x), g(y))^2 > 0$. Then by A3(b) $g(x) \notin C(g(y))$ giving $x \notin C(y)$ by A2(a). This contradicts the hypothesis that $A$ is a null trajectory.

(b) $Q_{f,c}(g(x), g(y))^2 < 0$. Consider $g(x)$ (and hence $x$) fixed. By D3 $Q_{f,c}(g(x), g(y))^2$ is continuous in $g(x)$, $g(y)$. There is thus a ball $B$ about $g(y)$, such that for any $z \in B$, $Q_{f,c}(g(x), z)^2 < 0$. But $g(y) \in BDY \times (C(g(x)))$ by A5 and T6, so any ball (like $B$) about $g(y)$ must contain a point $w$ such that $w \in S - C(g(x))$. For such a point $w$, $Q_{f,c}(g(x), w)^2 > 0$ by A3(b).

Thus the relativistic interval vanishes for any two points on any null trajectory. This result confirms the intuitive appeal of A5, which I interpreted as affirming the existence of light-ray paths.

D5 introduces some terminology for the transformations of the coordinate functions induced by our causal automorphisms.

D5. Let $\chi = \langle S, K, G, \mathscr{F}, c \rangle$ be an MS, $f \in \mathscr{F}$, $g \in G$. Then a transformation $f \to f \circ g$ is a *multiplicative Poincaré mapping* (MPM) iff there is a real number $\Upsilon^{1/2}$, a nonsingular 4-×-4 matrix $[\Lambda.]$ (called a *Lorentz matrix*), and a four-component column vector $\langle B^1, B^2, B^3, B^4 \rangle$ such that for all $x \in S$

(a) $(f \circ g)_\mu(x) = Y^{1/2} \Lambda^\mu_\nu f_\nu(x) + B^\mu$ $(\mu = 1, 2, 3, 4)$
(b) $B^i (i = 1, 2, 3, 4)$ are real numbers
(c) $\Lambda^\mu_\rho \Lambda^\nu_\sigma \eta_{\mu\nu} = \eta_{\rho\sigma}$, where $\eta_{\mu\nu}$ $(\mu, \nu = 1, 2, 3, 4)$ are components of a *Minkowski matrix* given by

$$[\eta_{\mu\nu}] = \begin{bmatrix} 1 & 0 & 0 & 0 \\ 0 & 1 & 0 & 0 \\ 0 & 0 & 1 & 0 \\ 0 & 0 & 0 & -c^2 \end{bmatrix}.$$

I use, in D5 and below, the *Einstein summation convention*, by which if an index letter occurs two or more times in a single term of a formula (without summation over that index being indicated), then the formula as written is short for a summation over the numerical range of that index letter. In this context summation is from 1 to 4. If $Y = 1$ we have a *Poincaré mapping*. The group of all Poincaré mappings is the inhomogeneous nonorthochronous Lorentz group, and $Y^{1/2}$ is the scale factor mentioned in Section I. With this terminology the main result of this Section can be stated as follows:

T8. Let $\chi = \langle S, K, G, \mathscr{F}, c \rangle$ be an MS. Then if $f, f' \in \mathscr{F}$, they are related by an MPM. That is, for some unique $g \in G$, $g$ transforms $f$ into $f'$ by an MPM.

I prove T8 by adapting a proof due to Anderson (1967, pp. 184–187) concerning minimal conditions for the derivation of the Poincaré Mappings. With T1–T7 at hand, one can follow his method to show that the existence of light ray and particle, i.e., null and nonnull, trajectories in a 4-manifold of class $C^2$ ensures the truth of T8. The independent interest of my theory, however, lies in the fact that T1–T6 require no metrical assumptions, as was explained above. It is surprising that one can get this far on my primitive basis without such assumptions.[16] Anderson begins by assuming T7 and a form of A3(b), and derives the mappings that send trajectories into themselves, and in particular that send light-ray paths into themselves. The existence of such mappings is assumed. In contrast, T1–T7 *prove* that elements of $G$ (for a given MS) have this 'closure' property, and that null trajectories systematically represent light-ray paths.

PROOF: Consider any member $f = \langle f_1, f_2, f_3, f_4 \rangle$ of $\mathscr{F}$. Let $A$ be any trajectory, and let $x, x_0 \in A$. A4, A5, and A3(c) guarantee the existence of trajectories in $S$. Since $A$ is linear and one dimensional by D2, we may write

(1)  $\quad f_\mu(x) = f_\mu(x_0) + \lambda \Delta_f^\mu \quad (\mu = 1, 2, 3, 4)$

where each $\Delta_f^\mu$ is a nonzero constant (dependent upon $f$ and $x_0$) and $\lambda$ (more precisely $\lambda(x)$) is a real-valued path parameter for $A$. Since $T(A)$ we have $Q_{f,c}(x, x_0)^2 \leq 0$ for all $x \in A$, by D2 and A3(b). This requires (for all trajectories)

(2)
$$\sum_{r=1}^{3} \left(\frac{\Delta_f^r}{\Delta_f^4}\right)^2 - c^2 \leq 0$$
$$\sum_{r=1}^{3} (\Delta_f^r)^2 - c^2 (\Delta_f^4)^2 = \eta_{\mu\nu} \Delta_f^\mu \Delta_f^\nu \leq 0$$

where $[\eta_{\mu\nu}]$ is a Minkowski matrix.

By T7 we know that the equality in (2) holds iff $A$ is null. Moreover by T6 we know that the causal automorphisms of $\chi$ map trajectories into trajectories. So if $x \in S, f, f' \in \mathscr{F}$, and if $g \in G$ is such that $f' = f \circ g$, (1) and (2) give for $A$

(3)  $\quad (f \circ g)_\mu(x) = (f \circ g)_\mu(x_0) + \lambda \Delta_{f \circ g}^\mu \quad (\mu = 1, 2, 3, 4)$

(4)  $\quad \eta_{\mu\nu} \Delta_{f \circ g}^\nu \Delta_{f \circ g}^\mu \leq 0.$

By T6 and T7 the equality holds in (4) iff it holds in (2). From (3) we get (writing $f'_\mu$ for $(f \circ g)_\mu$, $f'$ for $f \circ g$)

(5)  $\quad \dfrac{d}{d\lambda} f'_\mu(x) = \Delta_{f'}^\mu \quad (\mu = 1, 2, 3, 4)$

(6)  $\quad \dfrac{d^2}{d\lambda^2} f'_\mu(x) = 0 \quad (\mu = 1, 2, 3, 4).$

(5) and (6) follow from (3) and the one dimensionality of $A$. By (1), (6), and the differentiability of the map $f \to f'$ guaranteed by A1

(7)  $\quad \dfrac{d}{d\lambda} f'_\mu(x) = \dfrac{\partial f'_\mu(x)}{\partial f_\rho(x)} \Delta_f^\rho \quad (\mu = 1, 2, 3, 4)$

(8)  $\quad \dfrac{\partial^2}{\partial \lambda^2} f'_\mu(x) = \dfrac{\partial^2 f'_\mu(x)}{\partial f_\rho(x)\, \partial f_\sigma(x)} \Delta_f^\rho \Delta_f^\sigma = 0 \quad (\mu = 1, 2, 3, 4).$

Since each $\Delta_f^\sigma \Delta_f^\nu$ is nonzero and dependent on the choice of $\lambda$, there is sufficient freedom to choose them so that

$$\frac{\partial^2 f'_\mu(x)}{\partial f_\rho(x)\,\partial f_\sigma(x)} = 0 \quad (\mu, \rho, \sigma = 1, 2, 3, 4), \tag{9}$$

Therefore the transformation $f \to f \circ g$ must be affine, i.e.,

$$f'_\mu(x) = (f \circ g)_\mu(x) = \mathscr{A}^\mu_\nu f_\nu(x) + B^\mu \quad (\mu = 1, 2, 3, 4), \tag{10}$$

where $[\mathscr{A}^\mu_\nu]$ is a nonsingular 4-×-4 matrix and $\langle B^1, B^2, B^3, B^4 \rangle$ is a real-valued four-component column vector (Birkhoff and Maclane, 1965, p. 237). The nonsingularity of the map follows from the group property of $G$.

From (5), (7), and (10) we get, changing dummy indices

$$\Delta^\mu_{f'} = \frac{\partial f'_\mu(x)}{\partial f_\nu(x)} \Delta^\nu_f = \mathscr{A}^\mu_\nu \Delta^\nu_f \quad (\mu = 1, 2, 3, 4). \tag{11}$$

Now let $A$ be a null trajectory. By (11) and the equality condition on (4)

$$(\eta_{\rho\sigma} \mathscr{A}^\rho_\mu \mathscr{A}^\sigma_\nu)\, \Delta^\mu_f \Delta^\nu_f = 0. \tag{12}$$

Multiply (2) by an undetermined nonzero constant $\Upsilon$ and subtract the result from (12). This yields

$$[\eta_{\rho\sigma} \mathscr{A}^\rho_\mu \mathscr{A}^\sigma_\nu - \Upsilon \eta_{\mu\nu}]\, \Delta^\mu_f \Delta^\nu_f = 0 \tag{13}$$

since the equality holds in (2) by virtue of our assumption that it holds in (4).

By the argument leading to (10)

$$\eta_{\rho\sigma} \mathscr{A}^\rho_\mu \mathscr{A}^\sigma_\nu = \Upsilon \eta_{\mu\nu} \quad (\mu, \nu = 1, 2, 3, 4). \tag{14}$$

We therefore have

$$\mathscr{A}^\mu_\nu = \Upsilon^{1/2} \Lambda^\mu_\nu \quad (\mu, \nu = 1, 2, 3, 4) \tag{15}$$

where $[\Lambda^\mu_\nu]$ is a Lorentz matrix. Thus finally

$$f'_\mu(x) = (f \circ g)_\mu(x) = \Upsilon^{1/2} \Lambda^\mu_\nu f_\nu(x) + B^\mu \quad (\mu = 1, 2, 3, 4). \tag{16}$$

Now $\Upsilon^{1/2}$ and $B^\mu$ ($\mu = 1, 2, 3, 4$) must be real, since $f$ and $f'$ are both homeomorphisms of $S$ onto $R_4$. This shows that the map $f \to f \circ g$ is an

MPM. Thus any two members of $\mathscr{F}$ are related by a member of the Poincaré group modulo $\Upsilon^{1/2}$.[17]

The physical significance of a transformation like (16) is that it can be viewed as the result of the action of a member of the Poincaré group, preceded by a coordinate-scale change of magnitude $|\Upsilon^{1/2}|$. For (16) can be rewritten as

(17) $\quad f'_\mu(x) = \Lambda^\mu_\nu(\Upsilon^{1/2} f_\nu(x)) + B^\mu \quad (\mu = 1, 2, 3, 4)$

and the nature of $\Upsilon^{1/2}$ as a scale factor made evident. If $\Upsilon^{1/2} < 0$ the scale change may reverse the signs for some or all of the coordinate values for certain $x \in S, f \in \mathscr{F}$. One must note, however, that this change is of no physical significance. For example, if $\Upsilon^{1/2} < 0$ for a given MPM, so that

(18) $\quad f'_4(x) = -|\Upsilon^{1/2}| \Lambda^4_\nu f_\nu(x) + B^4,$

we may have (for some $x \in S) f_4(x) > 0$ but $f'_4(x) < 0$. But this *sign-reversal* does not represent a *time-reversal* transformation. The latter is characterized, *for any real $\Upsilon^{1/2}$ in* (17), by the conditions (a) det $[\Lambda^\mu_\nu] = -1$, (b) $\Lambda^4_4 \leqslant -1$ (Anderson, 1967, p. 142). The sign reversal of (18) thus does not represent a change in the attributed temporal order of any pair of events with respect to the relation *later than*. We simply have a change in the numerical names of points in $S$. The sign change for the spatial coordinates seems nonproblematic. Thus the *name* changes in (17) do not impair the ability of an MS to give a description of the spatiotemporal structure of possible processes: for the scale factors can always be absorbed into the coordinate functions. With this intuitive interpretation of our MPM's, the TCS for an MS is seen to be adequate to describe the topological-causal structure of possible processes: for the components of a TCS are embedded in an MS, which characterizes Minkowski space-time in the manner demanded by special relativity, i.e., by coordinate functions related by members of the Poincaré group modulo $\Upsilon^{1/2}$. This completes the argument outlined in Section I.

*Vassar College*

## NOTES

[1] I wish to thank Z. Domotor, D. Malament, D. Micham, and P. Suppes for help on an earlier draft.
[2] I wish only to provide a schematic account of the structure of temporally symmetric

causal relations. Thus I will not discuss such fundamental questions as the ontology of space, time, or causality.

[3] For topology see Royden (1963). For relativity see Moller (1952), Weyl (1952), and the references in this paper.

[4] Technically, I should call the ordered triple $\langle S, \theta, \mathcal{O} \rangle$ a differentiable manifold, where $\langle S, \theta \rangle$ is a connected Hausdorff space with countable basis and $\mathcal{O}$ is a certain set of real-valued functions on $S$ (Trautman, 1965, pp. 69–73). But since my axioms provide a specification of $\mathcal{O}$, we may call $S$ itself a 4-manifold, and shall mean thereby that $S$ is a set such that there exists a well-defined class of functions from $S$ onto $R_4$ which renders $S$ a 4-manifold.

[5] I will assume the philosophical respectability of $K$ as a *physical relation* (between events or points of $S$). That $K$ is thus respectable is argued by van Fraassen (1970, pp. 191–198). Davidson (1967) gives a defense of the thesis that one can study the logical structure of sentences concerning causal relations without having at one's disposal an analysis of the *nature* of causality.

[6] Since my axiomatization uses only informal set-theoretical techniques, I speak freely about defining properties rather than symbols for them.

[7] For a similar result for asymmetric causal relations see Zeeman (1964, 1967). Zeeman (1967) shows that the group of homeomorphisms of a certain topology for $S$, which leave the relational structure of an asymmetric causal relation < invariant or reversed, is generated by the Poincaré group modulo a scale factor.

[8] The system of Robb (1914) uses a primitive asymmetric relation of temporal succession. It therefore attributes temporal asymmetries to space-time. I suppose that Grünbaum, Mehlberg and van Fraassen would consider that system to be philosophically relatively unilluminating, just because an adequate characterization of space-time can be given which attributes no such asymmetries to it. Similar remarks apply to Carnap's C-T system (1958, pp. 197–201).

[9] I do not wish to speculate on the possibility of giving causal accounts of some or all spatiotemporal relations in the context of general relativity, which allows, for example, 'worlds' containing closed but everywhere timelike world lines (Gödel, 1949a, 1949b).

[10] The need for this cautionary paragraph was forced upon me by David Malament, who has also given arguments suggesting that A5 is not independent of A1–A4.

[11] I am indebted to Z. Domotor for simplifications in D1–D4, which led to a more elegant axiom set, and simpler proofs.

[12] Zeeman (1964) can define predicates that pick out point-sets that intuitively correspond to null trajectories. His definitions, however, make essential use of the asymmetry of his <-relation. Thus I cannot use them here.

[13] It is debatable if this sense of 'priority' is of any logical, physical, or ontological interest. But the notion of priority has played a role in the study of causal theories of space and time (cf. Carnap, 1958, p. 197; Lewin, 1923; Reichenbach, 1949, Ch. 4; Robb, 1914).

[14] The phrasing of this last sentence is due in part to Anderson (1967, p. 187).

[15] Grünbaum (1970) has recently tried to specify a sense in which space-time (or its submanifolds) can be said to have 'intrinsic' metrical and nonmetrical $n$-adic properties. I will not comment on the relation of my notion of 'priority' to his concepts.

[16] This fact saves the system from the objection that it is totally uninteresting as a gloss on special relativity, since (it is argued) I have 'merely replaced' the locution '$x$ and $y$ are separated by a timelike or null interval' by the *new terminology* '$K(x, y)$'.

[17] Anderson, following Einstein, argues that $Y = +1$. I see no way to reproduce his informal argument in my system.

## BIBLIOGRAPHY

Anderson, J. L., *Principles of Relativity Physics*, Academic Press, New York, 1967.
Birkhoff, G. and MacLane, S., *A Survey of Modern Algebra* (3rd ed.), Macmillan, New York, 1965.
Carnap, R., *Introduction to Symbolic Logic and Its Applications*, Dover, New York, 1958.
Davidson, D., 'Causal Relations', *Journal of Philosophy* 64 (1967), pp. 691–703.
Gödel, K., 'An Example of a New Type of Cosmological Solutions of Einstein's Field Equations of Gravitation', *Reviews of Modern Physics* 21 (1949), 447–450 (a).
Gödel, K. ,'A Remark About the Relationship Between Relativity Theory and Idealistic Philosophy', 'in P. A. Schilpp (ed.), *Albert Einstein: Philosopher-Scientist*, The Library of Living Philosophers, Inc., Evanston, Ill., 1949, pp. 557–562 (b).
Grünbaum, A., *Philosophical Problems of Space and Time*, A. Knopf, New York, 1963.
Grünbaum, A., *Modern Science and Zeno's Paradoxes*, Wesleyan University Press, Middletown, Conn., 1967.
Grünbaum, A., 'Space, Time and Falsifiability', *Philosophy of Science* 37 (1970), 469–588.
Hurewicz, W. and Wallman, H., *Dimension Theory*, Princeton University Press, Princeton, N.J., 1941.
Lewin, K., 'Die Zeitliche Geneseordnung', *Zeitschrift für Physik* 13 (1923), 62–81.
Mehlberg, H., 'Physical Laws and Time's Arrow', in H. Feigl and G. Maxwell (eds.), *Current Issues in the Philosophy of Science*, Holt, Rinehart, and Winston, New York, 1961, pp. 105–138.
Mehlberg, H., 'Relativity and the Atom', in P. K. Feyerabend and G. Maxwell (eds.), *Mind, Matter and Method*, University of Minnesota Press, Minneapolis, 1966, pp. 449–491.
Menger, K., 'What is Dimension?', *American Mathematical Monthly* 50 (1943), pp. 2–7.
Moller, C., *The Theory of Relativity*, Oxford University Press, Oxford, 1952.
Reichenbach, H., *The Philosophy of Space and Time*, Dover, New York, 1958.
Reichenbach, H., *Axiomatization of the Theory of Relativity*, University of California Press, Berkeley, 1969.
Robb, A. A., *A Theory of Time and Space*, Cambridge University Press, Cambridge, 1914.
Roman, P., *Theory of Elementary Particles*, North Holland, Amsterdam, 1960.
Royden, H. L., *Real Analysis*, Macmillan, New York, 1963.
Suppes, P., *Introduction to Logic*, Van Nostrand, Princeton, N. J., 1957.
Suppes, P., 'Axioms for Relativistic Kinematics With or Without Parity', in L. Henkin, P. Suppes, and A. Tarski (eds.), *The Axiomatic Method with Special Reference to Geometry and Physics*, North-Holland Publ. Co., Amsterdam, 1959, pp. 291–307.
Suppes, P., 'The Desirability of Formalization in Science', *Journal of Philosophy* 65 (1968), 651–664.
Trautman, A., 'Foundations and Current Problems of General Relativity', *Lectures on General Relativity* (Brandeis Summer Institute in Theoretical Physics, 1964), Prentice-Hall, Englewood Cliffs, N. J., 1965, pp. 1–248.
van Fraassen, Bas C., *An Introduction to the Philosophy of Time and Space*, Random House, New York, 1970.
Weyl, H., *Space-Time-Matter*, Dover, New York, 1952.
Zeeman, E. C., 'The Topology of Minkowski Space', *Topology* 6 (1967), 161–170.
Zeeman, E. C., 'Causality Implies the Lorentz Group', *Journal of Mathematical Physics* 5 (1964), 490–493.

# NOTES ON THE CAUSAL THEORY OF TIME

## I. INTRODUCTION

The causal theory of time is a special type of a relational theory of time. A relational theory holds that it is not necessary to postulate the absolute existence of instants of time, to think of instants as part of a being which is absolute in that it is a kind of event container whose existence is independent of the existence of the events it contains; rather, time is said to be nothing over and above (to be constituted by, to be reducible to) the structure of temporal relations between events. The causal species of this theory says that temporal relations can be defined in terms of 'physical relations', relations which, whatever else they are, are not 'specifically temporal'.

To put it politely, modern causal theorists set rather modest goals for themselves; for example, Adolf Grünbaum contents himself with presenting a causal analysis of the relations of 'temporal betweenness' and 'temporal separation'. Despite their modesty, various versions of causal theory which have been put forward by Reichenbach (1956, 1958) and Grünbaum (1963) run into a number of technical difficulties (Lacey, 1968). Recently, new versions have been formulated which escape these difficulties (Grünbaum, 1968; van Fraassen, 1970). However, unless there is some prospect that the causal theorist can eventually go beyond these modest goals in such a way as to justify the relational view of time, the work of Grünbaum *et al.*, will have little more than a curiosity value; I believe that this prospect is dim. Moreover, there remains the suspicion that the new versions of the causal theory (so-called) are trivial, for all of the new versions employ the notion of causal connectibility, and there is the suspicion that 'causal connectibility' is just another name for a spatiotemporal relation, a relation which must be understood in terms of space-time structure. I do not propose to confront head on the question of whether the relation of causal connectibility can be defined in terms of 'physical relations', relations which are not 'specifically spatial or temporal'; rather, my approach will be more indirect. One of the things I will try to show is that even if we concede the relation of causal connectibility to the causal theorist, a number of cases are encompassed by the general theory of relativity in which even the most recent versions of the causal theory break down and in which there appears to be no hope of getting an adequate version.

Since the references cited above are widely available and widely read, I will not present here any systematic view of recent causal theories, and I will assume that the reader has some familiarity with the basic concepts used by Grünbaum and van Fraassen.

## II. SPACE-TIME AS THE BASIC SPATIOTEMPORAL ENTITY

Space-time is the basis spatiotemporal entity; space and time are simply the spatial and temporal aspects of this entity. This observation holds not only in the context of relativistic physics, but also in the context of classical physics where the neglect of spacetime structure has led to a number of philosophical errors and oversights (Earman, 1970a, b). In itself, this observation need not cause any trouble for the causal theorist, for his doctrines about time can be translated into doctrines about space-time. But it is worth noting that such a translation cannot always be performed in a trivial manner. For example, Reichenbach's causal theory is said to be concerned with 'topological' properties of time; in particular, the notion of topological simultaneity plays an important role in his writings. But construed as a relation between space-time points, topological simultaneity is not a topological relation in terms of the manifold topology $T_{man}$ of space-time, i.e., this relation is not preserved by all homeomorphisms of $T_{man}$.[1] This raises two questions. First, can the causal theorist supply an analysis of $T_{man}$? I will argue below in Section V that the answer is no. Second, when we view the doctrines of the causal theorist as doctrines about space-time, just how interesting and plausible do they seem? From the discussion below, I will conclude that they do not seem very interesting or very plausible.

The above observation does raise a more immediate difficulty for the causal theorist when it is combined with the fact that in the usual formulations of relativity theory, space-time is taken as part of the basic ontological furniture, e.g., one quantifies over space-time points, and the notion of space-time point is used to explain other notions like spatiotemporal coincidence. Perhaps there is no need to postulate the existence of space-time points; for example, the causal theorist may be able to give an analysis of the notion of spatiotemporal coincidence without appealing to the notion of space-time point. Before examining this matter, it is first necessary to go into more detail on the causal structure of space-time.

### III. ELEMENTS OF THE CAUSAL STRUCTURE OF RELATIVISTIC SPACE-TIMES

Throughout, I shall concentrate on relativistic space-times since the modern proponents of the causal theory of time usually have such space-times in mind and since they often claim that relativity theory provides motivation for a causal analysis of time. And in any case, the causal theory would not command much interest unless it could cover such space-times.

By a relativistic space-time I shall mean a four-dimensional differentiable manifold $M$ equipped with a Lorentzian metric. In what follows, the metric will be used only to define the null-cone structure of $M$. Because of the signature of the metric, there is defined in the tangent space $T_x$ of any point $x \in M$ an object $C_x$ called the null cone at $x$. A tangent vector $v \in T_x$ is said to be timelike (respectively, null, spacelike) just in case $v$ lies inside (on, outside) $C_x$. $M$ is temporally orientable (intuitively, $M$ can be given a globally consistent time sense) just in case there exists on $M$ an everywhere defined continuous timelike vector field (see Appendix 1 of Earman, 1970a, for more details). A connected and temporally orientable $M$ has two temporal orientations, and the choice of one amounts to a choice of 'the future direction of time'. I will not deal here with the problem of how this choice is made – this is part of the cluster of problems often referred to as 'the problem of the direction of time'. Once a choice is made, an orientation is induced on certain curves of $M$: a smooth parameterized curve $C$ of $M$ is a future directed timelike (causal) curve just in case the positive tangent to $C$ at every point $x$ lies in the future lobe of $C_x$ (lies inside or on the future lobe of $C_x$). Let $x <\cdot y$ and $x \ll \cdot y$, $x, y \in M$, mean, respectively, that $x$ is causally earlier than $y$, and $x$ is chronologically earlier than $y$, i.e., $x <\cdot y$ if and only if there is a future directed causal curve of nonzero (affine) length from $x$ to $y$, and $x \ll \cdot y$ if and only if there is a future directed timelike curve of nonzero length from $x$ to $y$. The causal past $Ca_-(x)$ of $x$ is then defined as $\{y : y \in M$ and $y < \cdot\, x\}$. The causal future $Ca_+(x)$ of $x$, the chronological past $Ch_-(x)$ of $x$, and the chronological future $Ch_+(x)$ of $x$ are defined in analogous ways.[2] $S(x) \equiv (Ca_-(x) \cup Ca_+(x))$ is the set of points causally connectible with $x$ in the sense that it is physically possible for a causal signal of nonnull length to connect $y$ with $x$ just in case $y \in S(x)$. No doubt there are other senses of causal connectibility, but it is sufficient for our purposes to consider this sense. Also, causal theorists usually have causal signals in mind when they speak of causal connectibility.

As the above definitions indicate, I am assuming that propagation of causal signals does not take place outside the null cone. This assumption has been challenged in recent years, but a careful analysis seems to support it (Fox, *et al.*, 1970). In any case, this assumption is justified within the context of the present inquiry since a causal analysis of relativistic spacetimes becomes much more difficult or even impossible if it does not hold (e.g., if there is no upper bound on the velocity of causal signals, a causal analysis would not seem possible).

### IV. SPATIO-TEMPORAL COINCIDENCE

Can the causal theorist supply a causal criterion of spatiotemporal coincidence? The events to which the modern causal theorist refers are assumed to be 'unextended' – if space-time is taken as primitive, they are events which are localized at a space-time point. Thus, I will use capital letters $X$, $Y$, $Z$ to indicate events which occur, respectively, at space-time points $x$, $y$, $z$. An obvious, and seemingly the strongest, criterion of spatiotemporal coincidence which can be formulated in terms of the relation of causal connectibility is the following:

(C)  $X$ and $Y$ are spatiotemporally coincident just in case: for every $Z$, $Z$ is causally connectible with $X$ if and only if $Z$ is causally connectible with $Y$.

Van Fraassen (1970, p. 184) adopts (C) as a definition. (C) is adequate only for those space-times which satisfy the following condition:

(1)  For every pair of points $x, y \in M$, if $Ca_-(x) = Ca_-(y)$ and $Ca_+(x) = Ca_+(y)$, then $x = y$.

And (1) is *not* satisfied in many of the space-times encountered in the general theory of relativity.

The causal theorist might reply that this is not a very strong criticism of his theory, for, he might argue, any space-time in which (1) fails is physically unreasonable. For instance, (1) will fail if $M$ contains a point $x$ for which $Ca_-(x)$ and $Ca_+(x)$ are not disjoint; but a space-time which contains closed causal curves violates our everyday notions about causation. But $M$ need not contain closed causal curves in order for (1) to fail.[3] It is true, however, that if (1) fails, $M$ will contain 'almost closed' causal curves; that is, the following condition will not hold for every $x \in M$:

(2)  Every open neighborhood (in $T_{man}$) of $x$ contains an open neighborhood which is not entered twice by any causal curve.

But precisely what reasons, other than the question begging one that the causal analysis fails, is there for thinking that such a space-time is physically unreasonable? There are indications that we must learn to live with space-times which fail to satisfy (1) unless we are willing to reject or drastically modify the general theory of relativity; for some of the most fundamental solutions of the field equations violate (1) (Carter, 1968). And even if we should agree that a space-time which fails to satisfy (1) is 'unreasonable', should not an adequate theory of time be broad enough to handle 'unreasonable' temporal structures so that the degree of reasonableness or unreasonableness of a given structure can be discussed *within* the theory? Moreover, the presently active causal theorists want to be able to account for 'closed times', and viewed in terms of space-time structure, such a temporal structure would seem to imply closed causal curves (see Section VI).

The causal theorist still has two courses open to him. He could present some other analysis of spatiotemporal coincidence. I suspect that other approaches will be no more successful, but nothing final can be said until some concrete proposal is made. Alternatively, the causal theorist could simply take the relation of spatiotemporal coincidence as a primitive. Reichenbach (1958) followed this course in an early version of his theory, but this move considerably weakened the interest of his theory – for one thing, it left him no way to reply to the challenge that space-time points are assumed to be part of the basic ontology in the usual formulations of relativity theory – and subsequent causal theorists have sought to analyze this fundamental relation. I have argued that they have not been successful.

The implication of the above remarks extends beyond the issue of the soundness of the causal analysis of time. For instance, certain cases in which (1) fails can be used as counterexamples to Davidson's (1969) criterion of event identity – assuming, of course, that a necessary condition for $X = Y$ is $x = y$. By a criterion of event identity, Davidson means a 'satisfactory filling' for the blank in

(3)     If $E$ and $F$ are events, then $E = F$ if and only if ———.

Examples of the type of filling he finds satisfactory are 'Classes are identical if and only if they have the same members' and 'Physical objects are identical if and only if they occupy the same places at the same times'. Davidson's own filling for the blank in (3) is

(4)     If $E$ and $F$ are events, then $E = F$ if and only if for every event $G$, $G$ is a cause of $E$ just in case $G$ is a cause of $F$, and $E$ is a cause of $G$ just in case $F$ is a cause of $G$.

Consider a world in which causal influence is transmitted only via causal signals and in which (1) fails for some pair of points $x$, $y$ in the space-time background of this world. If either (a) the only causal curve of $M$ which is realized by a causal signal is a closed curve joining $x$ and $y$, or (b) every causal curve of $M$ is realized by a causal signal, then although $x = y$, we can choose $X$ and $Y$ so that every cause of $X$ is a cause of $Y$ and vice versa and every effect of $X$ is an effect of $Y$ and vice versa. (There are also counterexamples to (4) involving cases where no causal curve of $M$ is realized by a causal signal. In such a world, potential as well as actual causal relations can be taken into account; but at best, identity of potential causal relations can guarantee identitiy of spatiotemporal location of events, and not always that.)

I suspect that we cannot reach a satisfactory criterion of event identity without the help of a sufficiently deep analysis of the concept of event. One of the conclusions of this paper is that taking events and their causal relations as a primitive basis is not sufficient for getting at the subtle and complex spatiotemporal relations which can obtain between events set in a relativistic space-time background. This suggests that we should go the other way round and take space-time as primitive and use it to get an analysis of events – perhaps physical events can be analyzed in terms of ordered pairs consisting of a space-time region and a physical property which applies to that region – but this is a matter I cannot pursue here.

## V. TOPOLOGY AND CAUSAL STRUCTURE

Under what conditions is there hope of giving a causal analysis of the manifold topology of space-time? Define the chronological topology $T_{\text{chron}}$ of a space-time $M$ to be the coarsest topology in which $Ch_-(x) \cap Ch_+(y)$ as $x$ and $y$ range over $M$ are open sets. If $M$ satisfies a condition somewhat weaker than (2), to wit:

(5)     $M$ contains no almost closed timelike curves.

then $T_{\text{chron}} = T_{\text{man}}$. Conversely, if $T_{\text{chron}} = T_{\text{man}}$, then (5) holds (Kronheimer and Penrose, 1967). $T_{\text{man}}$ can also be approached by using the concepts of causal past and future rather than the concepts of chronological past and future, but the approach is more complicated (Kronheimer and Penrose, 1967).

Can the causal theorist make use of these relations between $T_{\text{man}}$ and the null-cone structure of $M$ to give a causal analysis of $T_{\text{man}}$? The answer is no. Leaving aside the limitation imposed by (5), the approaches mentioned above depend upon the existence of a globally consistent time sense for $M$ (though not on the distinction between the past and the future in the which-is-which sense) and on the notion of a temporally oriented space-time curve (though not on the distinction between past and future oriented curves in the which-is-which sense); but if we have the notion of a temporally oriented curve, there is no need for the causal theory. For example, if the curve is open, what the causal theorist hopes to do is to tell us which events on the curve are 'temporarily between' other events, but temporal betweenness falls out of tempral orientedness. Moreover, it seems to me that the discussions of Grünbaum et al., implicitly assume the property of temporal orientability.

Here the causal theorist might reply that indeed temporal orientability is a presupposition of his theory, but that once he has achieved his goal of defining 'temporal betweenness' and/or 'temporal separation', he can give an account of temporal orientability. But such a reply shows a confusion between the property of temporal orientability and temporal order properties. For we can have orientability without order: in particular, we can have space-times which are temporally orientable but for which neither the relation of 'temporal betweenness' (understood on the model of the triadic relation '$Q$ is between $P$ and $R$' for the points of the real line) nor the relation of 'temporal separation' (understood on the model of the tetradic relation '$P, Q$ separate $R, S$' for the points of a circle – intuitively, to get from $R$ to $S$ by going around the circle, one must either pass over $P$ or over $Q$) make sense.[4] This fact is directly related to the fact that we can have temporally orientable space-times which can neither be considered open nor considered closed in their temporal aspects. This matter is taken up in the following section.

## VI. THE CAUSAL THEORY AND A 'CLOSED TIME'

The most recent and systematic version of the causal theory is that of van Fraassen (1970). Throughout his book, van Fraassen emphasizes the logical and conceptual possibility that time might be 'closed', and he says that his version of the causal theory is designed to accommodate this possibility.

From the point of view taken in Section II, the first question we must ask is, what does a closed time come to in terms of space-time structure? Suppose that the space-time $M$ (which is assumed throughout to be temporally orientable) can be partitioned by a family of spacelike hypersurfaces. This allows us to project out a time $T$: $T$ is the quotient of $M$ by the equivalence relation $Rxy$ which holds between $x, y \in M$ just in case $x$ and $y$ lie on the same hypersurface, i.e., an instant of $T$ is an $Rxy$-equivalence class of points of $M$. Now if no future (or past) directed timelike curve of $M$ intersects more than one any of the hypersurfaces corresponding to the instants of $T$, we can define a mapping $f: M \to \mathbb{R}$ such that (i) the fibers of $f$ correspond in a one-to-one manner to the instants of $T$ and (ii) for any $x, y \in M$, if there is a future directed timelike curve of nonzero length from $x$ to $y$, then $f(x) < f(y)$. In this case, $T$ can be said to be open. Analogously, we can say that $T$ is closed if there is a mapping $g: M \to C$ ($C =$ circle) such that (i) the fibers of $g$ correspond in a one-to-one manner to the instants of $T$ and (ii) for any $w, x, y, z \in M$, if these points are distinct and there is a future directed timelike curve from $w$ to $x$, from $x$ to $y$, and from $y$ to $z$, then either this curve intersects all of the hypersurfaces of $M$ which correspond to the instants of $T$ or else the points $g(w)$, $g(y)$ separate the points $g(x)$, $g(z)$ on the circle $C$.

There are two problems here for the causal theorist. Some interesting and important solutions to the field equations of general relativity cannot be partitioned by spacelike hypersurfaces (Carter, 1968); and other solutions do not possess a single global spacelike hypersurface or 'time slice' (Hawking, 1968). In these cases a time $T$ cannot be defined, and the question of openness or closedness does not even arise. Even when it does arise, the answer may be neither 'Open' nor 'Closed'; at least there are obvious examples of relativistic space-times from which we can project out a time $T$ but which satisfy neither the above criterion for openness nor the criterion for closedness. Thus, those causal theorists who envision the possibilities of open and closed times as being mutually exclusive and exhaustive are working with a false dichotomy; and, in effect, all recent causal theorists make this assumption: they assume that either time is open, in which case they seek to analyze the relation of temporal betweenness, or else that it is closed, in which case they seek to analyze the relation of temporal separation.

Second, the most obvious examples of space-times which by the above criterion are closed in their temporal aspects are such that for every $x \in M$, $(Ca_-(x) \cup Ca_+(x)) = M$. Van Fraassen's theory does not apply to such a case (e.g., 1970, p. 185, Postulate VI fails); and it is hard to see how any causal theory remotely like his could apply.[5]

There are two other points worth noting. If the presumption of the above construction (i.e., if $M$ can be covered by a family of nonintersecting spacelike hypersurfaces) holds, then no closed timelike curve of $M$ is homotopic to zero (Carter, 1968). Therefore, we can always find another space-time $M'$ which contains no closed timelike curves and which forms a covering space for $M$ (intuitively, $M'$ is a larger space which is locally like $M$ but which has different global features). As a result, it could be argued that it is never necessary to admit that time is closed since there always exists the alternative interpretation, which cannot be ruled out by any local observations, that time is open and the history of the universe is periodic, that instead of encountering the same events in a closed time we encounter similar events over and over agin in an open time. Finally, note that Grünbaum and van Fraassen rule out by fiat certain topological structures for time, e.g., that of a figure eight. It seems to me that it may be possible to use the four-dimensional point of view to explain why such structures cannot occur. For example, it is hard to see under what circumstances it would be reasonable to assign to $T$ the topology of a figure eight; for the cross point of the eight must correspond to a spacelike hypersurface of $M$, and, seemingly, there would have to be two null cones at every point of the surface.

### VII. MOTIVATIONS FOR THE CAUSAL THEORY

Viewed as an autonomous being, time does seem mysterious and does seem to call for a philosophical analysis. If one balks at treating time as a substantial entity, then a relational theory of time becomes attractive, and a causal theory is a type of relational theory. But time is not an autonomous being; rather it is an aspect of the more fundamental entity, space-time. Presumably, the modern causal theorist sees space-time as a mysterious entity standing in need of analysis. This attitude is in direct contrast to the present attitude of the scientific community which is relatively comfortable with the concept of space-time but which sees the concept of causation and the concept of a localized particle, concepts which the causal theorist takes as basic, as being in need of analysis.[6]

The most sophisticated version of the causal theory is that of van Fraassen (1970). He sees space-time as an abstract mathematical construct used to represent the relational structure of events that constitutes world history. This representation need not be by way of an isomorphism; that is, there is a distinction drawn between the actual structure of world history and space-time, and "the structure of world history is set in [space-] time, and we conceive world history to be set in the same [space-] time regardless of the form it takes" (van Fraassen, 1970, p.106). This picture is not consistent with general relativity. In the first place, different world histories cannot always be set in the same space-time, for according to the general theory, the structure of space-time is influenced by the behavior of matter-energy; in particular, the space-time metric becomes a dynamical object which is related to the distribution of matter-enery through the field equations. Second, the notion that space-time is an abstract mathematical entity used to represent the relational structure of events does not square with general relativity; for the interaction between space-time and matter-energy pointed to above is a two-way street, and the behavior of physical objects is explained in part by the structure of space-time, e.g., the deviation of the world lines of a system of test particles is caused by the curvature of space-time. Solutions to the field equations in which the curvature of space-time is not due to the presence of gravitational matter prevent the causal theorist from replying that such explanations can be recast in a form which talks only about the direct gravitational interaction of bodies and which avoids mention of space-time curvature; this is especially clear when a nonzero cosmological term is present in the field equations, for in this case, empty space-time creates its own gravitational field. Finally, the geometrodynamical interpretation of relativity theory (Wheeler, 1962) sees space-time not as an arena but as everything – everything is made out of curved empty space-time. This view turns the relational theory on its head. Though geometrodynamics is in a very programmatic stage, it is clearly one of the more exciting programs to emerge from modern physics, and any philosophical theory which cannot accommodate it is *prima facie* too narrow as well as being in danger of becoming passe.

## VIII. CONCLUSIONS

I have argued that the most recent versions of the causal theory are subject to serious limitations. The causal analysis of spatiotemporal coincidence considered in Section IV does not apply to space-times in which (1) fails. And current versions of the theory collapse altogether for typical cases of relativistic space-times which are closed in their temporal aspects. Second, I have pointed out that the program of recent causal theorists is based on a false dichotomy – open vs. closed times; for only a small subclass of relativistic space-times can be said to be either open or closed in their temporal aspects, and the causal theory seems incapable of handling the cases which fall in between. Third, I have argued that the general theory of relativity does not provide motivation for the causal theory; on the contrary, general relativity promotes the view of space-time as a substantial entity.

As a result, I do not see that the causal theorist has a convincing argument against the position which holds that in order to understand the subtle and complex temporal structures encompassed by relativity theory, one must accept space-time as an entity which cannot be analyzed away as an abstract mathematical construct used for representing the 'physical', i.e., causal, relations between events. And I cannot agree with van Fraassen (1970, p. 140) that

> Philosophers were not long in appreciating this development [i.e., relativity theory], and the consequent construction of the causal theory of time and space-time must be considered one of the major contributions of twentieth-century philosophy of science.

For it seems to me that causal theorists have failed to appreciate this development and that the construction of the causal theory has served to obscure important and interesting facts about the temporal aspect of space-time.

## NOTES

[1] This point is made by Stein (1970).

[2] These concepts are adapted from Kronheimer and Penrose (1967), but the definitions given here differ somewhat from theirs. If $M$ contains closed, future directed timelike curves, then $\ll$ will not have the features that the relation of being temporally earlier than is usually assumed to have, eg., it will not be irreflexive.

[3] The example in Kronheimer and Penrose (1967, p. 487) can be used to demonstrate this fact.

[4] See van Fraassen (1970, pp. 66–70) for a discussion of 'temporal betweenness' and 'temporal separation'.

[5] As an example, let $M$ be Minkowski space-time and $(x, y, z, t)$ a pseudo-Cartesian coordinate system for $M$. Obtain $M'$ by identifying two points $(x_1, y_1, z_1, t_1)$ and $(x_2, y_2, z_2, t_2)$ just in case $x_1 = x_2$, $y_1 = y_2$, $z_1 = z_2$ and $t_1 = t_2$ modulo some positive integer. $M'$ is temporally closed by the above criterion, but every pair of points of $M'$ can be joined by a causal curve. Van Fraassen (private communication) has argued that he is not committed to the view that if $C$ is a causal curve, then it is physically possible for $C$ to be the path of some causal signal. But the usual interpretation of relativity theory does contain this view. Moreover, van Fraassen's analysis will not work for even the simplest relativistic space-times unless he grants that if $y \in (Ca_-(x) \cup Ca_+(x))$ then *some* causal path joining $y$ with $x$ is the path of a possible causal signal; and this is sufficient to make the above point.

[6] See Fox *et al.* (1970) and the references given there for some of the problems about causation currently under discussion in the physics literature. Kalnay (1970) reviews some of the problems in defining a satisfactory position operator in relativistic quantum theory.

## BIBLIOGRAPHY

Carter, B., 'Global Structure of the Kerr Family of Gravitational Fields', *Physical Review* **174** (1968), 1559–1571.

Davidson, D., 'The Individuation of Events', in N. Rescher (ed.), *Essays in Honor of Carl G. Hempel*, D. Reidel, Dordrecht, Holland, 1969, p. 216.

Earman, J., 'Space-Time, or How to Solve Philosophical Problems and Dissolve

Philosophical Muddles Without Really Trying', *Journal of Philosophy* **67** (1970), 259–277. (a)

Earman, J., 'Who's Afraid of Absolute Space?', *Australasian Journal of Philosophy* **48** (1970), 287–319. (b)

Fox, R., Kuper, C. G., and Lipson, S. G., 'Faster-Than-Light-Group Velocities and Causality Violation', *Proceedings of the Royal Society (London) A* **316** (1970), 515–524.

Grünbaum, A., *Philosophical Problems of Space and Time*, Alfred A. Knopf, New York, 1963.

Grünbaum, A., *Modern Science and Zeno's Paradoxes*, Wesleyan University Press, Middletown, 1968.

Hawking, S. W., 'The Existence of Cosmic Time Functions', *Proceedings of the Royal Society (London) A* **308** (1968), 433–435.

Kalnay, A. J., 'Lorentz Invariant Localization for Elementary Systems', *Physical Review D* **1** (1970), 1092–1104.

Kronheimer, E. H. and Penrose, R., 'On the Structure of Causal Spaces', *Proceedings of the Cambridge Philosophical Society* **63** (1967), 481–501.

Lacey, H. M., 'The Causal Theory of Time, a Critique of Grünbaum's Version', *Philosophy of Science* **35** (1968), 322–354.

Reichenbach, H., *Space and Time*, Dover, New York, 1958.

Reichenbach, H., *The Direction of Time*, University of California Press, Los Angeles, 1956.

Stein, H., 'A Note on Time and Relativity Theory', *Journal of Philosophy* **67** (1970), 289–294.

van Fraassen, B. C., *An Introduction to the Philosophy of Time and Space*, Random House, New York, 1970.

Wheeler, J. A., *Geometrodynamics*, W. A. Benjamin, New York, 1962.

BAS C. VAN FRAASSEN

# EARMAN ON THE CAUSAL THEORY OF TIME*

There is an important point behind Earman's (this volume) criticisms of the causal theory of time and space-time. This point has been made perspicuously in a recent paper by Glymour (1970). It concerns the novel problems raised for a theory of space-time by the *general* theory of relativity (henceforth GTR), and I shall explain it briefly in Section II below. Section I briefly states my own view of the status of the causal theory, and Sections III and IV deal with Earman's specific criticisms.

## I. THE STATUS OF THE CAUSAL THEORY

The efforts by Leibniz, Kant, and Lechalas, which today can be classed as attempts to justify a causal theory of time vis-à-vis classical physics, were not successful (see van Fraassen, 1970, Ch. II, Section 3). Unlike Earman, I do not believe that success was impossible given that classical physics set no upper limit to signal velocity. But the main obstacle was exactly the difficulty of displaying a physical basis for simultaneity. The situation was changed radically by Einstein's critique of simultaneity, and this is what made possible a revival and a large measure of success for the causal theory in this century.

In my opinion, this success is essentially a justification of the causal theory vis à vis the *special* theory of relativity (henceforth, STR). Moreover, this success has been attained through a self-critical process of the formulation and successive revisions of axiomatic theories, whose scope was limited through certain simplifying assumptions (van Fraassen, 1970, Ch. VI). The use of these assumptions was justified, because the aim was a demonstration of the consistency of certain theses. It was intended to show that the causal theory was compatible with principles of contemporary physics (indeed, that the causal theory could provide the foundations for its kinematics), and also compatible with certain cosmological possibilities such as that time, or space, is closed.

Earman refers specifically to the axiomatic theory displayed in Chapter

*P. Suppes (ed.), Space, Time and Geometry, 85–93. All rights reserved*
*Copyright © 1973 by D. Reidel Publishing Company, Dordrecht-Holland*

V, Sections 4–5 of van Fraassen (1970). At the time of writing, I echoed Grünbaum's conviction that a radical philosophical inquiry into the foundations of the *general* theory of relativity and relativistic cosmology was still in its first stages (as when I emphasized the restriction to the STR on pages 140 and 191). Reichenbach had been too sanguine about the extension of his ideas (and adaption of his axiomatic formulations) to the context of the GTR. Careful analyses of special problems concerning space-time in the GTR are to be found in Grünbaum (1963, Chs. 2 and 14; 1968, pp. 7–73, 196–242, 309–323; 1970).

Earman has exhibited a number of aspects of cosmology in the context of the GTR that might prevent a straightforward adaptation of the causal theory to that context. In the next section I shall briefly describe what I take to be the central problem. Whether it is justified to conclude with Earman that "there is little hope of getting an adequate version" of the causal theory in the light of this, I leave to the reader to judge.

## II. LOCAL AND GLOBAL GEOMETRY[1]

Space is locally Euclidean; what constraint does this place on the structure of space? This question was already discussed in depth by Felix Klein, and the answer is: *surprisingly little*. Indeed, space may be locally Euclidean and have the topological character of a sphere, a torus, or a Klein bottle.

The GTR began with the assumption that *locally*, the STR holds.[2] What constraint does this place on the structure of space-time? Again, surprisingly little. This is so even in the context of the other usual assumptions made in relativistic cosmology (discussed in Glymour, 1970, Sec III, but ignored by Earman; they do not follow from the GTR but limit its cosmological models). The field equations specify a local relation between the components of the metric and the energy-momentum tensors in any coordinate system. Accordingly, the GTR has a number of distinct cosmological models, in which the global geometric character of space-time varies greatly.

It must be realized that the usual presentations of classical physics and the STR are cosmologically naive, in that the multifarious possibilities opened up by Riemannian geometry are ignored. Cosmological questions are settled by default, as it were, through the assumption that time is open and space Euclidean (the simplest generalization of local structure). In the nineteenth century some attempt was made to show that the basic principles of classical physics are compatible with closed time, and this attempt was made (in a much more thorough manner) for the STR in this century. But this is only a small step compared to the problem posed for philosophy by the GTR which grants only local validity to the world picture of the STR.

Our experience and the relation between the STR and the GTR suggests that the causal theory can give a philosophical account of the local geometry of space-time. The questions we face at this point are:

(1) Can the causal theory be formulated in such a way that its main principles explicitly pertain to local geometric structure only?

(2) Can the causal theory be then extended so as to place constraints on the global geometric structure of space-time?

(3) If so, how are these constraints related to those commonly assumed in relativistic cosmology?

It has been suggested specifically by Glymour that topologically distinct cosmological models may be empirically indistinguishable. In that case it may have to be admitted that conventionality enters the (global) topology as well as the metric of space-time. At least, that would be my conclusion; a realistic philosophy of space-time, such as Earman's, would presumably hold that, in that case, real distinctions outrun the empirically distinguishable.

### III. CLOSED TIME

According to Earman, "the program of recent causal theorists is based on a false dichotomy – open vs. closed times." Given Reichenbach's fascination with time-travel (local closed causal chains) and Grünbaum's attack on a priorism with respect to topology (1963, p. 201), Earman cannot possibly have believed this to be a strong or deep-seated conviction. (It is true that Grünbaum explicitly ruled out a "figure eight" topology for time, but not a priori: he points out that this is impossible in the deterministic framework of his discussion; see 1963, p. 203.)

In my book I ruled out consideration of other possibilities for purely tactical reasons. I wished to show the compatibility of the causal theory of time with the basic principles concerning space-time of the STR and *either* open *or* closed time. When the aim is to show logical compatibility, one needs to construct a specific model or class of models, and it is advisable to keep the construction as simple as possible. However that may be, Earman is correct in pointing to the fact that no other alternatives have been explicitly discussed; though how he came to conclude that "the causal theory seems incapable of handling the cases which fall in between," I cannot tell from his paper.

The same tactical considerations led me to restrict attention to the *simplest* modification of the usual principles that would allow time to be closed. This is relevant to the next point raised by Earman, who argues that my characterization of closed time does not fit "the most obvious examples of space-times which... have a closed temporal aspect." In his note 6 he gives a precise example and discusses my reaction to that example (in correspondence). His response to my reaction begs the question; to show this I will briefly reiterate the technical notions used by Earman.

Earman considers, for the purpose of modeling space-times, four-dimensional differentiable manifolds $M$ with Lorentzian metric, and (in the present context) assumes $M$ is oriented. Then each point $x$ has a causal past $Ca_-(x)$ and a causal future $Ca_+(x)$, the union $S(x)$ of which is the set of points $y$ such that a *causal curve* links $x$ and $y$. Now a *causal curve* is defined to be a timelike curve of a certain kind (defined in an obvious way with reference to the null cone at $x$). In the usual presentation of Special Relativity, it is asserted that these curves, the causal curves, are the paths of possible signals, and indeed the only such paths because there is a finite limit to signal velocity. Hence, in the usual context, events at points $x$ and $y$ in space-time are causally connectible if and only if $y$ is in $S(x)$. It must be recalled that in the usual presentation of Special Relativity it is also assumed that time is open.

Now, suppose that we want to construct a mathematical model of space-time in which time is closed. There is an obvious procedure, the *identification* procedure, by which noncompact spaces can be turned into compact spaces. We introduce an equivalence relation on the first space, let the equivalence classes be the points of the second space, and define the basic geometric notions on the second space in an obvious way. This is exactly what Earman does to turn a Minkowski space-time into a space-time with closed time (calling $(x, y, z, t)$ and $(x, y, z, t')$ *equivalent* exactly if $t = t'$ modulo positive integer $m$, and no other pairs of points equivalent).

Looking now at what happens to $S(x)$ when $x$ is identified with all points of the original space-time thus equivalent to it, we see that every pair of the new points $x, y$ is such that $y$ is in $S(x)$. But it was a basic assumption of the post-relativity causal theory (introduced to give an account of the space-time of Special Relativity in its usual presentation) that for every event there are some events not causally connectible with it.

I discussed this problem explicitly in van Fraassen (1970, pp. 186–187). Nonmathematically, the difficulty is this: if time is closed, in what sense can there be a fastest signal? It would seem that a signal could be so slow that it would go 'all the way around time' and connect two events which cannot be connected by a signal which does not go faster than light, but whose path is restricted to a small region of space-time containing both events.

This is a genuinely new conceptual problem that arises when closed time is considered, for it appears that the principle that light is the fastest signal is not unambiguous but admits two possible interpretations. Under the first interpretation, the principle holds only locally: light signals are the fastest among signals in restricted spatiotemporal regions. Under this interpretation, any two events are causally connectible, but may not be 'locally causally connectible'. In other words, if this interpretation is adopted, then the principles about first signals introduced into the causal theory by Reichenbach must be given restricted ('local') validity. It would seem that this is the correct approach if one wants to provide a philosophical foundation for the General Theory. This is the moral that can be drawn straightforwardly from Glymour's and Earman's discussion of the qualitative variety of cosmological models of the General Theory.

In van Fraassen (1970), however, I chose the second interpretation, a global interpretation of the 'first-signal' principle, which is that arbitrary pairs of events are not causally connectible. That rules out actual trajectories 'all the way around time'.

If this choice is made, the relation between causal connectibility (a relation between events) and geometrically definable relations on space-time must be revised. The interpretation of 'causal curve' must now be that, if this curve is given by a function $f(t) = (x_t, y_t, z_t, t)$, for all $t$, then only its *proper* segments can be the paths of possible causal signals. In other words, the causal relational structure of events can be embedded in the geometric space such that causal signals travel along 'causal curves', but the structure of geometric relations is an extension of the structure of causal relations.

Earman comments in his note 6 that "van Fraassen's analysis will not work for even the simplest relativistic space-times unless he grants that if $y \in (Ca_-(x) \cup Ca_+(x))$ then some causal path joining $y$ with $x$ is the path of a possible causal signal; and this is sufficient to make the above point." It is indeed sufficient, for the premise in question is exactly that events are causally connectible if and only if they are located at points in space-time linked by a causal curve. But I see no justification for this comment. The abstract principle that there are pairs of events which are not causally connectible does not rule out that in the *usual* case of open time, it is exactly the events located in $S(x)$ that are causally connectible with events located at $x$. So there is no warrant for Earman's dismissal of my analysis.

## IV. SPACE-TIME AND WORLD HISTORY

It is crucial to the above argument that I do not regard the relation between the mathematical space-time and the actual structure of events (world history) as isomorphism. My interpretation of a cosmological theory is that it asserts that world history, whatever it turns out to be, can be embedded in the mathematical model(s) it provides. This is what is conveyed briefly by the assertion that space-time is a logical space. To accept a certain cosmological model is tantamount to delimiting the physically possible worlds to those that can be embedded in this model.

Earman asserts that this picture is not consistent with General Relativity. His first argument is that different world histories cannot be set in the same space-time, because the space-time metric is related to the distribution of matter-energy through Einstein's field equations. For this argument to be correct, two further premises must be added: (a) everything that happens is uniquely determined by the matter-energy distribution, and (b) the relation given by the field equations is one-to-one.³ But I shall not quibble; I am quite willing to conceive of a theory in which space-time structure uniquely determines world-history. Such a theory would presumably only specify abstract conditions $C$ on space-time structure, and assert only that space-time is a '$C$-space' (space satisfying conditions $C$). With only a little wiggling I could therefore point to some abstract structure of which the theory asserts, exactly, that any physically possible world can be embedded in it. But why bother? God, presumably, could then specify conditions $C'$ so closely that there remained only a single possible world. It is a limiting case, proper to God's limiting nature, and of course we must admit limiting cases.

Earman's second argument is this: "the interaction between space-time and matter-energy... is a two-way street, and the behavior of physical objects is explained in part by the structure of space-time – e.g., the deviation of the world lines of a system of test particles is seen to be caused by the curvature properties of space-time." Now this is not physics, but a particular philosophical interpretation of physics. I do not grant the interpretation; not every explanation is a causal explanation; the interpretation is naive; the assertion that space-time causes deviations in world lines is a category mistake. Anticipating such a reaction Earman says that such explanations cannot be recast in a form which talks only about the direct interaction of bodies and which avoids mention of space-time curvature, because there are solutions of the field equations in which the curvature is not due to the presence of gravitational matter. This remark can be understood with reference to an earlier argument by Earman already commented on by Grünbaum (1970, pp. 566–567). The argument is that the GTR admits space-times which have well-defined metrical structures which are devoid of matter-energy. Hence the metric cannot be explained with reference to actual extrinsic standards such as rigid bodies. In addition, Earman (forthcoming) argues we cannot appeal to possible test particles or counterfactual judgments about how a test particle would behave if it were present,

for since the space-time metric is a dynamical object, the theory implies that if these extrinsic standards were present, the metrical structure would be different....

The reason is not adequate. For the counterfactuals that have the form of the counterfactual assertion in the passage just cited from Earman would single out the space-time metric. To be precise, let $P$ stand for the perturbations envisaged here by Earman (introduction of a rod or clock) and $G$ stand for the resultant metric structure. Then the set of true counterfactuals of form $P > G$ determines the metric of the space-time in question (up to empirical undistinguishability). The argument Earman gave has roughly the form of: that this water has temperature $T$ cannot be explained in terms of counterfactuals about the behavior of thermometers immersed in the water, for this immersion will (generally) alter the temperature of the water, and you cannot know the extent of this alteration without knowing the original temperature.

Finally, Earman's third argument appeals to Wheeler's geometrodynamics. In Wheeler's proposal "everything is made of curved empty space-time. This view turns the relational theory on its head. Though geometrodynamics is in a very programmatic stage, it is clearly one of the most exciting programs to emerge from modern physics, – and any philosophical theory which cannot accommodate it is *prima facie* too narrow as well as being in danger of becoming passé." Earman has mentioned the program in several papers, and Grünbaum (1970, pp. 487, 522–524, 555–567) has offered a number of detailed criticisms of the conclusions. Reading Earman, I am amazed to find that there are such straightforward entailments between physical theories and philosophical conclusions, which apparently can be seen at once by anyone who understands the theories. In the past, philosophers have thought the connection less simple and more difficult to uncover; but as Earman indicates in his closing paragraph, our historical judgments are rather divergent.

## V. CONCLUSION

I have so far ignored Earman's Section IV in which spatiotemporal coincidence is discussed. The answer will be clear from the preceding: the exact definitions and principles of the exact theories we have displayed are to be discussed with reference to the special and not the general theory of relativity. But moreover, Earman's transition from (C) to (1) assumes what we do not grant: that events are causally connectible *exactly if* the points in the mathematical space-time at which they are located are linked by a causal curve.

This captures in a nutshell my own conclusions. The first is that the causal theory, after its success vis-à-vis the STR, must now provide a detailed analysis of spatiotemporal concepts in the GTR. The second is that the points raised by Earman do not provide substantive reasons for doubting the adequacy of the causal theory to this task, because Earman insists in his extrapolations on a much closer relation between the empirical structure of events and the mathematical structures that model it than ought to be assumed.

*University of Toronto*

## NOTES

\* I wish to acknowledge gratefully my debt to Dr. A. Grünbaum, University of Pittsburgh, to Dr. C. Glymour, Princeton University, and to the support of the John Simon Guggenheim Memorial Foundation.
[1] This section explains briefly some main points raised by Glymour (1970).
[2] For the precise cash value of this remark, see for example Pauli (1958, pp. 145–149).
[3] This is the relation between the components $g_{\mu\nu}$ and $T_{\mu\nu}$ given by $R_{\mu\nu} - \tfrac{1}{2}g_{\mu\nu} R + \Lambda g_{\mu\nu} = \kappa T_{\mu\nu}$, where $\kappa$ and $\Lambda$ are universal constants, and $g_{\mu\nu}$ determines $R_{\mu\nu}$ and $R$ locally.

## BIBLIOGRAPHY

1. Farman, J., 'Are Spatial and Temporal Congruence Conventional?' (forthcoming).
2. Earman, J., this issue, p. 72.
3. Glymour, C., 'Cosmology, Convention, and the Closed Universe', Mimeographed manuscript, Princeton University, 1970.
4. Grünbaum, A., *Philosophical Problems of Space and Time*, Knopf, New York, 1963.
5. Grünbaum, A., *Geometry and Chronometry in Philosophical Perspective*, University of Minnesota Press, Minneapolis, 1968.
6. Grünbaum, A., 'Space, Time, and Falsifiability, Part I', *Philosophy of Science* **37** (1970), 469–588.
7. Pauli, W., *Theory of Relativity*, Pergamon Press, London, 1958.
8. van Fraassen, B., *An Introduction to the Philosophy of Time and Space*, Random House, New York, 1970.

ROBERT PALTER

# KANT'S FORMULATION OF THE LAWS OF MOTION*

## I. THE METAPHYSICAL FOUNDATIONS OF NATURAL SCIENCE

Kant's interpretation of the mathematics of motion is to be found in his *Metaphysical Foundations of Natural Science* in the form of one fundamental principle of kinematics and three laws of mechanics. None of these propositions is especially original with Kant so far as the sheer mathematics goes, but the selection of just these four propositions and Kant's proof for each of them are at the very least strongly influenced by the special features of his critical philosophy. Before turning to details I should like to consider what Kant takes to be the significance and usefulness of the subject of metaphysical foundations of natural science (which in its strictest, and for Kant, most proper sense, is the metaphysical doctrine of corporeal nature). Kant's claims are, as usual, in one sense very modest and in another sense magnificently presumptuous: modest in that he thinks his achievement "no great work" [kein grosses Werk]; presumptuous in that he believes himself to have once and for all exhaustively investigated his subject. In any case, Kant does claim that all mathematical physicists, however much they may seek to 'repudiate any claim of metaphysics on their science', must, at least implicitly, make use of the principles which he is about to elucidate. Kant adds that these same mathematical physicists, in view of the apodictic certainty they recognize to be their goal, prefer to *postulate* their fundamental (metaphysical) principles rather than appear to be merely deriving these principles from experience. But it is important for the progress of science to be able to distinguish *limitations* of one's fundamental principles from mere *mistakes in the application* of one's principles; and to do this one must obviously first know exactly what these principles are. Besides this pragmatic value of the metaphysical foundations of natural science for natural science, the subject also possesses a pragmatic value for metaphysics itself – understanding here by 'metaphysics' the knowledge [*Erkenntnis*] of God, Freedom, and Immortality – namely, that of furnishing examples [*Beispiele*] in the form of intuitions [*Anschauungen*] to illustrate the pure concepts of the understanding [*reine Verstandesbegriffen*] of universal metaphysics. Kant remarks on the fact that it is indeed *only* the metaphysical doctrine of corporeal nature which can furnish such concrete instances, *via* the forms and principles of outer intuition, to universal metaphysics. Hence the desirability of developing the subject independently, and this apparently for both of the following reasons: (a) general metaphysics is inhibited in its 'regular growth' by the 'shoot' of metaphysics of corporeal nature, which had therefore best be lopped off and planted separately; and (b) the latter 'shoot' can itself develop better if it is not 'encumbered' by its parent stem.[1]

The architectonic of Kant (1883) is determined by two considerations: first, the table of categories (Quantity, Quality, Relation, and Modality), which, for Kant, exhausts the possible *a priori* determinations or specifications of any concept of matter in general; and second, the content of that initial general concept of matter, which is required in order for there to be something definite to determine or specify. The evaluation of the central role which the table of the categories plays here (as elsewhere) in Kant's philosophy is beyond the scope of this essay (though it is worth calling attention to the long footnote in which Kant, anticipating his about-to-be-published second edition of *Critique of Pure Reason*, defends his use of the categories against a contemporary critic).[2] Considerably more germane to my intended theme is Kant's view that *motion* is that characteristic of matter which the metaphysical doctrine of corporeal substance must take as its point of departure. Why motion (instead of, say, force, which would, for example, be Leibniz's (1695, p. 712) choice)?[3] Kant's answer is brief but unfortunately far from self-evident. Since matter is before all else an object of outer sense, and since this sense can only be affected by motion, it follows, Kant argues, that the 'fundamental determination' [*die Grundbestimmung*] of matter must be motion. But why, commentators on Kant (1883) have asked, should we accept Kant's unanalyzed and undefended assertion here? How do we know that the outer sense can be affected solely by motion? Surely this is not merely a question of definition; and if not, then it can only be an empirically derived and hence thoroughly contingent assumption dependent for whatever plausibility it possesses on the latest results in the physiology of sense perception. And indeed Kant (1883, p. 152; IV, 482) does hold that the movability of bodies is an empirical concept because it 'presupposes the perception of something movable' (Kant, 1933, p. 82; A41 = B58).

Does not the movability of bodies seem to follow, however, almost immediately from the mere fact that they exist in space and time, for is not motion simply the change of a body's place as time elapses? Presumably, what Kant would find lacking in this line of thought is an *a priori* proof that motion is *fundamental* to bodies. However, when one thinks of bodies as essentially located in a four-dimensional space-time manifold – which is customary in the post-Minkowski era – it becomes a bit difficult to see why the state of motion of a body is any less *a priori* than its mere location in space-time. It is true that the use of the four-dimensional framework is not as compellingly natural in classical (Newtonian) kinematics as in relativity kinematics. (This is because the temporal 'fibering' of classical space-time into instantaneous spaces is unique, i.e., classically, time is absolute. More precisely, each instant of time defines a unique three-dimensional spatial 'stratum' so that we can represent the totality of these 'strata' in the form of a *single* permanent or enduring space – a form of representation which is not possible in relativity kinematics.) Nevertheless, even for classical kinematics the four-dimensional approach can be both elegant and conceptually illuminating (Stein, 1967).

In any case, Kant (1883, part 1) does take up the study of motion as such – i.e., in abstraction from all the other properties of the movable. The three succeeding parts consist in the study of what happens to the concept of matter as merely movable when it receives successive increments of content: first, impenetrability (i.e., the ability of a movable body to resist motion by other bodies); then, the ability of a movable body to exert force (i.e., to produce motion in other bodies); and finally, the ability of a movable body to be experienced as such (i.e., to be known as objectively characterized by a certain definite state of motion). The study of impenetrability is concerned with more or less conjectural hypotheses about the internal *dynamical* structure of matter (attractions and repulsions). The study of force is concerned with the *mechanical* laws governing the motions and interactions of bodies. The study of the cognitive status of possible states of motion is concerned with how to distinguish among *phenomenologically equivalent motions* the true from the merely illusory.

It is worth noting that all four of Kant's subject-matters – phoronomy,[4] dynamics, mechanics, and phenomenology – can be found in Newton (1934), though the arrangement and emphasis there is, of course, very different from Kant's. Thus, Newton in his Preface alludes to the forces exerted on each other by the ultimate particles of matter; he explains in the Scholium to the Definitions how to distinguish absolute from relative motions by their 'properties, causes and effects'; immediately following he formulates his three Laws of Motion, or mechanical laws; and finally, Newton formulates Corollary I to his laws of motion, the parallelogram rule for adding motions which Kant takes as the fundamental proposition of phoronomy.

For all the differences between Newton (1934) and Kant (1883), on one central doctrine of both works there is complete agreement: the fundamental principles of natural science must be mathematical in form; or, in Kant's terms, in a natural science there is only so much science as there is applied mathematics.[5] For Newton the point is so obvious as to require little discussion beyond his bare assertion in the very first sentence of the *Principia* (1934, p. xvii) that "the moderns, rejecting substantial forms and occult qualities, have endeavored to subject the phenomena of nature to the laws of mathematics". For Kant on the other hand, the remarkable success of the mathematical physicists, and more especially of Newton himself, requires explanation in terms of certain fundamental features of the human cognitive faculties of intuition and understanding. Briefly, Kant's explanation is that mathematical physics has discovered that the synthetic *a priori* necessity of any discipline worthy of the name of natural *science* arises out of a process of mathematical construction which is ultimately grounded in pure intuition. (Natural science in this sense is to be distinguished both from empirical natural *history* concerned with mere particulars of the natural world and from pure natural *philosophy* not concerned with particulars at all.) It follows that the initial problem of the metaphysical foundations of natural science must be to demonstrate that motion is a genuine *magnitude*, which means that motions may, like spatial and temporal magnitudes, be added and subtracted.

## II. COMPOSITION OF MOTIONS

We have now arrived at the threshold of phoronomy, which, as Kant (1883, p. 167; IV, 495) points out, is not a "pure doctrine of motion" but a "pure doctrine of the quantity of motion". The single fundamental proposition [*Lehrsatz*] of phoronomy deals with the so-called composition of motions, specifically, the *composition* of two rectilinear motions of a material point into a third motion (usually called the *resultant*) of that material point. We must not, however, make the mistake of thinking that Kant is interested merely in providing some formal principle for compounding and resolving motions from which other important phoronomic (or kinematic) propositions may be deduced. In the formal approach to the study of magnitudes characteristic of modern algebra it is sufficient in defining any given species of magnitude, Q, to lay down axioms which express the formal properties of the laws of composition for the elements of Q. Q might consist, for example, of the positive integers, or complex numbers, or vectors. In the case of vectors the rules for 'addition' and 'multiplication' are rather different from those for ordinary numbers (thus, the 'sum' of two vectors is defined by the so-called parallelogram rule, and two kinds of 'product' – scalar and vector – are defined). Vectors are physically important, of course, because they possess the same formal properties as motions or velocities.[6]

At the other extreme, perhaps, from the formal algebraic approach is Newton's derivation of the composition of motions principle from his first two laws of motion. From Kant's (1883, pp. 158, 165; IV, 486–87, 493) point of view Newton's demonstration is unsatisfactory because it makes use of the concepts of inertia and force which have no place in phoronomy. What Kant wishes to do is to show how the composition [*Zusammensetzung*] of two motions can be constructed [*construiren*] *a priori* in intuition. But the means for solving this problem are already implicit in the formulation of the problem: *a priori* construction requires a pure form of intuition and the only such forms available to the human mind [*Gemüth*] are those of space and time. But since time itself can only be represented spatially,[7] the pure form of spatial intuition assumes a peculiarly central role in the study of motion.[8] Kant's solution to the fundamental problem of phoronomy, then, is to show how two rectilinear motions of a material point may be in a certain sense *added in*

KANT'S FORMULATION OF THE LAWS OF MOTION 99

*space*. This solution is formulated in the first (and only) proposition of phoronomy (Kant, 1883):

The composition of two motions of a single point can only be conceived [gedacht werden] by representing [vorstellen] one of the motions in absolute space, and the other by the equivalent motion of a relative space with equal and opposite velocity [p. 161; IV, 490].

Before examining Kant's proof of this proposition I shall attempt to explain briefly Kant's idea of absolute space (Palter, 1971). For Kant, every description of motion requires a (materially defined) kinematic framework, or *relative space*. Some kinematic problems may require for their solution the introduction of several such relative spaces. In every problem, however, there will be one relative space whose state of motion is left unspecified and this is, for the problem in question, the 'immovable' relative space, or *absolute space*. Thus, Kant's absolute space is, unlike Newton's, an indeterminate framework which may be thought of as shifting from problem to problem; absolute space functions, in Kant's view, as a mere regulative idea of reason.

In his proof of the fundamental proposition of phoronomy, Kant considers three cases, depending on the relative directions of the two original motions. In the first case the two motions are collinear and in the same direction; in the second case the two motions are collinear but in opposite directions; and in the third case the two motions are not collinear. For my purposes the essential features of the proof are all found in the first case – that '*Erster Fall*' of which Coleridge complained that even "after twice seven readings" he was "still in the Dark."[9] Central to Kant's (1883) proof is the Principle of the Relativity of (Uniform) Motion, which he formulates as follows:

Every motion as object of a possible experience, may be viewed, at pleasure, as motion of a body in a space that is at rest, or as rest of the body, and motion of the space in the opposite direction with equal velocity [p. 158; IV, 487].

We consider two motions (with velocity $v$) of the same material point, each represented in some relative space (say $S$) by a line segment of appropriate length (see Figure 1). Call these line segments $AB$ and $ab$; for simplicity and without significant loss of generality Kant makes them equal. Now, to add the motions represented by $AB$ and $ab$ it obviously will not do, argues Kant, simply to add the line segments themselves, for this addition would produce a line segment (call it $AC$)

which is not traversed in the same time as each of the two original line segments $AB$ and $ab$.[10] What Kant wants is for the representations of the two original motions to be still recognizably present in the representation of the resultant composite motion. The two original motions – i.e., the spaces traversed together with the corresponding time lapses – must

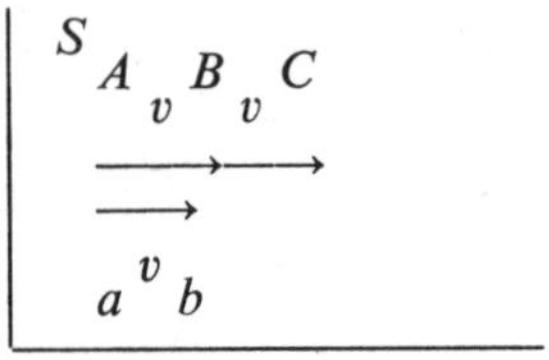

```
   S
    A   B   C
      v   v
    ───▶───▶
    ───▶
    a   b
      v
```

Fig. 1.

not be, so to say, swallowed up in the resultant motion but are somehow to remain accessible to pure (spatial) intuition alongside of the resultant motion. To accomplish such a ·phoronomic synthesis, as shown in Figure 2, Kant begins by letting $AB$ represent the motion of the point $A$ in absolute space (call it $v$), and then assumes that a certain relative space (call it $S$) is moving, also with respect to absolute space, with the same velocity but in the opposite direction. The velocity of $S$ (call it $V$) will be represented by $-ab$. But now, by the principle of relativity of motion, motion of the point with a certain velocity in some given (relative) space is equivalent to an equal but opposite motion of that space. It follows – or rather, it is *assumed* – that the velocity of the point $A$ in the space $S$ (call it $v_S$) is the difference of the velocity of $A$ in absolute space and the velocity of $S$ in absolute space, i.e., $v_S = v - V$, so that $v_S = AB - (-ab) = AB + ab$. Thus, the resultant motion of the point is now conceivable as containing [*enthalten*] the two original motions; a composite motion has been constructed in space, or more specifically, addition of motions has been shown to follow the same rules as addition of directed line segments or distances (at least for this one case).

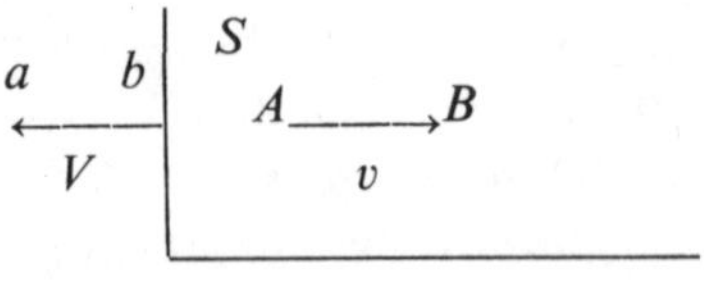

```
    a    b      S
    ◀────        A────▶B
      V            v
```

Fig. 2.

Corresponding to its somewhat elusive role, absolute space, though presupposed in the diagram, is not represented.

That there is something very convincing about Kant's procedure – what we might call the 'reversal of relative space procedure' – is apparent from the fact that Christiaan Huygens uses essentially the same procedure in his study of the impact of bodies.[11]

Kant's proofs in his second and third cases again make use of the reversal of relative space procedure together with the relativity of motion principle. It is important to see that in all three cases what has been proved is merely that the addition of motions can be grasped intuitively, in the same way, at bottom, as the addition of distances. Viewed as a *proof* of the rules for adding motions, the argument would have to be pronounced circular. To see this, consider for a moment the third case (of which the first and second cases are, in fact, just special instances). What Kant's proof does is merely to reduce the problem of *adding* two oblique motions of a single point to the problem of calculating the relative motion of a moving point and a moving coordinate system, that is to say, the problem of *subtracting* two motions. Given the relativity of motion principle, the two problems are easily seen to be equivalent. And indeed, if one looks closely at the proof of Kant's third case (the conclusion of which is the parallelogram rule for adding motions), one will find at a critical step in the proof a covert appeal to precisely this same rule for the determination of where the moving point is located in the moving coordinate system. Of course, now that we are aware of an alternative to the parallelogram rule, namely, the relativistic analogue of this rule, we find it a lot easier to detect such covert assumptions.

My interpretation of Kant's phoronomic proof has as an interesting consequence that it should not be difficult to formulate a relativistic analogue of the proof. There is, first of all, nothing incompatible about Kant's idea of absolute space and Einstein's (special) relativity principle – which is indeed formally identical to the classical relativity principle to which Kant appeals. We consider, then, just as before (see Figure 2) a material point $A$ and a relative space $S$, moving with the respective collinear velocities $v$ and $V$ with respect to some unspecified spatial framework (Kant's absolute space). The resultant velocity of $A$ with respect to $S$, $v_S$, must be some function of $v$ and $V$.

Now in classical kinematics it is implicitly assumed (as mentioned above) that there exists just a single universal time, the same for all spatial frameworks regardless of their states of relative motion. From this assumption it follows directly that velocities – distances divided by times – must follow the same rules of addition as distances, the times simply cancelling out (and, of course, not even being represented in such diagrams as Kant's). In relativistic kinematics, on the other hand, times and distances, as determined with respect to different moving spatial frameworks, are inextricably interrelated according to the so-called Lorentz transformation equations. On the basis of these equations a four-dimensional space-time can be constructed in which the *interval* between any two space-time points is given by the expression:

$$c^2 dt^2 - (dx^2 + dy^2 + dz^2),$$

where $x$, $y$, and $z$ are the three space coordinates, $t$ is the time coordinate, and $c$ is the velocity of light.

The metric of this space-time is termed pseudo-Euclidean; it is very similar to that of Euclidean space except that the interval can vanish even for pairs of distinct space-time points.

To return now to the question of finding how $v_S$ is related to $v$ and $V$: our recourse is naturally to the Lorentz transformation equations for distances and times, on the basis of which the relation between $v_S$ and $v$ can be calculated. The result of this calculation is as follows:

$$v_S = \frac{v - V}{1 - vV/c^2}.$$

Furthermore, on considering a velocity-space in which the rule for adding and subtracting velocities is given by the above expression, one finds that the space in question is Lobachevskian (or hyperbolic). Such a space possesses a constant negative curvature; by comparison, the curvature of Euclidean or pseudo-Euclidean space is zero[12]. We see then that Kant's reversal of relative space procedure would provide an intuitive representation of relativistically composite motions only if we were capable of grasping intuitively the addition of distances in Lobachevskian space. This question of the possibility of an intuition of non-Euclidean space has been often discussed but will not be considered here. It is worth mentioning, however, that recent experimental studies of binocular visual space suggest that it possesses precisely a constant negative curvature.[13]

III. THE LAWS OF MECHANICS

Kant (1883, Part 3) formulates three laws of mechanics. The mathematical character of the first two of these – the principle of mass conservation[14] and the principle of inertia[15] – requires no comment. The third of Kant's laws – the action-reaction principle – will presently be discussed in detail. First, however, an historical puzzle presents itself. Anyone today might be forgiven for wondering how Kant could have omitted Newton's second law of motion, which is universally understood in modern textbooks, usually in the form $F = ma$, as the heart of 'Newtonian mechanics'. Unable to give a definitive solution to the puzzle, I simply call attention to the fact that in the eighteenth century the equating of force to the product of mass and rate of change of velocity was by no means universally taken as a fundamental principle of mechanics, and this for numerous reasons (Hankins, 1967). Some of these reasons may have influenced Kant (though the demands of his architectonic can hardly be overlooked).

Kant's (1883, p. 223; IV, 544) third law of mechanics states that "In all communication [*Mittheilung*] of motion, action [*Wirkung*] and reaction [*Gegenwirkung*] are always equal to one another". In the proof of this principle – which Newton, it will be recalled, had adopted as an axiom – Kant uses once again the procedure I have called reversal of relative space. In addition, Kant appeals to universal metaphysics for the proposition that all external action is reciprocal action. The essence of this proof, as we shall see, is a straightforward application of the relativity of motion principle, but the motivation for this application seems to me somewhat obscure, as are certain details in the proof itself. What Kant *says* is that any interaction of two bodies in which motion is communicated from one to the other is inconceivable unless we think of each of the two bodies as having been initially in motion. Without being at all certain, I believe that Kant's reasons for this assertion are as follows. Since the interaction must be reciprocal, and since *a body can only move another body insofar as it is itself moving*,[16] it follows that each of the interacting bodies must be initially in motion. But why is motion necessary to produce motion? That a body should be capable of imparting motion to another only, so to speak, as its own stock of motion is depleted, may perhaps be made plausible for interactions taking the form of impacts; and, indeed, as we shall soon see, Kant's proof is restricted to just this form of interaction. The trouble is that Kant also holds that the action-at-a-distance force of attraction and the contact force of repulsion are the fundamental forms of interaction between bodies, and it is difficult to see why the former at least can only be exerted by a *moving* body. It must be admitted that Kant does seem to make precisely this assumption about *both* attractive and repulsive forces (see note 16); the only question is whether the assumption is warranted *given Kant's theory of the dynamics of such forces*. To complicate matters further, Kant (1883, p. 225; IV, 546) seems to adopt two different attitudes toward the distinction between impact, on the one hand, and attractive-repulsive forces, on the other, in the course of his discussion of his third mechanical law. At one point the difference is not very important so that the application of the law to attraction [*Zug*] follows directly from the proof for the impact case (where impact is presumably to be identified with mutual repulsion) – the only difference lies in the direction in which the bodies resist one another in their motion. On the other hand, Kant (1883, p. 227; IV, 548–549) subsequently draws a distinction between the *mechanical* law of the equality of action and reaction (which applies to impact) and the *dynamical* law of the equality of action and reaction (which applies to attraction and repulsion); and Kant even sketches what appears to be an independent proof for the dynamical version. In any event, all these are relatively minor matters given what I take to be the main point of Kant's proof in the impact case, to which we now turn.

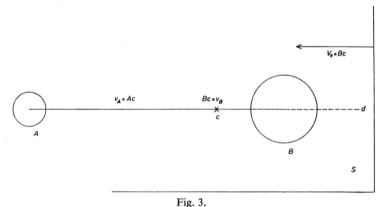

Fig. 3.

Consider two bodies of different masses, $A$ and $B$; $A$ is assumed to be moving with velocity $AB$ toward $B$ in some relative space $S$, and $B$ is assumed to be at rest in $S$ (see Figure 3). Now, in order to conceive an intuitively acceptable interaction when $A$ and $B$ collide, we must be sure (for the reason already discussed) that neither body is absolutely at rest before the impact. Here is where the idea of absolute space comes in: without, of course, knowing which of the relative spaces corresponds to absolute immobility, we simply consider an indeterminate relative space (= absolute space) and imagine the original velocity of $A$ in $S$ to be divided between the two bodies in such a way that neither is at rest in absolute space. Now, there are an unlimited number of ways in which to make this division; which shall we choose? Kant's (1883, p. 224; IV, 545) answer is that "each of the two bodies must have an equal share of the motion attributed to the one in relative space, since there is no ground for ascribing more to one of them than to the other", a conclusion which thus follows from the 'reduction' of the motion "to absolute space" [*auf den absoluten Raum reducirt*]. Here 'motion' must be understood as 'quantity of motion', or mass times velocity, so we are to suppose that the masses of $A$ and $B$ are inversely as their velocities in absolute space,[17] which is to say, $Ac/Bc = B/A$, where $Ac + Bc = AB$, and $AB$ is the original velocity of $A$ in $S$. $B$, then, is to be conceived as moving (in absolute space) toward $A$ with the velocity $Bc$, and the relative space $S$ must be conceived as moving with this same velocity $Bc$ (in absolute space). (In making these calculations we have, of course, been implicitly appealing to the principle of relativity of motion.) Kant now introduces the crucial assumption that the respective quantities of motion of $A$ and $B$, being equal and opposite, will 'reciprocally destroy one another' [*einander wechselseitig aufheben*]. It follows that both $A$ and $B$ will be at rest (in absolute space) after the impact, and hence the velocities of $A$ and $B$ in $S$ after impact will both be $-Bc = Bd$. Calculation of the action of $A$ in $S$ and the reaction of $B$ in $S$ during impact yields the following:

$$\text{action of } A = B(-Bc - 0) = B \cdot (-Bc),$$
$$\text{reaction of } B = A(Bd - AB) = A \cdot (-Ac),$$

which, in view of the inverse relation between masses and initial velocities, leads to the conclusion that action equals reaction. (Kant ignores the direction of the action and reaction, though in fact his proof suffices to show that the two are opposite in direction.)

Critical study of the above proof should quickly convince one that Kant has tacitly restricted his consideration to the special case of what are today called 'perfectly inelastic' bodies; and this is confirmed by comparison with an earlier but essentially identical version of the same proof in which Kant explicitly mentions 'abstracting entirely from elasticity' [*von aller Federkraft abstrahirt*].[18] It is indeed only for this special case that the two bodies would possess zero velocities after impact. (For a proof of this see the Appendix.)

What then is the significance of Kant's proof of the action-reaction principle? In the first place, we can see that by singling out the case of perfectly inelastic impact, Kant has in fact made a particularly felicitous choice; for – unlike the case of perfectly elastic impact, which is an ideal never fully realized by actual material bodies – perfectly inelastic impact can be readily realized by simply placing a bit of some sticky substance like pitch or wax at the point of contact of the colliding bodies. Kant may perhaps be understood here as deliberately narrowing the empirical basis of his proof to a simple and near-unmistakable judgment, namely, the recognition by direct inspection that two colliding bodies remain stuck together after impact (and hence possess the same velocity). If this is what Kant had in mind, one only wishes that he had been more outspoken about the grounds for his choice of this particular kind of impact. In the second place, by ascribing equal though opposite quantities of motion to his two unequal masses before impact, Kant has succeeded in formulating a situation which at once transcends the too special case of equal masses and yet possesses a high degree of symmetry. The point of the symmetry is that it enables Kant to conclude, without any further empirical appeal, that the two bodies must have equal and opposite quantities of motion also after impact (why should one body rather than the other have a larger quantity of motion?); from which it follows immediately that the bodies must be at rest after impact (since only zero velocity is consistent with equal and opposite quantities of motion and equal velocities). Thus, with a minimal appeal to empirical data, Kant has been able to show that action and reaction are equal during the impact of the two bodies.

## APPENDIX

### *The Equality of Action and Reaction in Impact*

In trying to see the full import of Kant's procedure (see above, pp. 103–6), we will find it helpful, I believe, to reformulate his basic impact situation in more general terms. Also, in the discussion which follows, I shall ignore Kant's idea of absolute space – as he himself did in an earlier version of the proof (cited in note 18).

Consider, then, two bodies $A$ and $B$ of unequal masses $m_A$ and $m_B$ moving toward each other with velocities $u_A$ and $u_B$; and after impact let their velocities be $v_A$ and $v_B$. Let us now introduce the so-called coefficient of restitution, $k$, a constant for any specified material composition of $A$ and $B$, defined as follows:

$$k = \frac{v_B - v_A}{u_A - u_B}.$$

We see that $k$ is a kind of measure of the fraction of the relative velocities of $A$ and $B$ which is conserved during the impact, $k = 1$ corresponding to the perfectly elastic case and $k = 0$ to the perfectly inelastic case. For $A$ and $B$ to stick together and move as a single body after impact – as they do in Kant's proof – it is clear that $k$ must be zero (so that $v_A = v_B$). This verifies my assertion (in the text, p. 106) that Kant is tacitly presupposing that special type of impact called perfectly inelastic. Indeed, it is easy to derive for this case the formula:

$$v_A = v_B = \frac{m_A u_A + m_B u_B}{m_A + m_B}.$$

However, the velocities $v_A$ and $v_B$ in Kant's case are not only equal but they are also equal to zero (that is, $A$ and $B$ are at rest after impact). This shows that some further assumption is necessary (in order to deduce $v_A = v_B = 0$). We appeal, therefore, to Kant's explicit assumption concerning the respective masses and initial velocities of $A$ and $B$:

$$\frac{m_A}{m_B} = -\frac{u_B}{u_A},$$

where the negative sign is required to take account of the opposite direction of $u_A$ and $u_B$. But since only $u_A$ and $u_B$ are involved in Kant's assumption, one clearly cannot deduce from it alone that $v_A = v_B = 0$.

We obviously need some principle relating the masses and the initial and final velocities; this principle can only be the conservation of momentum. Thus we have:

$$m_A u_A + m_B u_B = 0,$$

and

$$m_A u_A + m_B u_B = m_A v_A + m_B v_B,$$

from which it follows that $m_A v_A + m_B v_B = 0$. And this, together with our earlier result that $v_A = v_B$, leads immediately to the conclusion that $v_A = v_B = 0$.

We can now see that, while Kant's proof holds for the special case of perfectly inelastic impact, it is certainly not valid in general. In particular, Kant's key assumption that the equal and opposite motions of the two bodies will 'reciprocally destroy one another' is definitely not true in general. For, consider any impact for which the condition $m_A u_A + m_B u_B = 0$ holds, and $k \neq 0$; then, after impact, $v_B - v_A = k (u_A - u_B)$, so that $v_B \neq v_A$ (since we assume $u_A \neq u_B$), and so at least one of the two bodies is moving after impact, and the equal and opposite initial motions of $A$ and $B$ have not destroyed each other. Indeed, for $k = 1$, the case of perfectly elastic impact, $A$ and $B$ simply retain their velocities after impact but with reversed directions. The proof:

$$u_A - u_B = v_B - v_A, \qquad (1)$$

$$m_A u_A = - m_B u_B, \qquad (2)$$

$$m_A v_A = - m_B v_B \qquad (3)$$

Solving (3) for $v_A$ and substituting in (1), one gets:

$$v_B = u_A - u_B - m_B v_B / m_A,$$

or,

$$m_A v_B = m_A u_A - m_A u_B - m_B v_B.$$

Hence,

$$v_B = \frac{- m_A(- u_A + u_B)}{m_A + m_B}.$$

Now, using (2):

$$v_B = \left[-m_A\left(\frac{u_B m_B}{m_A} + u_B\right)\right]/m_A + m_B$$
$$= -m_B u_B - m_A u_B/m_A + m_B = -u_B.$$

*University of Texas at Austin*

## NOTES

* This work was supported in part by the National Science Foundation (Grant GS 2413). I am indebted for several helpful discussions to Dr. Jürgen Ehlers of the Max Planck Institute in Munich.

[1] The preceding discussion is based on the Preface (Kant, 1883, pp. 137–49; IV, pp. 467–79). (References to the orginal texts of Kant's writings other than the *Kritik der reinen. Vernunft* are to the volume and page of the Akademie edition. In quoting Kant (1883), I have occasionally altered the translation).

In Kant (1933) there is an amusing if not very instructive application of a metaphysical principle of natural science to the discussion of a topic in transcendental dialectic. Kant is considering transcendental illusion and he begins by arguing that error always results from "the unobserved influence of sensibility on the understanding". Therefore, he continues,

"In order to distinguish the specific action of understanding from the force which is intermixed with it, it is necessary to regard the erroneous judgment as the diagonal between two forces – forces which determine the judgment in different directions that enclose, as it were, an angle – and to resolve this composite action into the simple actions of the understanding and of the sensibility [p. 298; A295 = B351]."

What Kant has in mind here is, of course, the parallelogram rule for adding forces, which is formally equivalent to the rule for adding motions that we shall soon encounter in Kant (1883).

[2] This note (1883, pp. 11–4; IV, 474–7), has been taken a marking a turning point in the evolution of the critical philosophy by H.-J. de Vleeschauwer (1962, p. 93), who says "Kant suddenly and completely altered the situation and... substituted for the problem of objectivity [central to the first edition of *Critique of Pure Reason*] that of the limitation of pure reason to phenomena as the real Critical demonstrandum." Further, in this note "added at the last minute... Kant succeeded in bringing the function of judgment right into the foreground of the deduction [de Vleeschauwer, 1962, p. 98]."

[3] "...there is nothing real in motion itself except that momentaneous state which must consist of a force striving toward change. Whatever there is in corporeal nature besides the object of geometry, or extension, must be reduced to this force."

[4] The term derives from the Greek φορά (motion), and was used earlier by Leibniz for the study of abstract motion (as opposed to the study of force, or dynamics). See Leibniz's essay, 'The Theory of Abstract Motion', in which the term 'phoronomy' occurs (Leibniz, 1671, p. 222). (The term in common use today is, of course, 'kinematics'.) A treatise with the title *Phoronomia* (Hermann, 1716), by a pupil of James

Bernoulli is not restricted to kinematics but deals with such mechanical problems as the tension in a string and the kinetic theory of gases.

[5] It follows for Kant that neither chemistry nor empirical psychology can be a genuine natural science – chemistry, because we are unable to construct *a priori* in space the laws regulating chemical reactions in terms of the motions of the interacting particles; and empirical psychology, because mathematics is inapplicable to the phenomena of the inner sense and its laws. (1883, p. 141; IV, 470–1.) Kant's prediction that chemistry must forever remain merely empirical has, I think, been falsified by the contemporary reducibility in principle of chemical to physical phenomena. No doubt what Kant had in mind was the impossibility of accounting for such properties as color, odor, and, more especially, chemical 'reactivity' and 'affinity' in terms of the mechanical interactions of the particles of different substances. But precisely such an account *is* given by quantum mechanics even to the extent of explaining why there exist just so many (stable) chemical elements. To be sure, the basic concepts of quantum mechanics are very different from Kant's 'mechanical' concepts, the nature of which Kant (1933) alludes to in a chemical context in the following interesting passage:

...in order to explain the chemical interactions of bodies in accordance with the idea of a mechanism, every kind of matter is reduced to earths (*qua* mere weight), to salts and inflammable substances (*qua* force), and to water and air as vehicles (machines, as it were, by which the first two produce their effects) [p. 534; A646 = B674].

[6] For an elementary discussion of these matters, see Felix (1960, Chap. V).

[7] "Even time itself we cannot represent, save in so far as we attend, in the *drawing* of a straight line (which has to serve as the outer figurative representation of time), merely to the act of the synthesis of the manifold whereby we successively determine inner sense, and in so doing attend to the succession of this determination in inner sense [Kant, 1933, p. 167; Bl54]."

[8] Cf. what Kant (Zweig, 1967), pp. 170–171; XI, 245–247) says about the role of space as the "general and sufficient ground" of both the principle of inertia and the "reciprocal opposition" of the actions and motions of bodies, in his letter of January 3, 1791 to C. F. Hellwag.

[9] I have run together two of Coleridge's comments in his copy of the *Metaphysische Anfangsgründe der Naturwissenschaft*; see Schrickx (1959, pp. 167–8).

[10] What makes it easy to suppose at first glance that the desired composite representation has been achieved simply by juxtaposing the two initial line segments is the fact that no explicit temporal representation is associated with the spatial representations of the motions; one is supposed simply to remember that time is lapsing as the respective distances are traversed. If one attempts to remedy this lack by introducing a time-axis, one obtains something like the accompanying diagram (cf. the following page). But here, once again, there is no directly intuitable relation between the representations of $v$ and $2v$, though, of course, we know that $\theta_2 = \tan^{-1}(\frac{1}{2} \tan \theta_1)$.

[11] Huygens takes as his two axioms the relativity of motion principle and the assumption that when perfectly elastic and equal bodies collide with equal speeds, they will rebound with equal speeds. In his proof Huygens asks us to imagine a man standing in a moving boat and holding pendulums with equal bobs, one in each of his hands. If the bobs are moved toward one another by the man in the boat with equal speeds, $v$, and if the boat is also moving with the speed $v$, relative to the bank, then the speeds of the bobs before impact will be respectively 0 and $2v$, relative to the bank, and the

# KANT'S FORMULATION OF THE LAWS OF MOTION

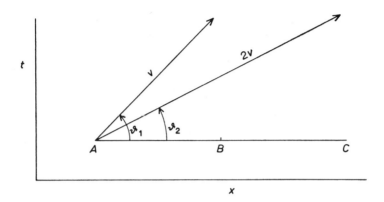

speeds of the bobs after impact will be exactly reversed, again relative to the bank. Thus a case of nonsymmetrical impact has been reduced to the symmetrical case by Huygens' device. See Dugas (1955, pp. 177–78).

[12] The mathematics of this result may be sketched as follows. We begin with the Lorentz transformations for distance and time:

$$x' = \frac{x - Vt}{\sqrt{1 - V^2/c^2}},$$

$$t' = \frac{t - \frac{Vx}{c^2}}{\sqrt{1 - V^2/c^2}}.$$

Here, for simplicity, we are considering two coordinate systems, $S$ and $S'$, in uniform motion $V$ along their mutual $x$-$x'$ axes; and $c$ is the velocity of light. Consider now a particle moving parallel to the $x$-$x'$ axes. If we designate the velocity of the particle in $S$ by $v$ and in $S'$ by $v'$, then $v$ and $v'$ will be related as follows:

$$v' = \frac{dx'}{dt'} = \frac{\frac{dx}{dt} - V}{1 - \frac{V}{c^2}\frac{dx}{dt}},$$

$$v' = \frac{v - V}{1 - \frac{Vv}{c^2}}.$$

Suppose now that we evaluate $dv'^2$, the squared increment of $v'$, by substituting $V = v + dv$ in the formula for $v'^2$. The result is as follows:

$$dv'^2 = \frac{dv^2}{\left(1 - \frac{v^2}{c^2}\right)^2},$$

where the term $vdv/c^2$ in the denominator has been ignored as very small compared with $v^2/c^2$. Since it can be shown that an identical formula holds for $dv^2$ as a function of $dv'^2$ and $v'$, it follows that $dv'^2$ is invariant (like the interval defined for space-time), and hence may be thought of as defining a metric for velocity-space. Recalling the well-known expression for 'distance' in Riemannian space:

$$ds^2 = \frac{\sum\limits_{i=1}^{n} dx_i^2}{\left[1 + \frac{K}{4} \sum\limits_{i=1}^{n} x_i^2\right]^2},$$

we see that $K$, the measure of curvature, is $-4/c^2$ in velocity-space, thereby confirming the fact that the curvature is constant and negative. Of course, we have been considering only the special case in which the velocities $v$ and $V$ are parallel, so that the resulting velocity-space is one-dimensional. For the general case, see Fock (1964, pp. 45–53).

[13] On the intuition of non-Euclidean space, see Reichenbach (1958, pp. 48–58). An older discussion in the form of an exchange between von Helmholtz (1876, 1878) and a Kantian named Land (1877) is still worth reading. Grünbaum (1963, pp. 154–57) analyzes the significance of the Luneburg-Blank studies of binocular visual space, with references.

[14] It is interesting to compare Kant's version of the principle of conservation of mass with that of Lavoisier, published three years later in 1789: "We may lay it down as an incontestible axiom, that, in all the operations of art and nature, nothing is created; an equal quantity of matter exists both before and after the experiment [Lavoisier, 1790, p. 130]." Lavoisier had actually applied the principle but without formulating it some nine years earlier (see Meyerson, 1930, p. 169). Kant's version reads as follows: "With all changes of corporeal nature, the quantity of matter remains, on the whole, the same, unincreased and undiminished [Kant, 1883, p. 220; IV, 541]."

[15] About the principle of inertia Kant (1883) makes the striking statement that "the opposite of [this principle] would be *hylozoism*, and therefore the death of all natural philosophy" [p. 223; IV, 544 (1)]. Kant goes on to distinguish the inertia of matter from the very different inert states of living organisms, which result from positive efforts at self-maintenance (what we today call 'homeostasis'?).

[16] I am interpreting Kant (1883): "one [body] cannot act on the other [body] by its own motion, unless, through approach by means of repulsive force, or at a distance by means of attraction. As now both forces always act equally and reciprocally in opposite directions, no body can act by means of it, through its motion, on another, except precisely in so far as the other reacts with equal quantity of motion. Thus no body can impart motion through its motion to an *absolutely resting* [body]... [p. 226, note; IV, 547]."

[17] Velocities satisfying this condition were termed *proper* velocities (*proprias velocitates*) by Wren (1668) in his study of perfectly elastic impact. See Dugas (1955, pp. 175–76). On the other hand, in his study of perfectly inelastic impact Wallis (1668) deduced the conclusion that two bodies with equal quantities of motion will be at rest after impact (this being precisely Kant's case).

[18] In this early essay Kant (1758, II, p. 23) is also careful to specify that the impact must be 'direct' [*gerade*] – a point which he omits to mention in Kant (1883) (though it is clear from his diagram). One must be careful to distinguish elasticity from hardness, which, as Kant himself points out (1883, p. 228, note; IV, 549), are distinct properties of bodies. Thus, when Kant says (*ibid.*) that his treatment of the action-reaction principle is independent of whether the bodies in question are absolutely hard [*absolut-hart*] or not, he is not saying – which would indeed be false – that his proof would hold for bodies of *any* degree of elasticity (see Appendix). The question of how bodies of varying degrees of hardness and elasticity behave during impact was much debated during the eighteenth and nineteenth centuries (see Scott, 1959). Newton (1934, pp. 24–25) incidentally, like Descartes, equated absolute hardness with perfect elasticity, at least in the context of a search for the laws of impact. However, subsequently, until about the middle of the nineteenth century, the prevalent view, even among Newton's disciples such as Maclaurin, was that atoms, at least, are both hard and inelastic.

## BIBLIOGRAPHY

de Vleeschauwer, H.-J., *The Development of Kantian Thought* (translated by A. R. C. Duncan), T. Nelson, London, 1962.

Dugas, R., *A History of Mechanics* (translated by J. R. Maddox), Central Book, New York, 1955.

Felix, L., *The Modern Aspect of Mathematics* (translated by J. H. and F. H. Hlavaty), Basic Books, New York, 1960.

Fock, V., *The Theory of Space, Time and Gravitation* (2nd ed., translated by N. Kemmer), Macmillan, New York, 1964.

Grünbaum, A., *Philosophical Problems of Space and Time*, Knopf, New York, 1963.

Hankins, T., 'The Reception of Newton's Second Law of Motion in the Eighteenth Century', *Archives Internationales d'Histoire des Sciences* 20 (1967), 43–65.

Hermann, J., *Phoronomia, sive de viribus et motibus corporum solidorum et fluidorum* R. & G. Wetstenios, Amsterdam, 1716.

Kant, I., *Neurer Lehrbegriff der Bewegung und Ruhe* (1758), Akademie ed., II, 13–25.

Kant, I. *Prolegomena and Metaphysical Foundations of Natural Science* (translated by E. B. Bax), Bell, London, 1883.

Kant, I., *Critique of Pure Reason* (translated by N. K. Smith), Macmillan, New York, 1933.

Land, J., 'Kant's Space and Modern Mathematics', *Mind* 2 (1877), 38–46.

Lavoisier, A-L., *Elements of Chemistry* (translated by R. Kerr), William Creech, Edinburgh, 1790. Reprinted by Dover, New York, 1965.

Leibniz, G.W., 'The Theory of Abstract Motion (1671)', in L. Loemker (ed.), *Philosophical Papers and Letters*, Vol. 1, The University of Chicago Press, Chicago, 1956.

Leibniz, G. W., 'Specimen Dynamicum (1695)', in L. Loemker (ed.), *Philosophical Papers and Letters*, Vol. 2, The University of Chicago Press, Chicago, 1956.

Meyerson, E., *Identity and Reality* (translated by K. Loewenberg), George Allen & Unwin, London, 1930.

Newton, I., *Principia* (translated by F. Cajori), University of California Press, Berkeley, Calif., 1946.

Palter, R., 'Absolute Space and Absolute Motion in Kant's Critical Philosophy,' *Synthese* **23** (1971), 47–62. Reprinted in L. Beck (ed.), *Proceedings of the Third International Kant Congress*, Reidel, Dordrecht, Holland, 1972, pp. 172–87.

Reichenbach, H., *Philosophy of Space and Time* (translated by M. Reichenbach and J. Freund), Dover, New York, 1958.

Schrickx, W., 'Coleridge Marginalia in Kant's Metaphysische Anfangsgründe der Naturwissenschaft', *Studia Germanica Gandensia* **1** (1959), 161–187.

Scott, W. L., 'The Significance of Hard Bodies in the History of Scientific Thought', *Isis* **50** (1959), 199–210.

Stein, H., 'Newtonian Space-Time', *Texas Quarterly* **10** (1967), 174–200. Reprinted in R. Palter (ed.), *The annus mirabilis of Sir Isaac Newton 1666–1966*, The M. I. T. Press, Cambridge, Mass., 1970, pp. 258–284.

von Helmholtz, H., 'The Origin and Meaning of Geometrical Axioms', *Mind* **1** (1876), 301–321.

von Helmholtz, H., 'The Origin and Meaning of Geometrical Axioms (II)', *Mind* **3** (1878), 212–225.

Wallis, J., 'A Summary Account Given by Dr. John Wallis, of the General Laws of Motion', *Transactions of the Royal Society* **3** (1668), 864–66.

Wren, C., 'Theory Concerning the Same Subject... lex naturae de collisione corporum', *Transactions of the Royal Society* **3** (1668), 867–8.

Zweig, A. (ed.), *Kant: Philosophical Correspondence*, University of Chicago Press, Chicago, 1967.

ROBERT WEINGARD

## ONTRAVELLING BACKWARD IN TIME

In his recent article 'It Ain't Necessarily So', Hilary Putnam (1962, p. 668) remarks that "in the last few years I have been amused and irritated by the spate of articles proving that time travel is a 'conceptual impossibility'."

Consequently, he attempts to remove this irritation by showing that time travel is, after all, a 'conceptual possibility'. Unfortunately, this attempt is not entirely successful. While I think Putnam has overcome some of the problems connected with showing that time travel is a 'conceptual possibility', other problems remain. By examining both the difficulties he overcomes and the ones he does not, I hope that in this paper we will get a clearer idea of what is required in order to show that time travel is a 'conceptual possibility'.

### I. PUTNAM'S TIME TRAVEL EXAMPLE

To start with, Putnam agrees that one is easily led to absurdities if one tries to describe what time travel would be like in ordinary language. However, he thinks that this difficulty can be gotten around if one can find a mathematical technique of representing all the phenomena which fall under a particular notion of time travel, because then a way of speaking corresponding to the mathematical representation can be worked out. The mathematical technique he finds for representing phenomena that might reasonably be called 'time travel' is that of world lines and Minkowski space-time diagrams. He then uses this 'mathematical technique' to present what he believes is a convincing example of time travel.

Let us, then, look at Putnam's proposed case of time travel. In Figure 1, we have the world lines of Oscar Smith and his time machine (with respect to an observer's frame of reference $t - x$). At $t_0$ the observer sees Oscar at $A$. At $t_1$ he sees Oscar at $A$, and at $B$ the creation of two objects: a person who looks and acts like an older Oscar (Oscar 3 in the Figure) and a machine containing a creature seated in it who looks like Oscar (Oscar 2 in the figure) going from being biologically older to being biologically younger, with all processes going in reverse. (Oscar 3 is the being whose world line is segment 3 of the Oscar Smith world line, Oscar 2 is the being whose world line is segment 2 of the Oscar Smith world line.) In other words, if a motion picture of the machine

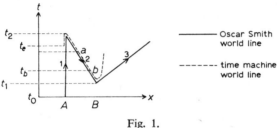

Fig. 1.

and its contents are taken from $t_1$ to $t_2$ and played backwards, all events look normal. Thus, entropy in this system is decreasing rather than increasing with time. Then, at $t_2$, this system collides with the time machine that Oscar has stepped into and they both disappear.

During a certain period of time there are three Oscars: Oscar$_1$, Oscar$_3$, and Oscar$_2$ living his strange life. According to Putnam, any reasonable scientist would interpret these phenomena in terms of the above world line diagram so that instead of three Oscar Smiths there is really only one. At a time $t_e$ a little before $t_2$ Oscar gets into a time machine located at $A$ which at $t_2$ begins to run backward in time. At $B$ and time $t_1$ Oscar gets out of the time machine and begins living forward in time again for the rest of his life, while moving away from the empty time machine. Thus, Putnam has drawn the arrows on segments 1 and 3 pointing upwards to agree with the arrow on $t$, while the arrow on segment 2 points downwards.

To see now why the technique of world lines is essential to Putnam's time-travel interpretation of the three Oscars, we must remember that in classical physics we have only the notion of time simpliciter, i.e., there is simply a single universal time. Interpreting the diagram in Figure 1 classically, the ordering of events in time is determined with respect to the universal time axis $t$ which represents the 'flow' of time. Referring to the $t$ axis, then, events at $a$ in the time machine occur after events at $b$ in the time machine because $t_a > t_b$. Yet in order for the time machine and its contained Oscar$_2$ to go back in time, it must have, in some sense, developed from $a$ to $b$, events at $a$ occurring before those at $b$. But this can not make sense if events at $a$ occur after those at $b$.

This problem does not arise, however, if instead of using the notion of time simpliciter to interpret the Oscar diagram, we replace it with the notion of time respect to a frame of reference. To do this, we interpret the world line (space-time path) of a physical system as being the time axis of a frame of reference whose space axes are at rest with respect to the physical system (the physical system being at the origin). The distance between two events on a world line is thus the time (called 'proper time') interval between the occurrence of these events registered by a clock that is carried along as part of the system.

Therefore, the $t$-axis in Figure 1 is the world line of the observer (or his laboratory, say) while the world line of Oscar$_2$ and his machine is the time axis of a frame of reference which is at rest with respect to the time machine. Given this, if we now wish to suppose that Oscar has travelled back in time, we can say that although events at $a$ occur after those at $b$, with respect to the observer's frame of reference, events at $a$ occur before those at $b$ in the time machine's proper time.

In classical physics, although it would seem to Oscar$_2$ that events at $b$ occur after those at $a$, as judged by his experiences, the behavior of clocks and other physical systems in the time machine, he would be wrong since events at $a$ are simply after those at $b$. However, by dropping the idea of a single, universal time independent of any frame of reference, we have seen that Putnam can say the following: If, with respect to Oscar$_2$'s proper time, events in the time machine are temporally ordered one way (with respect to earlier-than) while they are ordered the reverse way with respect to the observer's frame of reference, then with respect to the observer's frame of reference, Oscar$_2$ is going back in time.

Usually, the mathematical techniques of world lines and Minkowski space-time diagrams are associated with the special theory of relativity. Consequently, we might take Putnam's example, if correct, as implying that time travel is compatible with special relativity. As we have just seen, however, only the ideas of proper time and time with respect to a frame of reference are essential to his example. While Putnam probably derived these ideas from special relativity, I think it is clear that he can divorce them from the particular space-time structure of special relativity. Therefore, one thing that I am not going to be concerned with in assessing Putnam's example is whether or not it is compatible with special relativity.[1]

## II. WHY PUTNAM'S EXAMPLE IS NOT CONVINCING

By using the notion of world lines and time with respect to a frame of reference, I think Putnam has overcome some of the problems of showing that time travel is a 'conceptual possibility'. But some problems remain. In particular, even supposing we are reading the Oscar diagram (or interpreting the phenomena that are diagrammed) in terms of world lines and proper time, we would still normally consider the diagram as showing the world lines of three different individuals rather than showing the zig-zag world line of one individual.

Thus, Putnam considers the following objection to his example. The phenomena just described as imaginable but the correct description of them is that $Oscar_1$ lives normally until $t_2$, when he collides with a strange system and both disappear. At $t_1$, this strange system somehow was created along with a person who resembles Oscar $Smith_1$ very closely. Therefore, the arrow on segment 2 of the Oscar diagram should also point upwards to agree with the arrows on segments 1 and 3. If you do not describe the situation in this way you will be using language with a change in usage or meaning from the ordinary way of speaking, and then this would not show that time travel is a conceptual possibility because what we want, of course, is a case that can be described as 'time travel' in terms of our present concepts of time, travel, and change.

Putnam's (1962, p. 667) reply is that rather than his description of the example involving a change in usage or meaning, it is the objector's description that is involved in "a host of difficulties which make us doubt whether to speak in this (the objector's) way is to go on using language without any change of usage or meaning." Some of the difficulties Putnam mentions are, for example, $Oscar_1$ murders someone and is not caught before he vanishes. Do we punish $Oscar_3$ for $Oscar_1$'s crime or not? On the objector's account we shall have the 'strange' situation of $Oscar_3$'s thinking he did what someone else did, and so we should console him, not punish him. And what if $Oscar_1$'s wife lives with $Oscar_3$

before $t_2$ or after $t_2$? Is it adultery or unlawful cohabitation? Putnam (1962, p. 668) remarks, "In this kind of case, to go into court and tell the story as my friend (the objector) would tell it would be to use language in a most extraordinary way." It seems to me, however, that the objector's description of Putnam's example does not involve using language with a change in usage or meaning such that we are forced to accept Putnam's description because it avoids these changes. To show this I will describe devices for the creation and annihilation of Oscars. Then, by means of these devices, I will fill in some details to Putnam's case of the three Oscars so that in this new case the objector's description is the correct one and Putnam's time-travel description is clearly ruled out.

To begin with, consider a matter transmitter that works as follows: When an object is placed in it, the machine analyzes the object's molecular and atomic structure and stores this information in its electronic memory. In the course of the analysis, the object is decomposed into a pile of its basic elements in an uncombined state. It then transmits this information by radio waves to another matter transmitter which, on the basis of the received information, reconstructs the original object out of supplies of basic elements which it contains for this purpose. If, for example, a person steps into one such transmitter in New York which then beams a directed radio signal to another one in Paris, the being that then steps out of the Paris transmitter would, I think, be naturally described as the person who stepped into the New York machine and not just as a copy of that person. For example, this might become a popular mode of travel, not just between points on earth, but between planets.

But not all situations involving matter transmitters are so routine or simple. Thus, suppose the radio beam from the New York transmitter is picked up, not only by the Paris station, but also by ones in London, Moscow, and some other cities. It seems clear that the beings stepping out of these transmitters cannot all be the person who entered the New York transmitter. And since nothing distinguishes all these 'people' from each other when they step out of the transmitters, except their position on earth, none of them is the person who walked into the New York transmitter before the transmission.

However, to develop my example we must complicate things even more. We need to further imagine that once a person's structure is recorded in the memory of a transmitter, we can affect that memory and vary the structure recorded there. For example, if Oscar$_1$ (as he was at $t_1$) went into the New York machine, we could then adjust the memory of the New York machine (or the receiving Paris machine) so the signal sent to London is Oscar$_1$ (as he was at $t_1$) but the signal received at Paris is Oscar$_3$'s (as he was at $t_1$).

The situation I propose can now be described, with reference to Figure 2. Up until $t_2$, a person named Oscar is waiting at $A$ for what he thinks will be a time machine that will take him back to $t_1$. Meanwhile, at

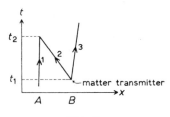

Fig. 2.

time $t_1$, at $B$, a special matter transmitter creates two beings. One is just like Oscar is at $t_1$, except that he is just enough biologically older, and he has just the right beliefs and memories, as if he had been Oscar and had waited until $t_2$ at $A$ for a time machine and then gone back to $t_1$ at $B$. This being, who is like Putnam's Oscar$_3$, lives a normal life along world line 3. The second being created at $t_1$, at $B$, is identical in every way to the first except that it is composed entirely of the antimatter counterparts of the normal particles of the first. This antimatter Oscar is created inside (and along with) the so-called time machine, which then moves off in the $-x$ direction toward $A$. We need to suppose further that in the so-called time machine are antioxygen, antimatter seats, etc., but that the outside of the machine is normal matter, and somehow inside and outside are kept from reacting.

Now the interesting thing about antimatter, as Feynman (1962) has taught us, is that it behaves as if it is normal matter going back in time. For a concrete example, consider a positron (positive electron or antielectron) travelling from $x_1$ at $t_1$, to $x_2$ at $t_2$, as shown in Figure 3. We see from the relativistic analogue of the Lorentz force law, $f_i = e(dz^j/ds) F_{ij}^2$, that changing the sign of the proper time $ds$ leaves the force $f_i$ unchanged if the sign of the charge $e$ is also changed. This means that a positron going forward in time will look just like an electron going back in time. In general, the antimatter Oscar will behave just like Putnam's Oscar$_2$. At $t_1$, when Oscar enters the so-called time machine, he and his antimatter double will be annihilated and disappear (the burst of radiation will also destroy the ordinary matter shell of the so-called time machine).

In this case, too, Oscar$_1$ can murder someone, and Oscar$_3$ can get involved with Oscar$_1$'s wife, and thus situations can arise that lead to

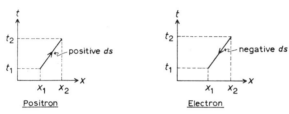

Fig. 3.

what Putnam (1962, p. 668) refers to as "a host of difficulties which make us doubt whether to speak in this way is to go on using language without any change of usage or meaning." Concerning his original example, Putnam's idea was that if we did not describe it in terms of time travel, such change of meaning difficulties would arise, which shows the appropriateness of the time-travel description. But in my modification of Putnam's example, it is clear that the objector's description is the correct one because I have provided a mechanism for the creation and annihilation of the Oscars that is, I believe, philosophically unimpeachable, being based only on current scientific theories. In this case, to use a time-travel description would be simply false. Therefore, *even if we* have to use language involving changes of usage and meaning in describing the actions of the matter transmitter's creations, there is no question of replacing our description in terms of creation and annihilation of Oscars with a time-travel description.

Thus, if such difficulties arise, the reason why they arise is not because the description of our case in terms of the creation and annihilation of Oscars is somehow faulty or objectionable. In Putnam's example, there were no matter transmitters, so the objector talked instead of men simply being created and appearing and disappearing. But if the "host of difficulties" Putnam mentions arises, we have just seen from my modification of Putnam's case that the reason they arise is not because there is something philosophically objectionable to the objector's description of the case in terms of being created and then disappearing suddenly.

What does happen, I think, is that there are many sentences that we think do not make sense because normally we cannot or do not think of circumstances where they can be used to make true statements. But then, because of unexpected discoveries and changes in the world, such circumstances come about and suddenly these sentences gain a use. Because what did not make sense does make sense now, we might say that there has been a change in the relevant concepts, or that there has been a change in meaning. But far from this kind of meaning change (if it is indeed that) being damaging to Putnam's objector, these so-called 'difficulties' arise from Putnam's description also. For example, suppose that $Oscar_3$ takes something from $Oscar_1$, who then goes to the police for assistance. Or what if $Oscar_3$ assaults and injures $Oscar_2$? According to the objector's view, $Oscar_3$ has assaulted another person and can thus be punished for this, while in Putnam's view, we have the 'strange situation' of someone assaulting himself (albeit an earlier version of himself) and thus not being guilty of a crime against another person. Rather, he should be consoled for his illness. In such cases, I do not see that Putnam's account helps to resolve the problem of what to do any better than the objector's account does.

We see from the matter transmission case that if Putnam's "host of difficulties" arises from a description, it does not mean that the description is somehow faulty or itself involved in changes in language. It is simply that on either Putnam's or his objector's account, our every-day notions of ethics, identity, etc., will be strained because the very unusual events involved are not the kind of events we are used to dealing with in terms of these notions.

### III. ENTROPY AND CAUSATION IN TIME TRAVEL

Putnam's argument from linguistic change has thus failed to convince us that his example must be described as one of time travel, rather than in his objector's terms. In another response to his objector's alternative account, Putnam remarks that the objector's account has the disadvantage of countenancing strange occurrences like someone vanishing into thin air, and someone appearing out of thin air (assuming no matter transmitter) – occurrences which, on the time-travel view, are not strange at all, for they are accounted for by the time-travel hypothesis. But now, is someone's disappearing or appearing out of thin air so much more strange or mysterious than someone's going backward in time? On the surface, I would think not. In the context of Putnam's example, having people disappearing or appearing out of thin air is strange or mysterious because there is nothing in the example to explain how such events could happen. But then, to claim that instead of having people appearing and disappearing we have a person going backward in time is to substitute one set of strange events for another, unless some idea is provided in the example of how someone could go back in time. We must ask, then, what explanation does Putnam give for how a person can go back in time?

About the time machine containing $Oscar_2$, Putnam (1962, p. 666) writes, "if we take a moving picture of this physical system during its entire period of existence, we will find that if the movie is played backward then the events in it are all normal. In short, this is a system running backward in time-entropy in the system is decreasing instead of increasing." We have two things here, then. One is that if we take a motion picture of a system and the events photographed appear normal when the picture is played backwards, then the system is running backward in time. And the second is that in a system where the entropy is decreasing instead of increasing (with respect to our time) time is running backwards because the direction of time is determined by the direction of entropy increase. In other words, the second suggestion here is that if $S_1$ and $S_2$ are temporally successive states of a (relatively) closed system, then if the entropy of $S_2$ is greater than the entropy of $S_1$, $S_1$ occurs before $S_2$ in the time of that system, i.e., in that system's proper time (where here, as before, the length of a proper time interval between $S_1$ and $S_2$ is the distance between $S_1$ and $S_2$ along the world line of $S$.)

But clearly, the fact that in a motion picture of a system, when the picture is played backwards the system appears normal, does not by itself mean that time is running backwards in that system. Consider, for a moment, a clock that is running counterclockwise, between $t_1$ and $t_2$, $t_1 < t_2$, so that a motion picture of it, when played backwards, shows the clock running normally, i.e., in the clockwise direction. This does not mean that the clock system is running backward in time because the clock may simply be made to run counter clockwise at $t_1$. That this is so is connected with the following considerations.

Suppose that an event $E$ is the cause of an event's $F$ occurring during an interval of time $\Delta t$ but that $E$ is not simultaneous with $F$ (the time of occurrence of $E$ does not overlap the time of occurrence of $F$). We would ordinarily suppose that $E$ is able to so cause $F$ to occur only because the influence of $E$ is propagated, by some process or structure, through time to the time of occurrence of $F$. For example, if I break a window by throwing a ball, the influence of my ball throw is propagated through time (and space) by the ball which broke the window.

We can think of this supposition as simply expressing the temporal version of the action-by-contact doctrine, which in this case becomes the view that there can be no action at a temporal distance. This view rules out such things as, in the above example, the ball's bouncing off the window and the window's breaking 10 minutes later (because of the bounce), unless the impact of the ball sets up vibrations or some such disturbance in the glass. Since I believe the action by temporal contact view is supported by our scientific practice, I will accept it in the discussion that follows.[3] But it should be kept in mind that the most important points to be made there can be made independently of the doctorine by using as examples *particular* events that can be non-simultaneous causes of events only by means of processes which propagate their influence through time.

Thus, if to some observer, i.e., with respect to his frame of reference $t - x$, $E$ is the cause[4] of $F$ but $E$ occurs after $F$, this must involve a time reversed process (with respect to $t$) since the influence of $E$ must be propagated by this process back through time to the time of occurrence of $F$. Just as in our discussion of a man's traveling back in time (with respect to some frame of reference $t - x$), if the influence of an event $E$ is propagated back in time to the time of occurrence of an event $F$, then along the world line of the system's propagating this influence from $E$ to $F$, $E$ is before $F$.[5] And similarly, if to some observer, i.e., with respect to his frame of reference, $E$ occurs before $F$, and $E$ is the cause of $F$, then the process or structure connecting the two through time is not a time reversed system with respect to that observer but is running in the same direction as his (proper) time. Thus, in the case of the clock's running counterclockwise, the observer making the motion picture does not consider it a time reversed system because he knows that the partial cause of the clock's running counterclockwise occurred at $t_1$ before the clock started to run.[6]

Clearly then, the time machine containing Oscar$_2$ is not supposed to be a time reversed system in virtue of the fact that when a motion picture of it is played backward, it appears normal, but this is true because the entropy in the time machine is decreasing, rather than increasing, with respect to the observer's time axis. But if the entropy of a system is decreasing rather than increasing with respect to an observer's frame of reference $t - x$, does this mean that the system is running backward in the time of $t - x$? That is, does this mean that the ordering of events in the system's proper time is the reverse of the ordering of the system's events with respect to the frame of reference $t - x$? I think we can see that it does not.

For example, suppose that with respect to our observer's frame of reference $t - x$, the entropy of a (relativity) closed system $C$ is changing between $t_1$ and $t_2$, $t_1 < t_2$, such that the entropy of the system at $t_1$ is greater than at $t_2$. Further, suppose that the entropy of the system at $t_2$ is less than at $t_1$ *because* of an event that occurs in the system $C$ at $t_1$. For the event at $t_1$ to be the cause of the entropy decrease from $t_1$ to $t_2$, it follows from our earlier discussion that its influence must be propagated through time from $t_1$ to $t_2$. This means that the ordering of events in the proper time $T$ of $C$ is the same as in the observer's frame of reference. Consequently, we do not have a system that is time reversed with with respect to the $t - x$ frame of reference, even though the entropy of $C$ is decreasing with time.

This conclusion parallels the conclusion we reached in the case of the clock. Both the clock and the above system $C$ are 'running backwards' with respect to $t - x$. But this does not make them time reversed systems with respect to $t - x$ because the cause of their unusual behavior between $t_1$ and $t_2$ occurred at $t_1$, $t_1 < t_2$. And this also explains how we were able to use matter transmitters to construct a case like Putnam's but that was clearly not one of time travel.

The crucial feature in the above argument against the entropy criterion of time ordering is that the entropy decrease in $C$ between $t_1$ and $t_2$ is caused to occur by an event $E$ at $t_1$. As a result, the direction of causal propagation is coupled to the entropy decrease such that the framework about which this entropy decrease in $C$ occurs is the same structure that propagates the influence of $E$ from $t_1$ to $t_2$. The entropy decrease, as it were, is itself propagated, or carried along, with the influence of $E$ from $t_1$ to $t_2$. This coupling prevents Putnam from replying to my argument by claiming that instead of the propagation of the influence of $E$ being a criterion of the direction earlier-to-later-than in the proper time of $C$, the entropy criterion shows that $E$'s causing $F$ is a case of backward causation with respect to $C$'s proper time $T$.

Another objection that might be raised against the above argument depends on the fact that usually it is believed that the existence of the second law of thermodynamics is what gives entropy its presumed significance in determing a time ordering with respect to the relation earlier-than. Therefore, to show that the direction of increasing entropy does not determine such a time ordering, we must use a means that does not violate the second law. It might be objected, then, that since causing the entropy of $C$ to decrease from $t_1$ to $t_2$ by means of an event $E$ at $t_1$ violates the second law, the above argument against the entropy criterion of time ordering does not work.

Clearly, causing the entropy of a closed system to decrease with time is ruled out by the phenomenological or nonstatistical form of the second law because any entropy decrease with increasing time in a closed system is so ruled out. However, as is well known, the correct form of the second law is given by statistical mechanics. And in this form, the second law is compatible with the entropy decreasing with increasing time in a closed system. Yet it might still be thought that the statistical form of the second law rules out such an entropy decrease being (at least partially) caused to occur.

However, I think that by employing a variation on the idea of Maxwell's demon, we can see how (in a closed system) a decrease in entropy from $t_1$ to $t_2$, $t_1 < t_2$, can be caused or brought about by an event at $t_1$, in a way that is compatible with the laws of thermodynamics.[7] The situation originally imagined by Maxwell was a gas at uniform pressure and temperature within a rigid and adiabatic container. Dividing the container into two parts $A$ and $B$ is a partition with a little hole in it that can be opened and closed by a little intelligent demon. While the gas is at a uniform pressure and temperature, the velocities of the individual molecules are not uniform, as Maxwell discovered when he derived his velocity distribution function. When the demon perceives a fast molecule in $A$ coming towards the partition, he opens the hole and lets the molecule into $B$. Thus, says Maxwell, the demon, without doing work, raises the temperature in $B$ and lowers the temperature in $A$ and thus apparently violates the second law. And what is relevant to our purpose, the demon brings about, or causes the entropy of the container to decrease.

Unfortunately, the idea of such a demon, as presented above, can be objected to on a number of grounds. These are mainly that the entropy decrease brought about by the demon partitioning the molecules would be nullified by the increase in entropy involved in the operation of the demon himself and in the operation of observing the molecules to see which are fast and which are slow.

But suppose that instead of an animated demon we have some sort of automatic device that does not increase entropy in maintaining and operating itself. For example, the hole in the partition might be covered by a simple shutter on the $B$ side that would allow molecules to pass from $A$ to $B$ when a molecule in $A$ collides with it, but would not allow molecules to pass in the other direction from $B$ to $A$ because the shutter is larger than the hole. Having such a shutter results in a pressure difference and so a decrease in entropy.

Unfortunately, the shutter will not work as I described it above because it cannot be more massive than the molecules hitting it or it would not open on collision. But then the behavior of the shutter would be subject to random fluctuations which, in general, would not be correlated with the fluctuations it makes use of in decreasing the entropy. This means that the shutter will open once in a while to allow molecules to pass from $B$ to $A$ which would tend to keep the pressure equalized on both sides by offsetting the movement of molecules from $A$ to $B$.

However, we do not have to make use of an automatic mechanism. All we need is some phenomena whose fluctuations involve large decreases in entropy (even if they are low probability ones) and a mechanism whose operation involves relatively small entropy increases (low friction, low power). Thus, suppose we have a pile of heavy rocks and above them a platform with a sliding trap door that can be set to open at preset times. Further, the platform is raised a small distance above the rest of the platform. There is a very small, but finite, probability that one of the rocks in the pile will fly up towards the trap door. There is an even smaller probability that at 5-second intervals the rocks of the pile will take off, one after another, towards the trap door, and in such a way that each rock's velocity is zero just above the trap door.

Now suppose further that we set the trap door to open at 5-second intervals over a period of time $\Delta t$ and then seal the platform and the rocks in an adiabatic chamber. And now, the highly improbable fluctuations happen to occur and the rocks fly up one after another, each time just as the trap door is opening and reaching zero velocity just above the door as the door is closing.[8] This keeps each rock on the platform, the rocks rolling away from the door, so the individual decreases of entropy due to each rock's flying up is stored. Thus, during the time period $\Delta t = t_2 - t_1$, the entropy decreases (partially) because of an event that occurred at $t_1$ – the event's triggering the mechanism of the door – and so the influence of that event must have been propagated from $t_1$ to $t_2$.[9]

If the trap door had not been set properly at $t_1$, and if the influence of this event had not been propagated from $t_1$ to $t_2$, the individual fluctuations that we are supposing occurred during $\Delta t$ would not have been stored up. Thus there would have been no net entropy decrease from $t_1$ to $t_2$. It is this net decrease the triggering event at $t_1$ (and events the triggering event triggers after the chamber is sealed) helped bring about, and not the individual fluctuations themselves.[10]

IV. CONCLUSION

We must conclude, from the above discussion, that Putnam has not satisfactorily explained how a person can go back in time and thus has not offered any compelling reason why we should accept his description of Oscar rather than his objector's description. However, earlier in our discussion, a possible way to show that Oscar did go back in time came to light: namely, if it could be shown that $Oscar_2$ was at $B$ at $t_1$ *because* $Oscar_1$ entered the time machine at $t_2$. Thus, if backward causation were possible, such backward causes could be used to send a person back in time. But whether backward causation is a conceptual possibility or not is the topic for another paper.

*Rutgers College*

NOTES

[1] For some discussion of this point see Earman (1967) and Graves and Roper (1965).
[2] $F_{ij}$ is the electromagnetic field tensor.
[3] Perhaps the view that there cannot be action at a temporal distance rests partly on the idea that if the time of occurrence of the cause, $E$, does not overlap with the time, $\Delta t$, of occurrence of the effect, $F$, then there is no reason for $F$ to occur during $\Delta t$ rather than at some other time. If $E$ has already occurred and is over before $F$ occurs, why should $F$ occur 10 minutes after $E$ rather than 10 hours? To put it simply, how can something that does not exist or is not occurring have any influence on what does exist to bring about the occurrence of an event – in this case $F$?
[4] Unless it makes a difference, I am referring to total and partial causes as just 'causes'.
[5] Hans Reichenbach also uses the direction of causal propagation as a criterion of time ordering. However, I differ from Reichenbach in that I use it as a criterion of time ordering only along the world line of the processes, or structure, that propagates the influence of the cause to the time of the effect. See Section 21 of Reichenbach (1958).
[6] A consequence of this discussion that is worth noting is that backward causation is not compatible with classical physics. That is, the occurrence of backward causation requires, just as time travel does, a world where the appropriate concept of time is time with respect to a frame of reference.
[7] About the example I am going to present it might be claimed that instead of being a case where the entropy is caused to decrease from $t_1$ to $t_2$ by an event $E$ at $t_1$, it is one where the event $E$ at $t_1$ causes a necessary condition of the decrease to be fulfilled. Even if this is correct, I do not think it hurts my argument because it is still true that the influence of $E$ must be propagated through time from $t_1$ to $t_2$ in order that the entropy decrease.
[8] Of course, the rocks cannot simply take off without there being an opposite and equal reaction in the ground, otherwise momentum would not be conserved. But again, I suppose the improbable circumstance that the entropy gained in the interaction between ground and rock would be much less than the entropy decrease when the rocks are stored.
Furthermore, in this case the adiabatic container must contain some of the ground or have very thick walls. To avoid such complications we could have used the more ordinary case of the partitioned container of gas, instead of the rocks and platform. In that case the shutter over the hole would be preset to open after a specified interval.
[9] It might be objected that my example violates the second law because according to this law causation is essentially a statistical concept such that cause and effect are distinguished by which has the higher or lower entropy. But this is wrong since classical mechanics is, of course, compatible with statistical mechanics and causation in classical mechanics is not statistical (for example, as in collision phenomena).
[10] I want again to point out that the above criticism of the entropy criterion of time

order does not, strictly speaking, depend on the correctness of my view that *all* nonsimultaneous causes must have their influence propagated through time in order for them to be causes. It only depends on using in the example particular events that can be nonsimultaneous causes only by means of processes which propagate the influence of these events through time.

BIBLIOGRAPHY

Earman, J., 'On Going Backwards in Time', *Philosophy of Science* **34** (1967), 211–22.
Feynman, R. P., *Theory of Fundamental Processes*, W. A. Benjamin, New York, 1962.
Graves, J. C. and Roper, J. E., 'Measuring Measuring Rods', *Philosophy of Science* **32** (1965), 39–55.
Putnam, H., 'It Ain't Necessarily So', *Journal of Philosophy* **59** (1962), 658–71.
Reichenbach, H., *The Philosophy of Space and Time*, Dover, New York, 1958.

P. J. ZWART

# THE FLOW OF TIME*

## I. THE MEANING OF THE TERM 'THE FLOW OF TIME'

Examining the way in which we usually speak about time, the impression is strongly created that time is regarded as something in motion. We regularly talk about time 'flowing', 'passing', 'flying', etc. Of course, we have an approximate knowledge as to the meaning of these terms, but what is their exact significance? When comparing the word 'flow' in the expressions 'the flow of time' and 'the flow of water', 'the flow of air', we wonder whether this word 'flow' has essentially the same meaning, or whether the resemblance of these expressions is purely superficial, that is, confined to the use of the same word 'flow'. This article aims at examining whether 'the flow of time' could be compared with 'the flow of a liquid or gas' in any real sense.

It stands to reason that we are to commence our investigations with the fundamental question: what is time? It is beyond the scope of this article, however, to examine and evaluate all the existing theories and views on time; for this I may refer the reader to other works on the subject of time (Whitrow, 1961; Zawirski, 1936). I hold Zawirski's opinion: only a moderately realistic view – as he calls it – is tenable. The idealistic view is unacceptable, since it boils down to a negation of time. For if time is something produced by the mind only, any change and becoming would be mere appearance; true reality would be immutable, i.e., timeless. This theory meets with the fundamental – and, in my opinion, insurmountable – objection that it makes the appearance of change and becoming entirely incomprehensible. For apparent change is nevertheless a change, even though it is a change of appearance only, and as such it is unthinkable without time (cf. Capek, 1961, pp. 164–165).

The purely realistic view on time is not tenable either, since it leads to Newton's absolute time. For if time exists in itself and not in something else, the consequence is that it "flows equably without relation to anything external," as Newton puts it. But even if it existed, this absolute time would be completely imperceptible and immeasurable. The only time we have access to is what Newton calls "relative, apparent and common time": the time measured by observing real processes, such as the motions of celestial bodies. The concept 'absolute time' is of no use whatoever to us, which means it has no sense, and therefore it is completely redundant. It is absolutely pointless to assume another time behind our common time.

In this way we come to the conclusion that time is relative. However, the meaning of this word 'relative' is by no means unambiguously clear. When we say that time is relative, we might for example mean that it is in some or in all respects dependent on the nature and/or attitude of the observing subject. In its extreme form this subjectivistic point of view closely approaches the idealistic view that time is appearance only. In a less extreme form it holds that time is dependent on the state the subject is in. According to modern physics, for example, the duration of a certain interval of time is dependent upon the state of motion of the observer or his clock. However, strictly speaking this is not a philosophical standpoint, but a purely scientific one, since it only relates to the *measuring* of time and not to its *nature*. Only when adopting the positivistic view that time is nothing more than what is measured by a clock, could the relativistic theory of time be considered a philosophical one. Then it could indeed be said that time in a system in motion flows more slowly than in a system at rest. However, when one is of the opinion that the scientific description of time as a quantity applied in certain formulas and measured by means of a clock is far from complete, one should restrict oneself to the statement that clocks (and processes in general) in systems in motion run more slowly than in systems at rest.

When we say that time is relative, we certainly do not mean that it is subjective in the sense that each individual has his own time. This may be true with regard to the so-called psychological time, which may creep for the one and simultaneously fly for the other, but this is certainly not true as regards the physical time measured by our clocks. Apart from its dependence on the state of motion of the observer, which only very seldom plays a part, physical time may certainly be called objective, since any observer having the disposal of a reliable clock will obtain exactly the same measuring results. But, as I already indicated, we do not observe or measure time itself; the only things we observe are events and processes (changes of state). It then should be clear, however, that time does not exist in itself, but only in the events and processes that take place. Time thus is relative in this sense that the events are fundamental and time is nothing but a concept based on certain relations among events.

This theory, usually known as the relational theory of time, is the only acceptable one in my opinion. Time is no doubt more than mere appearance; on the other hand it is doubtlessly not an independent reality. All reality it has is derived from the events. Without events there would be no time; in a universe where absolutely nothing would happen, no time would 'flow' either. Events do not just have their places in time, like pieces of wood floating in a river, but events *constitute* time. There is no flow of time beside or beneath the flow of events, but *the flow of time is nothing but the flow of events*. Therefore we should not compare events in the flow of time with objects floating in a river, but with the molecules of water the river is composed of. As the passing molecules of water constitute the flowing river, so the passing of events, i.e., their occurrence, constitutes the flow of time.

The fundamental relation from which the concept of time is derived is the relation 'before-and-after' or 'earlier-and-later'. This relation should be considered primitive, i.e. irreducible to something else. It forms part of our perceptions and may not be considered the result of our interpretation of these perceptions. Psychological investigations have also led to the conclusion that the relation 'before-and-after' (among sensations) cannot be reduced to something else. Thus Fraisse (1957, p. 75) writes: "notre perception de l'ordre des sensations n'est pas, à sa naissance, réductible à un autre mécanisme. L'ordre est donné dans l'organisation même du successif." It also appears that the perception of a succession is highly elementary and that the power to perceive what is earlier and what is later is very seldom disturbed. Even mental patients can still tell whether a sound precedes a light signal or follows it, although in every other respect they entirely miss the capability to orientate themselves as to time.

So it is obvious that the most fundamental temporal relation must be found in the relation 'before-and-after'. Moreover, it is obviously the most general time relation, for it may be said of practically each pair of sensations which one is earlier and which one is later (if this should be impossible, the relation among the sensations is another one, viz., simultaneity). By means of the relation 'before-and-after' the sensations can be ordered, i.e., they can be placed in one series in which their order of succession is completely defined (simultaneous sensations together occupy one place in this series). This series could be called the subject's own time (it is also called his psychological time). When this method of ordering is extended to events in the external world, the concept of 'physical time' is formed. The end is reached, when this operation is extended to *all* events in the universe, i.e., when we assume that all events in the whole universe can be ordered into one unique series by means of the relation 'before-and-after'. In this way the concept 'universal time', in general simply called 'time', is formed. Thus time is to be defined as the succession of events in the hypothetical series of all events. The series of events which already took place forms the past, the series of events which are still to take place forms the future, and the events which are taking place form the present. In other words: *time is the generalized relation of before-and-after extended to all events.*

## II. THE RATE OF THE FLOW OF TIME

The relational theory of time originates from Leibniz (Alexander, 1956, p. 25), who, in his third letter to Clarke, defined time as "an order of successions." Clarke immediately raised an apparently obvious objection: the order of successions may continue to be the same, whereas nevertheless the quantity of time may increase or decrease (*op. cit.*, p. 52). Leibniz' reply is not very satisfactory. "If the time is greater," he writes, "there will be more successive and like states interposed; and if it be less, there will be fewer; seeing there is no vacuum, nor condensation, nor penetration (if I may so speak), in times..." (*op. cit.*, pp. 89–90). Apparently Leibniz had not succeeded in freeing himself completely from the absolute and realistic view on time; for if he had, he could have answered Clarke quite simply that, when the order of the phenomena continues to be the same, the quantity of time will continue to be the same as well, because time *is* that order and nothing but that order!

Even in our days Clarke's objection is still frequently raised. Thus Whitrow (1961, p. 40) writes that the relational theory of time can only tell us something about the order of succession of events, but nothing about the rate in which they succeed each other. Black and Smart use the same argument in a slightly different form, namely directed against the conception of time as flowing. Thus Black (1962, p. 185) writes: "if the claim that time always flows had a literal sense, we ought to be entitled to ask how fast time is flowing.... I cannot see that any sense attaches to this question." And Smart (1963, p. 136) thinks that "we should need to postulate a hypertime with reference to which our advance in time could be measured". In this form the argument rests on an analogy between the flow of time and the flow of a liquid or a gas, an analogy pressed much too far. It is by no means a foregone conclusion that the concept of 'the rate of the flow of time' is meaningless. However, what is certain is that its meaning (if it has one) will differ widely from the ordinary meaning of 'the rate of flow', when this term is applied to a gas or a liquid. Another certainty is that it is completely wrong to postulate some kind of hypertime behind our ordinary time (cf. the arguments against Newton's absolute time). For one thing, it would not help us at all, because it would only lead to an infinite regression.

Our analysis of the concept 'the rate of time flow' has to start with an examination of the methods we use in measuring time. Time is measured by means of a clock, but what is a clock exactly? Essentially a clock is nothing but a series of events, according to our feelings following each other regularly. In fact it is a series of events chosen from the hypothetical series of all events and considered to be representative for the universal series. Consequently any clock is ultimately based on some sense of regularity being present in ourselves and presumably connected with our sense for rhythm. And, in its turn, this sense of regularity is most probably based on a clock too: an internal or physiological clock (cf. Fraisse, 1957, pp. 20–50), probably consisting of a bodily process that takes the form of a constantly recurring event, like our breathing or our heartbeat. This physiological clock was man's first clock, which he could use to test all following clocks.

For of course the physiological clock was too subjective to be very useful. What was needed was a process easily observable for everybody and consisting of a phenomenon constantly recurring at a suitable rate. At first sunrise and sunset came up to these requirements best, but, possibly because of an observed discrepancy between this clock and the physiological clock, the measurement of time was based on the culmination of the sun later on. In its turn this event was replaced by the culmination of a fixed star, and quite recently the atom clock, in which the clock events consist of the vibrations in an atom, was introduced. However, the essential thing is this: man has introduced a standard clock and measuring time simply consists of counting the number of events of this standard clock. The time interval between two successive clock events forms the unit of time length; for the physiological clock this is the second, for the sunclock or sidereal clock it is the day. The most natural form of a time-measuring instrument would therefore be a counting mechanism that could register the number of clock events.

However, this view on time and time measuring does not agree with the widespread belief that time is closely connected with motion and that measuring time consists of measuring the distance covered by a body in motion. This Aristotelian view is, however, contradicted by the fact that simply by counting time can be measured very well. This method is still widely applied, particularly when small time intervals are concerned and no clock is available. As this method of measuring time by means of counting is very general, especially with children, it is hard to maintain the opinion that the other method (measuring time by way of the distance covered by a body in motion) is more fundamental. In my opinion it is just the other way round: man's inability to construct a counting device forced him to search for other methods to measure time intervals. One of the methods hit upon is based on the motion of the sun: it is the method of determining time by checking the position of the sun (directly or by means of a sundial). But other time-measuring instruments were constructed as well, based on completely different principles. For example, hourglasses and waterclocks measure the amount of a substance (sand or water, respectively) passing through a narrow orifice. In ancient times the burning knotted rope and the striped candle were frequently used methods. The general principle of all these methods of measuring time is that time is determined by how far some continuous and uniform process has proceeded. Any process of this sort can in principle be used, whether it be mechanical, physical, chemical or even biological (aging of persons). Merely due to the easy way of measuring, man has above all selected the motion of a body as the process to base his clock on.

Anyway, the essential point is that all clocks are based on one standard clock and that the duration of a time interval is nothing but the number of standard clock events that have taken place during that interval. If we assume for matters of convenience that the sidereal clock is our standard clock, the duration of the time interval between two events *A* and *B* then consists of the number of times the earth has revolved on its axis between *A* and *B*. Our ordinary clocks can be regarded as auxiliary instruments, inserted between the standard clock and the processes to be measured.

From the above definition of 'duration' it clearly appears that the concept 'real duration' has no meaning at all. For 'duration' implies that the process in question is compared with (the process of) a clock and without this comparison any duration whatsoever is out of the question. When the clock used is the standard clock itself, we find the 'true' duration of the process; there is no duration more real than this. The 'true' duration is nevertheless not the only one possible, for if we switch over to another standard clock, the duration in question will change accordingly. In this sense the duration of a process or the length of a time interval may be called relative.

This also applies to the rate of a process, for the concept of rate is derived from the concept of duration. What we call the rate of a process is nothing but the number of events in a certain time interval (as when we say that the rate of someone's heartbeat is 80 to the minute), or the magnitude of a change of state in a certain time interval (as when we say that the temperature in a system rises 2° per second). As the duration of a time interval is equal to the number of clock events in this interval, the rate of a process is nothing but a ratio: the proportion of the number of process events to the number of clock events in the same time interval, or the proportion of the magnitude of a change of state to the number of clock events in the same time interval.

Thus 'rate', like 'duration', is not an absolute, but a purely relational concept. In themselves these concepts have no meaning; they only make sense when they are connected with a certain clock. When we speak about the rate of a process we imply we are comparing this process with another, namely with the process of the standard clock. It is true that this last process is generally not mentioned explicitly, but it is implied in the use of the units of time. The rate of a process, like its duration, has no 'real' or absolute value, because it entirely depends upon the choice of the standard clock.

But how fast does the standard clock run itself? From the above it follows that the term 'the speed of the standard clock' implies the comparison of the standard clock with itself. Of course, this is not forbidden, but it will be clear that the only answer that can be given to the question put is a tautology: a second per second, or a day per day. Asking how fast the standard clock actually runs is exactly the same as asking after the real length of the standard meter. The only possible answer to such questions is a tautology, because in both cases the value of the quantity in question is fixed on the unit by definition. Thus the duration of the time interval between two successive culminations of the same fixed star is defined as one day = 24 hours = 1440 minutes = 86 400 seconds, and this interval can never increase or decrease so long as we adhere to the same standard clock. If, for example, the earth should rotate faster on its axis we should probably have the feeling that the days were shorter than before, but nevertheless the days would continue to last 24 hours.[1] All our clocks and watches would suddenly be slow, but once we had adjusted them to the new situation we could continue in the same way as before. The only consequence would be that all kinds of processes would prove to proceed more slowly than before, and a number of constants of nature, such as the velocities of sound and light and the constant of gravitation, would appear to have become smaller. Of course, in such a case it would be simpler to decrease the number of seconds in a day by definition, so that the value of the constants of nature could be maintained. This amounts to a transition to another standard clock: one that is based on the constancy of the constants of nature.

We are now in a position to consider what should be understood by such terms as 'the rate of the flow of time'. As the flow of time is in fact the stream of events, the rate of this flow can only be the number of events in the unit of time (that is to say, per standard clock event). With the sidereal clock as our standard clock the rate of time flow is therefore the ratio of the total number of events in the whole universe in a certain period to the number of revolutions of the earth in the same period. This total number of events in the entire universe is, however, infinitely great and therefore indeterminable. It is also indefinite, because it is undefined what exactly belongs to the class of events and what does not.

So we see that the expression 'the rate of the flow of time' does have a meaning (although it is vague), but it does not have any significance. For what is the use of a quantity which cannot be measured and about which we do not know anything? The only thing we know about the rate of time flow is that it cannot be negative: events can happen, but they cannot *unhappen*. Neither is there any way as to knowing whether the rate of time flow is constant or not. If all events were to succeed each other faster or more slowly (cf. Clarke's argument against Leibniz' thesis), this would not make any difference in the rate of time flow at all, for the total number of events in the universe per standard clock event would continue to be exactly the same. This would also be in complete agreement with our observations, because we should not notice such a 'change' at all. Strictly speaking we may not even call this a change. Only an observer situated outside the stream of all events (which is a contradiction in terms, because observations are events as well) might become aware of any change here.

### III. THE DIRECTION OF THE FLOW OF TIME

If our thesis that the flow of time is nothing but the stream of events is correct, the direction of that flow cannot be anything but the order in which the events succeed each other. To put it more exactly: only when some kind of order can be discovered in the succession of events can we speak of a direction of the flow of time. In this way we arrive at a definition of the direction of time (as we shall call it henceforth for brevity's sake) that is the same as Leibniz' definition of time itself. Leibniz' definition, as a matter of fact, does not seem to me to be correct, for it implies that, if the order of succession of events should change, time itself would change too. However, I do not think that, if events were going to succeed each other in a most unusual and unfamiliar way, we should have the slightest inclination to say that *time* had become different. It is most improbable that we should drag in time at all, but if so, we should at most call it a change in the *direction* of time. In any case there would be no need to call it more than that, because in practically all respects time would have remained completely unchanged. The relation 'before-and-after' would, for example, still be transitive and asymmetrical; the relation of simultaneity would still be transitive and symmetrical; the measurement of time would hardly change, even if all clocks (including the earth) would rotate in opposite directions. For this reason a change in the accustomed order of succession of events could better be called a change in the direction of time than a change in time itself. Consequently the correct definitions are, in my opinion: time is the generalized relation of before-and-after among events; the direction of time is the order in which the events succeed each other. Thus, according to these definitions, the flow of time exclusively depends on the occurrence of discrete phenomena, whereas the direction of that flow is coherent to a certain order in their succession.

However, the relational conception of time is often defined somewhat differently, viz., with the help of the concept of 'state'. Leibniz himself seems to have had states in mind when he defined time as an order of successions, witness his reply to Clarke quoted above. However, in doing so the problem *what* states we have to consider and of what systems, is raised. In this respect the concept of event is much less problematic: every observable change in any system that is not too slow can be termed an event, and *all* events can be ordered into one series with the help of the relations 'before-and-after' and simultaneity. But we cannot use *all* states of *all* systems in our definition of time, because 'state' and 'system' are such vague concepts. Nobody can tell where one state ends and the other begins, and the same holds good for systems. Therefore the first thing to do is to specify what states of what systems are to be considered. Moreover these states and systems must be such that they can sufficiently clearly be defined and described. It is for example senseless to define time (or its direction) as the order of states of the whole universe. This definition is quite contrary to what a definition should be, for it is a reduction of the simple and known to the intricate and unknown. A sensible improvement is obtained when the universe is exchanged for some sort of small and surveyable system, as Reichenbach (1956, pp. 108, 127) for example does. Even in that case, however, the concept of state is still undefined. We are generally not concerned with the total states of the system (it is for that matter hard to see how these total states could be defined), but with certain aspects of these states only. Thus Reichenbach is merely concerned with the entropy in every state. But, of course, then it is much better to restrict oneself to the relevant aspects and leave all other state factors out of consideration.

Therefore Grünbaum's (1963, p. 259) explicit definition of the direction of time as the direction into which the entropy increases in a certain kind of system is, in my opinion, a significant improvement over Reichenbach's definition. Using a narrowly defined quantity or a combination of such quantities is always to be preferred over using the vague term 'state'.

To make our definition as general as possible, the best thing to do is probably to say that time is the generalized relation of before-and-after among phenomena, the term 'phenomena' comprising both events and states. The direction of time then becomes the order of succession of phenomena. Now when we say that time has a definite direction, one of two things can be meant only:

(1)  The events in nature succeed each other in a fixed, determinate order;
(2)  The states of a system in which a process is going on succeed each other in a fixed, determinate order. This is tantamount to saying that the process has a fixed determinate direction, or that it is irreversible.

*ad* 1. The only kinds of order in which the events in nature could conceivably succeed each other are the causal order and some kind of nomological order. The causal order comprises the following: every event has one or more causes and can have one or more effects; an effect always precedes its cause (in the limiting case they are simultaneous).[2] Events are ordered nomologically if there are laws according to which one event is always followed by another, whereas the reverse is not the case. For example, when there is a law that says that every event *A* is followed by an event *B* (under specified circumstances), but there is no law that says that an event *B* is followed by an event *A*, we may say that *A* and *B* are nomologically ordered. The nomological ordering of events is highly defective, for there are only a few laws of the kind mentioned above. Moreover, the few existing laws of this kind can usually be reduced to general causal propositions ('if an *A*, then a *B*' can be reduced to 'an *A* causes a *B*') (Zwart, 1967, pp. 95, 171). Consequently we can confine ourselves to the causal ordering of events and conclude that the only general order that can be discovered in the succession of events is this, that the effect always succeeds the cause.

*ad* 2. It is easy to ascertain that all processes in nature that proceed of themselves are irreversible. What is high falls down, what is warm cools down, what moves comes to a standstill and so on, but the reverse processes never occur spontaneously. In the middle of the last century several physicists started trying to formulate a law that would bring this obvious fact that all natural processes proceed into a certain direction under a general rule. The result of their investigations has become known as the second law of thermodynamics. This law can be formulated in several ways. A formulation that is very general, but rather vague, is the following: in nature a general trend towards leveling exists; that is to say, the differences occurring in a system (in temperature, pressure, composition, potential, etc., etc.) tend to disappear. This may be considered a generalization of the first formulation, which was given by Clausius, and reads as follows: heat flows of itself from a place of high temperature to a place of low temperature; without compensation the reverse process is impossible. In the same way we can say that gases and liquids flow from high to low pressures, that electricity flows from high to low potentials, etc. The effect in all these cases is the same: the originally existing difference becomes less and less and eventually disappears. Boltzmann, in an effort to explain this trend, showed that it is based on the (self-evident) tendency of a system to pass from a less probable state into a more probable one, or, what amounts to the same thing, to pass from a more ordered state into a less ordered one. Moreover, the probability of a state (or its degree of disorder) appeared to be connected with a certain quantity, also introduced by Clausius, the entropy. In this way Boltzmann arrived at the following formulation of the second law of thermodynamics (at present also called the entropy principle): a system has the tendency to pass from a state of low entropy into a state of high entropy. When, for example, a system in equilibrium is acted upon and some kind of process then takes place in the system, the entropy of the system will increase in this process. The successive states of the system are thus indeed ordered in accordance with a certain principle, namely, the principle of the increase in entropy.

So we see that there are two kinds of order in nature: a causal order applying to events, and an entropic order applying to the states of a system. The first question to be answered now is whether one of these orders can be identified with the temporal order, that is, with the order of before-and-after. It is obvious that the answer must be negative, for we simply do not call an event $A$ earlier than an event $B$ because $A$ is a cause of $B$, nor do we say that state $S$ is earlier than state $S'$ because its entropy is lower. Much more sensible is the view that temporal order is *based* on one of the other orders, causal or entropic, without being identical with it. In this view temporal order is a subjective, psychological reflection of an objective physical reality. In order to assess this point of view we compare the following three ways in which we can determine which of two phenomena $A$ and $B$ is earlier and which is later:

(1) We perceive both $A$ and $B$ directly, in addition to which we also perceive that one of them is earlier than the other. In the indirect form of this method we compare the points of time at which $A$ and $B$ occur, that is to say, the positions of the hands of the clock at the moments $A$ and $B$ occur.

(2) We examine whether a causal connection exists between $A$ and $B$, and if so, which of them is cause and which is effect.

(3) When $A$ and $B$ are two states of a closed system, we determine in which of them the entropy has the higher value.

Now it will be clear that the first method is far more general than the other two. For it can be applied to all possible events, whereas (2) could only be applied to events among which a causal connection exists, and (3) can only be applied to the states of a closed system. However, (1) is not only far more general than (2) and (3), but also much more fundamental. If (1) and (2) should, for example, yield different results, we might not believe our ears or eyes, but nevertheless we should have to, especially when repeated experiments would give the same results. Even if we are absolutely certain that $A$ is a cause of $B$, when we perceive that $B$ occurs first followed by $A$, we have no alternative but to change our opinion (unless another explanation is possible, for example, that the signal from $A$ requires more time to reach us than the signal from $B$). This applies to an even stronger extent to the case of (1) and (3) yielding contrary results. For example, if when stirring our coffee the homogeneous mixture should separate, the sugar crystallizing out and the cream beginning to float on the surface, we should be amazed to observe this, but still we should never suppose that the latter state was earlier than the former!

From the fact that the temporal relation of before-and-after is much more general and fundamental than the causal relation it clearly appears that the temporal ordering of the phenomena cannot possibly rest on their causal ordering. If temporal order could be reduced to causal order '*A* precedes *B*' would have to imply '*A* is a cause of *B*' which, of course, is not at all true. On the contrary, it is the reverse implication that holds: '*A* is a cause of *B*' implies '*A* precedes *B*'. This implication is true because it forms part of the meaning of the concepts of 'cause' and 'effect'; or to put it differently, it is an essential feature of the causal relation that the cause precedes the effect (although the time difference may be imperceptibly small). Neither may we conclude from this, however, that causal order could be reduced to temporal order. The relation of before-and-after and the causal relation are two completely different relations, neither being reducible to the other (cf. Zwart, 1967, pp. 92–95).

Anyway, it is clear that the temporal order is much more fundamental and primitive than the causal order. Only one condition has to be fulfilled for the former, viz., we must perceive discrete events. For the latter, however, we must also regard one event as a 'conditio sine qua non' for another (cf. Zwart, 1967, p. 29). If, for example, we should suddenly perceive that all the events we used to call the effects of other events were preceding these latter events, it is inconceivable that we should adjust the temporal order to the causal order, instead of the reverse. We should say that the world was upside down from a causal point of view, but not from a temporal one. We should therefore (if at least we survived this reversal) adjust the meaning of the terms 'cause' and 'effect' to the new situation, but the relation 'before-and-after' would remain completely unchanged.[3]

Even more absurd than the proposition that temporal order is based on causal order is the proposition that it is based on entropic order. In the former case we could at least say that the cause *must* precede the effect, but in the latter case even such a statement is impossible. If $S$ and $S'$ represent the entropies of a closed system at two moments $t$ and $t'$, the implications $t > t' \rightarrow S > S'$ and $S > S' \rightarrow t > t'$ are both incorrect. For the second law of thermodynamics may be considered either an empirical law or a statistical one, but in neither case could a logical implication be based on it. We can only say that if $t > t'$, it is highly probable that $S > S'$, but no more, and the same applies to the reverse implication. Should it appear that at an earlier moment the entropy of a system were greater than at a later moment, we should not begin to doubt the temporal order, but we should start searching for a physical explanation, for example, some action from the outside. If this decrease in entropy with time should be quite general, and if no causes could be found, we should simply have to adjust the second law of thermodynamics.[4]

One thing is quite inconceivable: that we could perceive a later event before an earlier one, or an earlier event after a later one. This kind of proposition is self-contradictory, that is to say, time reversal in this sense is *logically* impossible. Reversal of the entropic ordering is not logically impossible, however, though highly unlikely. But even if it occurred, we should only have to adjust our *laws*, not our *concepts*. As regards reversal of the causal order, this is not logically impossible either, at least if we drop the condition that a cause must precede its effect. From a logical point of view this does not need to encounter difficulties, for a cause, purely regarded as a conditio sine qua non, can in principle quite as well occur after its effect as before it. Thus Costa de Beauregard (1965, pp. 324, 331) considers the fact that causes precede their effects an empirical datum, which he calls the law of retarded causality. He even thinks that this law is essentially statistical. In any case, he does not preclude the possibility of a reversed causal order ('advanced causality'). If we wish to hang on to the condition that causes must precede their effects, however, causal reversal is logically impossible too. In that case and if the normal order of succession of events were to be reversed, the words 'cause' and 'effect' would simply have to be changed as well.

So the only direction we can attribute to time is the direction from before to after, or from earlier to later. It simply means that earlier events take place before later events. Of course this is only a tautology, but for this reason it is also completely irrefutable. We could also say that time flows from the past through the present to the future, which simply means that the past precedes the present and the present precedes the future. This is also a tautology: even if time were going to 'flow backward', we should not experience the future first and then the past, but by definition what has happened belongs to the past, and what has not yet happened belongs to the future.

There is no reason whatsoever to suppose that next to the direction of psychological time there is another direction of physical time. We only know one time, that is to say, we order all events in one single series, whether they are internal or external. Introducing a physical time apart from the ordinary or psychological time only creates new problems, such as the relationship existing between these two times. In my opinion this distinction is completely unfounded, since on the one hand all psychological events can be inserted in the 'physical' time of the standard clock, whereas on the other hand the 'psychological' terms 'past', 'present' and 'future' also have a meaning for physical systems. Nobody will deny that it is completely acceptable to say that I was glad at three o'clock and sad at four, but one often encounters statements like: "...the concepts of past, present and future have significance relative only to human thought and utterance and do not apply to the universe as such," (Smart, 1963, p. 132) or: "...nowness, and thereby pastness and futurity are mind-dependent" (Grünbaum, 1967, pp. 21–22). However, I do not see why it should be meaningless to speak about the past, present and future of a physical system like the moon, or a rock in the desert. For one thing, all actions of external things left their traces upon it, so how can it be denied that it has a past? Moreover, how could we speak of the age of the system (as is quite often done) if it had no past? Only if a system were completely unchangeable might it be said that it had no past. No such systems exist, however. The difference between the present on the one hand and the past and the future on the other hand is the same as the difference between being and not being. For something out of the present exists, whereas something out of the past or the future does not. The view taken by Smart and Grünbaum leads to a denial of the reality of time, that is, to eleatism. This view was already shown to the untenable in the first part of this article.

## IV. REVERSIBILITY AND IRREVERSIBILITY

The conclusion reached here that time does not have any direction (except from earlier to later, or from past to future – which is only a tautology) must not be confused with the – at first sight – identical conclusion of, among others, Mehlberg (1961, pp. 106–138). In my opinion all natural processes have a certain direction, which the second law of thermodynamics tries to formulate in as general a way as possible, but which may not be identified with the direction of time. Mehlberg's view is the exact opposite: the identification of the direction of time with the direction of natural processes is regarded by him as self-evident, but he denies that natural processes do have a direction. According to him such a direction must be based on laws; so we can only say that time has a direction (or that it is anisotropic) if there are events $E$ and $E'$ such that the laws of nature allow $E$ to precede $E'$, but preclude that $E'$ precedes $E$ (Mehlberg, 1961, p. 107). From this definition Mehlberg first concludes that prescientific time has no direction, because in ordinary life there are hardly any laws and therefore it is hardly ever possible to say that $E'$ cannot precede $E$. This conclusion is formally correct, but as it does not fit the facts at all, it simply proves that Mehlberg's definition is wrong. For practically all processes in everyday life are irreversible, life itself to begin with. Then there is aging and wear, the succession of the seasons, the irreversibility in one's personal life (what is done cannot be undone) etc., etc. It is a mystery to me how the isotropy of prescientific time could be defended seriously, seeing that for everybody time so conspicuously flows from his birth to his death.

With regard to scientific time, this would be thought to be clearly anisotropic (in Mehlberg's sense of the word), seeing that there is a scientific law (the second law of thermodynamics) that precisely indicates what is needed: some events can follow upon others, whereas the reverse is impossible. The validity of this law is, however, frequently disputed, among others by Mehlberg. The rejection of this law is based on two well-known objections, known as the reversibility objection of Loschmidt and the periodicity objection of Zermelo. Briefly the reversibility objection goes as follows: macroscopic processes are based on molecular motions governed by time-symmetrical laws. Consequently such processes must be completely reversible, i.e., they must proceed into one direction as often as into the other. For example, two gases would have to unmix as often as they mix. The periodicity objection runs as follows: any conceivable state of a system possesses a certain probability which is greater than zero. Consequently every state, even the most improbable, has to return regularly in the course of time, even if this should be at very, very long intervals. From these two objections it is concluded that the entropy of a closed system must decrease as often as it increases. Therefore a state of high entropy need not be later than a state of low entropy; it is quite as probable that it is earlier. From this it follows that processes in closed systems do not necessarily proceed into one definite direction; and so the second law of thermodynamics, which says that they always proceed into the direction of increasing entropy, is incorrect.

It is amazing how often this conclusion is accepted without further ado, although it is at complete variance with the facts. For there is no denying that all the processes we observe are irreversible. For example: never do gases of themselves separate from a mixture (life would be very hazardous if they did), but separate gases do always mix. Even mechanical processes, like the motion and collision of billiard balls – often regarded as reversible – turn out to be far from reversible on close inspection. For the conversion of kinetic energy (work) into heat through friction and collision always manifesting itself in such processes is quite irreversible. As a matter of fact the second law of thermodynamics in a different formulation says that complete conversion of work into heat is possible, whereas the reverse is not. As was already said in the preceding section, differences are always noticed to disappear, but never to appear spontaneously, whether it be a difference in composition or a difference in some quantity or another, such as temperature or pressure. It was exactly with the aim at bringing this tendency under a general rule that the second law of thermodynamics was formulated. And now should this law be abandoned, even though it is still in complete accordance with the facts observed? The periodicity objection does not take into account the fact that every system has only a limited life. Therefore it is almost completely certain for every system that it will long since have disappeared before it could return to a really improbable state. Before anywhere on earth for instance the oxygen and nitrogen in only one millilitre of air can separate spontaneously the earth itself is practically certain to have ceased to exist.

As regards the reversibility objection: the changes of entropy occurring in a closed system are very small, so small that in practice they are generally quite unobservable. In fact they represent nothing but extremely small fluctuations around a state of equilibrium. A thermometer, for example, put in a perfectly closed and isolated vessel containing a gas or a liquid, will finally always indicate one constant value. Even a highly sensitive thermometer would most probably never deviate from this one indication, even if it were read for years on end. Once equilibrium has been reached, no processes proceed any more. Only fluctuations take place, but in general they are quite surpassed by the measuring faults resulting from the restricted accuracy of the measuring instruments applied (which is sometimes a consequence of entropic fluctuations itself). In practice these fluctuations are generally not taken into account; in physics and chemistry the second law of thermodynamics is therefore generally accepted without restrictions. In lumping together fluctuations and processes the opponents against this law not only lose sight of the true proportions, however, they also disregard the fact that fluctuations are essentially different from processes. We shall return to this point in a moment.

As already stated, the reversibility and periodicity objections only touch certain formulations of the second law of thermodynamics, notably the one most frequently used: "in a closed system the entropy can only increase, not decrease." Several attempts were made to change this formulation in such a way that the objections mentioned could not affect it, so that it would still be suitable to base a direction of time on. Reichenbach (1956, p. 118ff.), for example, sought to accomplish this by changing over to another kind of system, the so-called branch system, defined as a system branching off from a comprehensive system and remaining isolated for some length of time. He then defined the direction of time as the direction of entropy increase in such a branch system. I fail to see, however, that this theory is in any way an improvement over the other. For entropic fluctuations occur in branch systems too, so in this respect the transition from closed systems to branch systems does not benefit us in the least. This transition has, on the contrary, several disadvantages, notably the introduction of the vague concept of branch system, and also the fact that in many branch systems (living beings for instance) the entropy decreases in time, which forces Reichenbach to define the direction of time as the direction of entropy increase in *most* branch systems. Especially objectionable in Reichenbach's treatment is the fact that the course of entropy in a branch system is connected with the course of entropy in the universe as a whole. For the entropy of the whole universe is a quantity that is absolutely unknown, that is not even definable, and about which really no sensible word can be said. Grünbaum (1963, p. 254ff.) is quite right when he leaves the entropy of the universe out of consideration, but now it becomes unclear why we have to consider branch systems. It suffices to simply state how entropy changes with time in closed or quasi-closed systems; whether these systems are branches of more comprehensive systems does not matter at all.

In this way we have arrived at our starting point again. Now it will be clear, however, that the solution of the problem should not be sought in the introduction of a different kind of system. In my opinion the solution should be sought in a differentiation between *processes* and *fluctuations*. In other words, fluctuations occurring in a system in equilibrium should not be considered changes of state; they form part of the state the system is in. We can thus accept the formulation of the second law of thermodynamics often found in text books: "processes proceeding spontaneously in a closed system are always attended by an increase in entropy," provided it is clearly understood that fluctuations around a state of equilibrium should not be considered processes. Only transitions from one state into another (generally a transition from a state of non-equilibrium into a state of equilibrium) may be considered processes. That once in a thousand years a somewhat greater fluctuation occurs, even so great that it tends to look like a change of state, may easily be taken into the bargain, particularly when we realize that such a deviation from the state of equilibrium will rapidly have vanished of itself. This formulation is certainly applicable in practice, for although completely closed systems are very rare, many existing systems can be considered quasi-closed, i.e., closed by approximation. Examples are our cup of coffee with sugar and cream and also Grünbaum's ice cube in a glass of water.

In this way the direction in which the processes proceed in nature is embodied in a general rule, which is sufficiently strict as well and which can be applied in practice. In case the use of the concept of entropy is impracticable, we can also fall back on somewhat vaguer terms, such as 'leveling', or 'increase in disorder'. This was already demonstrated above. In this way the second law of thermodynamics can, in my opinion, account for *all* irreversibility that occurs in nature.

Lately, however, several attempts have been made to find examples of a so-called nonentropic irreversibility, that is, an irreversibility which does *not* follow from the second law of thermodynamics. Best known are two examples advanced by Popper (1956a, b, 1957, 1958): (a) the wave motion resulting from a stone dropped into a pond, and (b) the expansion of a sufficiently thin gas in a vessel without walls, i.e., the infinite universe. As to the first example, I fail to see why this should be called a nonentropic irreversible process. Does not a leveling of the water surface take place in the wave motion following the original disturbance (denting of the water surface), perfectly in accordance with the second law of thermodynamics? The second example is, in my opinion, rather doubtful. Popper thinks the degree of order of the expanding gas increases, because eventually the fastest particles will all be found near the periphery of the expanding gas ball, and the slowest particles near its center. However, I am not so sure of this, because the direction of the motion of the particles (radial or tangential) also plays a part. There is also much to be said for the contrary opinion, namely that the entropy of the expanding gas increases in time, because the number of microstates increases.[5] For when the volume of the gas has become infinitely great, the number of microstates must also be infinitely great and, consequently, the entropy as well. Thus the entropy has increased from a finite to an infinite value.[6] However, I do not think it worthwhile to dwell further on this problem, because it is not only highly ambiguous (as we have seen), but it is extremely academic too. For the expansion of the gas is to take place in a vessel not only missing walls, but neither containing any other obstacle the molecules of the gas might collide with, owing to which they could return and again fill the space they had just left. Consequently the argument calls for a gas that expands in an infinite and completely empty space, which seems to me a bit hard to realize.

In this way all the arguments in favor of a nonentropic irreversibility can be met. Either the examples chosen are ambiguous, or they can be shown to fall under the second law of thermodynamics after all. And this is the way it should be, in my opinion: we should try to bring all observed irreversibility under the same law as long as possible, even if we should have to alter the definition of entropy or the formulation of the second law of thermodynamics to accomplish this. But, even if we succeed, this law remains either empirical or statistical and therefore it remains possible that the entropy of a closed system decreases instead of increases in time. Whether this happens or not, the mere possibility makes this law unfit to found the direction of time on. There is absolutely no sense in trying to convert a law of nature (and a purely empirical or statistical one at that!) into a definition of something as fundamental as the direction of time. I cannot see that it presents any advantages; on the contrary, it is bound to lead to all kinds of difficulties in connection with the direction of experienced or psychological time.

## V. TIME REVERSAL

Now we are also in a position to shed some light on a problem being frequently discussed nowadays: the possibility that the direction of the flow of time is reversed. This subject used to be considered fit for science fiction only, but since Stückelberg and Feynman suggested in scientific publications that a positive electron (or positron) might be regarded as a negative electron moving backward in time (Feynman, 1949; Margenau, 1954; Reichenbach, 1956, pp. 264–267), the possibility of time reversal (as it is mostly called) is seriously considered, and at the moment especially in scientific quarters. This idea is probably further facilitated by the use of expressions like 'the flow of time', for, if time is a flow, why should that flow not be able to reverse its direction, at least in principle? However, we have seen that we must be very careful with our analogies. The flow of time is quite unlike any other flow in that it only consists of the occurrence of events, and not of the motion of some sort of medium. We can easily understand what is meant by the reversal of (the direction of) a motion, but what are we to understand by a reversal of the direction of time?

The answer given to this question depends of course on one's conception of time. In Section III we saw that the direction of time is nothing but the order in which the events take place, and that this order is neither causal nor entropic, but simply tautological: the earlier events come first, followed by the later ones. Of course, this makes time reversal logically impossible: as the event perceived first by definition is earlier than the event perceived later, we cannot possibly perceive the later event before the earlier one (we leave the possibility that there is a significant difference in the traveling times of the two signals out of consideration). Time reversal in the strict sense of an exchange of before and after (or of earlier and later) is therefore simply a contradiction in terms.

It follows that what is usually called time reversal is not really time reversal at all. Sometimes reversal of causal order or reversal of the second law of thermodynamics is meant, but as we saw in Section III, these reversals may certainly not be called reversals of time. However, what is generally meant is that time flows back, in the sense that everything that has happened takes place again, but in the reverse direction of before. All processes reverse their directions, all systems run through the same series of states as before, only in the reverse order of succession. In this way all traces of what happened when time flowed forward are destroyed when it flows back, as if it had never happened at all. It would be like what we see when a motion picture is run off in the reverse: burning ashes become houses, pieces of glass spread over the floor jump up and join together to form a vase, etc. This notion of time reversal seems to be connected with the Newtonian conception of an absolute time flow. Consequently time reversal in this conception can only be general and universal; it cannot confine itself to part of the universe exclusively, for then it would be impossible that all traces of what had happened were obliterated. This conception of time reversal is, as far as I can see, the only one that is logically tenable.

It is obvious that it is not this kind of time reversal scientists talk about. It is a completely metaphysical concept, since such a reversal could never be demonstrated. It follows that something must be wrong with the scientific concept of time reversal. What is called time reversal is either no time reversal at all, or the concept contains some logical flaw. The latter is the case when we speak of time flowing backward in one system only, but not in the rest of the world. Thus Stückelberg and Feynman consider it possible that the time of one particle (the positron) would reverse its direction, whereas the time of its surroundings would not. This suggestion that two times could flow next to each other in reverse directions does not make sense. For 'time flows' just means 'events happen' or 'processes proceed' but how could the order in which this takes place be different for different observers? If, for example, an electron 'experiences' events in the order $A$, $B$, $C$, it is impossible that an observer would perceive first $C$, then $B$ and finally $A$. There is *one* series of events only and *one* order in which they can be perceived.

Another objection arises from the fact that the positron and the two electrons travel along different paths. In Feynman's interpretation this would mean that the electron in 'negative' time travels along a different path than it does in 'positive' time. However, this cannot be called time reversal any more. We could interpret this only as time flowing on for the electron, whereas it is standing still for its surroundings (and the world in general). But how could this motion of the electron then leave traces in its surroundings? How is it possible that the path of the positron can be photographed? In view of these difficulties it seems to me that the original interpretation in terms of pair production and pair annihilation is very much to be preferred over this interpretation in terms of time reversal.

There is one other important topic in physics in which the concept of time reversal is involved, namely, the problem of whether the laws of nature are invariant under time reversal. Physicists used to think that all laws of nature are time symmetric, except of course, the second law of thermodynamics. In the words of Eddington (1928, p. 80): "So far as physics is concerned time's arrow is a property of entropy alone." If this is true, the laws that govern the behavior of elementary particles must be time symmetric, for entropy only comes into play when we have to do with a great many particles. However, nowadays this principle of time-reversal invariance of the basic laws of nature is often challenged, for reasons that do not concern us here.[7] The only point we are interested in is: what exactly do physicists mean when they speak about time reversal in this context? As I see it they only mean that the direction of a process is reversed.[8] In a system consisting of a number of elementary particles this means that all particles reverse their motions. Now the question is whether in such a case the system runs through the same series of states as before, in reversed order of course. If this is the case the laws of nature involved are called time invariant; if not, it is said that time invariance is violated. However, no such violations have been found so far.

As has already been pointed out above, however, reversal of the directions of the processes proceeding in one system only may not be termed time reversal at all. Only if *all* particles and waves in the whole universe were to reverse their motions could we speak of time reversal. But a second condition would have to be fulfilled as well: the universe would have to run through exactly the same states as before, but only in the reverse order. And this is where the 'experiments in time reversal' come in. If time invariance were to be proved violated, even in one instance only, the second condition could not be fulfilled, which means that time reversal would be impossible.

Thus the experiments in question are not really experiments in time reversal, but they are experiments in the reversibility of elementary processes. From their results it can then be concluded whether in principle time reversal would be possible. As up to now no instances of irreversible elementary processes have been found, we have to conclude that at the present moment our knowledge of nature does not make time reversal a priori impossible. It goes without saying that this fact does not make time reversal any less improbable.

## NOTES

* The subject of this paper is dealt with more fully in Zwart (1971), which is, however, written in Dutch.

[1] So long as we have not changed over to another standard clock we cannot, strictly speaking, talk about the earth rotating on its axis slower or faster at all, for the revolving rate always remains the same by definition: one revolution in one day = 24 hours, etc.

[2] Here we are only concerned with so-called proximate causes, i.e., events, and not with ultimate causes, i.e., state factors. Cf. Zwart (1967, pp. 136, 140–141).

[3] For a more extensive treatment see Lacey (1968).

[4] Bridgman (1955, p. 251) is of the same opinion: "In no case is there any question of time flowing backward, and in fact the concept of a backward flow of time seems absolutely meaningless.... If it were found that the entropy of the universe were decreasing, would one say that time was flowing backward, or would one say that it was a law of nature that entropy decreases with time?"

[5] The entropy $S$ of a gas in a (macro) state comprising $W$ microstates is defined as follows: $S = k \ln W$. $W$ also indicates the number of ways in which the state in question can be realized.

[6] de Beauregard (1965, p. 319) arrives at approximately the same conclusion: "I believe that the expansion of a thin gas implies an *increase in disorder*, that is of a (non-Boltzmannian) *entropy*."

[7] For a recent exposé, see Overseth (1969).

[8] For the problem of the T-invariance of the laws of nature see Bunge (1968, pp. 379–381).

## BIBLIOGRAPHY

Alexander, H. B. (Ed.), *The Leibniz-Clarke Correspondence*, Manchester University Press, Manchester, 1956.

Black, M., *Models and Metaphors*, Cornell University Press, Ithaca, 1962.

Bridgman, P.W., *Reflections of a Physicist*, Philosophical Library, New York, 1955.
Bunge, M., 'Physical Time: The Objective and Relative Theories', *Philosophy of Science* **35** (1968), 355–388.
Capek, M., *The Philosophical Impact of Contemporary Physics*, Van Nostrand, Princeton, 1961.
de Beauregard, O. Costa, 'Irreversibility Problems,' in Y. Bar-Hillel (ed.), *Logic, Methodology and Philosophy of Science, Proceedings of the 1964 International Congress*, North-Holland, Amsterdam, 1965, pp. 313–342.
Eddington, A. S., *The Nature of the Physical World*, Cambridge University Press, Cambridge, 1928.
Feynman, R. P., 'The Theory of Positrons', *Physical Review* **76** (1949), 749–759.
Fraisse, P., *Psychologie du Temps*, Bibliothèque Scientifique Internationale, Paris, 1957.
Grünbaum, A., *Modern Science and Zeno's Paradoxes*, Wesleyan University Press, Middleton, Conn., 1967.
Grünbaum, A., *Philosophical Problems of Space and Time*, Knopf, New York, 1963.
Lacly, H., 'The Causal Theory of Time', *Philosophy of Science* **35** (1968), 332–354.
Margenau, H., 'Can Time Flow Backward?', *Philosophy of Science* **21** (1954), 79-93.
Mehlberg, H., 'Physical Laws and Time's Arrow', in H. Feigl and G. Maxwell (eds.), *Current Issues in the Philosophy of Science*, Holt, Rinehart and Winston, New York, 1961.
Overseth, O. E.,'Experiments in Time Reversal', *Scientific American* **221** (1969), 89–101.
Popper, K. R., 'The Arrow of Time', *Nature* **177** (1956), 538. (a)
Popper, K. R., 'Irreversibility and Mechanics'. *Nature* **178** (1956), 382. (b)
Popper, K. R., 'Irreversible Processes in Physical Theory', *Nature* **179** (1957), 1297.
Popper, K. R., 'Irreversible Processes in Physical Theory', *Nature* **181** (1958), 402–403.
Reichenbach, H., *The Direction of Time*, University of California Press, Berkeley, 1956.
Smart, J. J. C., *Philosophy and Scientific Realism*, Routledge and Kegan Paul, London, 1963.
Whitrow, G. J., *The Natural Philosophy of Time*, Nelson, London, 1961.
Zawirski, Z., *L'évolution de la notion du temps*, Académie Polonaise des Sciences et des Lettres, Krakau, 1936.
Zwart, P. J., *Causaliteit*, Van Gorcum, Assen, 1967.
Zwart, P. J., *Het Mysterie Tijd*, Van Gorcum, Assen, 1971.

# PART II

# GEOMETRY OF SPACE AND TIME

JULES VUILLEMIN

## POINCARÉ'S PHILOSOPHY OF SPACE

There is in the story of mathematics a problem, the so-called Riemann-Helmholtz-Lie problem of space, to which Poincaré gave an important and final contribution concerning two-dimensional space. Of this problem I shall only speak allusively, although mathematical methods are presupposed everywhere by Poincaré's philosophy of space. Even when concepts seem to be borrowed from experience, idealization has to transform them into geometrical tools.

I intend to distinguish three meanings of space's conventionality in Poincaré. The first one will be analyzed in the first and the second parts of this lecture, which respectively show how a new theory of the associations of ideas allows us to define displacements, and how these displacements form a group by which space is endowed with continuity and unlimitedness. The third part will be on a second sense of convention, which results from a higher-level abstraction. Joined to the inductive definition of dimension Poincaré introduced into topology, this allows us to give space its three dimensions. The fourth and last analysis will deal with the third and more popular sense of convention, which is wellknown under the name of the Duhem-Quine thesis; space will then be shown to be homogeneous and isotropic.

### I. HOW DO WE ASSOCIATE IDEAS BY COMPENSATION?

Psychologists and philosophers till now seem to be acquainted with only two kinds of association: by resemblance and by contiguity. Contiguity was further studied either by Pavlov's passive method or by Skinner's active one. Russian dogs are maintained stock-still in their harness, while American rats freely run around and press levers.

Poincaré's theory, although he insisted on activity, is completely original and had remained ignored. Surely, before Poincaré, people denied space had an a priori and intuitive character; they remarked that, being common to different sensations, it necessarily was a concept of an associative nature. But how the association worked, nobody explained (with the possible exception of Helmholtz). Poincaré was the first to clearly answer the question, "How can we construct space from spaceless impressions?"

He put the question as follows, "How do we distinguish, among our impressions, between those which correspond to changes of state and those which correspond to changes of position?"

Through resemblance we become aware of two kinds of impressions. Some are external and involuntary. Others are internal and voluntary. Being purely empirical, this classification does not always succeed: some impressions remain unclassified, and there is a borderline where belonging to one or another class is ambiguous. But we cannot expect more from an empirical classification: it allows us to say 'generally', not to say 'in all cases'. Moreover, imperfect as it is, it is enough and works quite well. We observe too that impressions which are classed as active and internal consist of sequences of kinesthetic sensations. On the other hand, as it is natural for our species, only tact and sight will detain us among our passive and external impressions.

Poincaré now makes a fundamental observation.

Among external changes (I mean changes of external impressions!), some can be compensated for or cancelled by an internal change, which restores our impressions to their primitive state. But we also experience other external changes which cannot be compensated for by an internal one. This is the ground for a new classification of our external impressions. We call changes of position those which can be compensated, and changes of state those which cannot. Moreover, we remark that among changes of our internal impressions only some of them are able to perfectly compensate the changes of position: they are those that we experience when we move our body in one block without change of attitude; moves which are accompanied by changes of attitude cannot compensate external changes or can do so only imperfectly. We thus separate among all our changes of impressions two distinguished and correlative classes: the class of external changes which can be compensated for by one of those internal changes occurring without any change of the body's attitude.

While our changes of impressions are subjective, this last classification allows us to define by abstraction new objective entities, which we call displacements and each of which correspond to a definite class of our distinguished changes of impressions. I do not take the word 'displacement' here in the Euclidian sense, the Euclidian group of displacements comprehending translations, rotations and symmetries, but in a more general sense, where as we shall see, a displacement is any element of the group of motions.

Let the Greek letters, $\alpha$, $\beta$, ... and the capital letters $S$, $T$, ... be variables, which respectively have as values external and internal changes of impressions. Succession of two changes will be designated by the succession of their two corresponding letters, and 1 will mean that a compensation occurs. Then, objective external displacements corresponding to changes of impressions will obey the following condition of identification:

(1) $\quad (\alpha)(\beta)\{D(\alpha) = D(\beta) . =_{Df} (\exists S)(S\alpha = S\beta = 1)\}.$

We thus refer two changes of external impressions to the same objective displacement if the same change of internal impressions is capable of compensating them.

Although not under our control, there are cases where two changes of internal impressions are compensated for by a same change of external impressions. We identify then the displacements of our body, we associate with them:

(2) $\quad (S)(T)\{D(S) = D(T) . =_{Df} (\exists \alpha)(\alpha S = \alpha T = 1)\}.$

Notice that these two definitions make sense only if empirical compensations exist. We can imagine worlds where they would not exist; that the world which we live in allows such compensations, we learn by experience. In this sense, geometry supposes experience.

Experience teaches us a third law too:

(3) $\quad (\alpha)(\beta)(S)(T)\{(S\alpha = S\beta = 1 . \alpha T = 1) \supset (\beta T = 1)\},$

which allows us to conclude that $D(S) = D(T)$. The two first laws disclose a reciprocal duality between external and internal impressions; the third class is self-dual. The first two suppose a world where compensations exist, and they make it possible to pass from impressions to displacements. The third law merely speaks about the world of displacements. It supposes commutativity of compensations, and says that if a member of an external displacement is compensated for by any internal change of impressions, compensation will be preserved by every member of the corresponding internal displacement.

Passing from impressions to displacements, from subjectivity to objectivity, involves among our impressions an entirely new principle of classification. Before taking notice of compensation, we considered two impressions as identical, in the sense of empirical indiscernibility, when we were unable to find between them a just noticeable difference. This identification I call a natural one. On the contrary, compensation theory puts into the same class of external changes, not only indiscernible ones, but all those which can be compensated for by the same internal change. The third law even makes it possible to eliminate any reference to this last particular impression, and, with it, to all natural identifications. To begin with, let $\alpha$ be the given external change in a first class $E$, assuming there is a compensatory change $S$ which will then belong to a second corresponding class $I$. Then, put with $\alpha$ in $E$ every $\beta$ which is compensated for by $S$, and with $S$ in $I$ every $T$ which compensates for $\beta$. You never need reference to sameness of impressions; belonging to one of the two classes merely means being compensated by or being able to compensate any member of the dual class. So, you experience that the same visual change can be cancelled by many internal changes (movement of the eyeball, movement of the head, movement of the body), and the dual statement is nothing but the most natural representation of an object's same change of position.

Now, displacements are conventional, as changes of impressions are natural. Here, the conventionality of Poincare's space acquires its first meaning. In a word, there would not be an objective world without the conventions which we agree to when we classify our impressions by the means of compensation.

## II. INTERNAL DISPLACEMENTS FORM A GROUP

What are then the laws which displacements obey?

Let us now consider only internal changes of impressions. We observe that for each such change there is a compensating one. This allows us to speak of inverse internal changes for two sequences of kinesthetic impressions $S$ and $T$, which compensate each other, although except for its capacity to compensate $S$ it would be highly dubious from the psychological point of view to interpret the equality

$$T = S^{-1}.$$

We observe too that there exists a neutral internal change we call 1, which is such that

$$1 = SS^{-1} = S^{-1}S,$$

and which, psychologically, corresponds as well to the absence of kinesthetic impressions as to the compensated sequences. We already experience with these two facts how conventional our classification is, but conventional as it is, it always makes sense from a psychological point of view. On the other hand, we cannot, generally, assert by what would be the law of closure that

$$(S)(T)(\exists U)\ U = ST.$$

The succession of two sequences of kinesthetic impressions $S$ and $T$ makes sense only if the final state of the first sequence $S$ happens to be psychologically indiscernible from the initial state of the second sequence $T$. Then and only then are we allowed to consider as a new continuous sequence of kinesthetic impressions the composition $U$ of $S$ and $T$. In this case, we shall say the two sequences are concatenated. In general, it is false that any two internal changes of impressions could be concatenated, but if three such changes can, then we notice that concatenation is associative: $S(TU) = (ST)U$.

Internal changes are therefore subject to the laws of what some mathematicians call a groupoïd[1]. The law of internal composition between members of the set of these changes is not always defined. In other words, you cannot take the Cartesian product of this set by itself to define a function whose value belongs to the set. So, Galois distinguished permutations, which can be associated only if the final state of one is identical with the initial state of the following one, and substitutions, which can always be combined with one another; and he insisted on the fact that permutations are states resulting from operations, while substitutions do not differ from the operations themselves. The same remarks apply to the difference between ordinal and cardinal numbers, and between fixed oriented segments and vectors.

Now, when the law of internal composition is always defined, we obtain not just a groupoïd but a group. And the set of internal displacements forms a group, while the set of internal changes only forms a groupoïd.

The importance of this introduction of displacements consists in yielding the geometrical concept of congruence. Suppose, indeed, we only dispose of our three laws. If, at two different times, two patches of different colors appear to us in two very similar sequences, and they will so appear provided there exists the same internal change of impressions which compensate them, compensation allows us to say that these two sequences correspond to the same displacement, that is, that these displacements are congruent. But suppose two pairs of stars are seen in two very different positions. A first eyeball's move $S$ will be necessary to look successively at the two members of the first pair, and a second one, $T$, will correspond to the second pair. Now, generally, not only $S$ and $T$ will subjectively differ, but they cannot be directly concatenated. On the other hand, setting forth the second law to put $S$ and $T$ into the same class by alledging the same external change which any of them should compensate would involve us in a vicious circle.

On the contrary, displacements are always comparable and composable and you can everywhere define congruence on displacements. According to mathematicians, to say that congruence is an equivalence relation between displacements amounts to saying in a precise language that displacements form a group. Now, our third law is easily seen to warrant the transitivity of the congruence relation between displacements, if displacements can always be composed. But from the three laws I was not able to infer the closure property for the composition of displacements. Now this amounts to a real difficulty. How do we know that two displacements are equal, when the two internal changes, which respectively compensate them, are not connected as in the third law by a common external change they compensate together?

Poincaré often blurs the distinction between impressions of internal changes and internal displacements. And he went so far as to speak of the group of our impressions. This could destroy the duality between internal and external changes and above all would entail our giving up the conventional definition of internal displacements. To avoid such disastrous conclusions, it is better to suppose the responsibility of Poincaré's confusions lies in his elegant, but sometimes loose style together with his propensity to a weak phenomenalism. On the other hand, his own firm distinction between representative and geometrical space and his constant opposition to empiricism give us systematic reasons to correct his hesitations and to frankly admit that, although displacements are defined by impressions, compensation being one of them, laws of displacements are not laws of impressions.

How can we now imagine the idealization which we need to enlarge the class of internal displacements as obtained from our three laws in order to cover all internal displacements as required by the law of group closure? Poincaré often refers the development of geometry to our handling of rigid bodies. That is another way of speaking about the peculiar features of appearances when referred to changes of position. But let us take him at his word and show by an example how such handling can be an idealization too. The problem pertaining to congruent pairs of stars is reminiscent of the problem of Thales, who wanted to find the distance of a ship from the shore. He is said to have constructed a rough compass made of a straight stick with a crosspiece fastened to it so as to turn about the fastening and to form any angle with the stick, while remaining where it was put. The observer stood in the top of a tower on the shore; the stick was fixed in an upright position and the crosspiece was directed towards the ship. The observer kept the stick vertical and the angle of the crosspiece constant, but turned the stick around, until the crosspiece pointed to some object on the land. The distance between the foot of the tower and this object would equal the distance to the ship (*Euclid* I, 26).

In any case, other idealizations have to be brought about in order to endow the group of internal displacements with its required properties. Space's infinity or, more exactly, illimitation[2] corresponds to our being able to iterate any internal displacement an unlimited number of times. Such an iteration can apply directly neither to our internal changes of impressions nor to their correlated displacements, inasmuch as those are conceived of as real behaviors of compensation. In other words, arithmetical thinking is presupposed by geometrical space. As Klein (1925, p. 193) remarked, Kant's error consisted in believing that intuition could embrace global space; but from the point of view of our representations, we are strictly unable to decide if, through a point taken outside a straight line, we can draw any or many parellels to this line as soon as we imagine the point distant enough from the line.

Illimitation called for abstract displacements in place of internal changes of impressions. In the same way, continuity demands abstract points in place of the minimal external impressions, and the complicated constructions of calculus in place of the minimal changes of external impressions, that is, the just noticeable difference of two point external impressions, which the group of our internal displacements leaves invariant. Owing to the laws of psychological thresholds, minimal impressions are only subject to physical continuity. Two of them can, at the same time, be discernible between themselves and indiscernible from a same third one. To avoid the logical contradictions which would result from taking indiscernibility, which is not a transitive relation, for identity mathematical continuity with its two orders was invented: density of rational numbers and completeness of the real numbers. The mind's creations never fit straight onto the impressions themselves, since by definition we cannot imagine, as Berkeley rightly objected to Newton, impressions smaller than the minimal ones. On the contrary, when all imagination has disappeared with the just noticeable differences, nothing prevents us from thinking of the new members of the mathematical continuum. Poincaré never tried, as Whitehead and Russell did, with I think only apparent success, to construct mathematical continuity from physical continuity by using a process such as that of extensive abstraction. For him, arithmetic and analysis are only developed by our thinking on the occasion of the experience and representation. We therefore do not claim to derive exact concepts from approximate data, although as Kant said, they begin with these. In any case, the solution we admitted for the law of closure is coherent with the general rationalism of Poincaré: the unlimitedness and continuity of space involve a new kind of idealization. Experience, he explicitly says, merely invites us to particularize the general concept of group, which "at least potentially, is preexistent in our minds," and "forces itself upon us not as a form of our sensibility, but as a form of our understanding" (Poincaré, 1902, p. 93).

It would be a pity to stop after such a good start. Up to this point we conceived of internal displacements as elements of a finite group. But we experience elementary kinesthetic changes as we experience minimal external impressions. And as we introduced the continuum of mathematical points in place of physical continuity by idealization, we can conceive infinitesimal displacements as elements to generate our group. Infinitesimals do not behave well, but by rewording them in terms of instantaneous velocities with Sophus Lie or more fashionably in terms of fibrated spaces, we can get them to do their work without committing ourselves to their difficulties. Our group of internal displacements then becomes a real continuous group of infinitesimal transformations.

Let us now compare the axiomatic construction of space by Lie with our own psychological derivation. We brought together a new theory of association by compensation and the requirements of the mind's irreducible idealizations. We thus got a part of Lie's first axiom, according to which space is a number-manifold (*Zahlenmannigfaltigkeit*), although we still ignore both how many dimensions there are and his second axiom, which says that motions of space form a real continuous group generated by infinitesimal transformations.

Relativity, conventionality and objectivity are correlative notions.

Space, being relational, must be constructed as an organization of successive data. Among the sequences of external impressions, we classify into the set of changes of position all those which we can compensate for by an internal change; here lies the convention. Since the displacements which correspond to these internal changes form a group, we look to the invariants of this group as being the objective geometrical properties of things.

An important consequence follows. Even supposing that experience would choose a determinate group, which we shall see is a false hypothesis, as soon as the relation between relativity, conventionality and objectivity had been understood, we should consider more remote and complicated relations between our impressions, and by imagining more comprehensive groups, that is, more general conventions, we would get new objects as their invariants.

In this sense, Poincaré (1916, p. 56) says that two worlds would be indiscernible if we could transform them one into the other, not only by the group of Euclidian transformations or by the fundamental group we obtain if we add dilatations, but also by the group of point transformations. Only general contact transformations are excluded, because they change space's elements. So broad, he says, is the true concept of the relativity of space.

Poincaré, unfortunately, did not develop the consequences of this widening of relativity for his conception of convention and compensation. I shall suggest a possible way out by using Klein's (1925, p. 74) distinction between active and passive interpretations of changes of coordinates. Till now, we considered them from a passive point of view. Objects were given by external impressions, our changes of internal impressions corresponding to changes of systems of coordinates. Objective geometrical relations are then interpreted as relations which are preserved by the group of such changes of coordinates. We say that the same object gives our body the same impressions, provided required compensatory moves are made. Now we can imagine our body and the object changed without changing our sensations, provided that this change retains their mutual relations. To the same tactile impression will then correspond all changes which it undergoes through the group of point transformations. By definition we cannot perceive them. But when we possess the idea of objectivity, we can abstract the group concept from its commitment to real compensation. We then take two spaces, one of which is the space of our unmoved coordinate system, the other being now itself subject to the active transformations.

This distinction perhaps gives a hint towards the ground of the difference between spontaneous representation and geometry. Poincaré's 'representative' spaces, which require the automatic activity of the nervous system, would fall under the passive point of view. On the contrary, 'geometrical' spaces would belong to the active point of view. The presence of a theory of similitudes inside the oldest parts of Euclid's *Elements* would suggest that even in order to conceive of a Euclidian space people have to free themselves from the bounds of the real compensations. Euclidian displacements indeed cease to result from such crude motions as soon as they are regarded as a subgroup of the fundamental group of the elementary geometry.

### III. THE TRIDIMENSIONALITY OF SPACE

We first have to complete Lie's first axiom regarding the dimensionality of space. A simple method would be to observe the group of internal displacements. It would seem to have six dimensions, which would correspond to the six infinitesimal transformations of the group of motions. The space so obtained would merely be 'akin' to our real space and the further reduction would have something artificial about it.

More natural is the second method, which fully uses the correlation between internal displacements and external configurations to lead us into a second and deeper understanding of compensation and convention. Moreover, let us exclude sight from examination, because visual space is too complicated with its combination of two-dimensional perspectives and binocular identifications.

Our external data are thus nothing more than the physical continuum of our tactile impressions. Our body has as many dimensions as there exist discernible parts on our skin. Indeed if two parts $A$ and $B$ of the skin are discernible, each of them will have its own succession of tactile impressions which are qualitatively completely different, so that there is no way to go continuously from one to another. As such places are very numerous; the group of displacements has a very great number of dimensions.

The construction of space's dimensionality will proceed as follows. First we shall only consider a given physical point of the skin, say, of one finger, and show that the group which leaves invariant the impression of this point has three dimensions. Then all the three-dimensional spaces which are bound to the diverse points of the skin will have to be identified.

The elements of space are points, which are only identified by the motions to which our body is subject in order to touch them. The space bound to a tactile point will then have as many dimensions as the group of internal displacements has, which leaves the contact invariant. Now, among all our internal displacements which freely alter the contact of a finger's point, we learn to distinguish three successive sets, each of which is contained as a part in the previous one, such that except for the last one, we experience them as being continuous (there always exist between two members of this set a chain of members belonging to the set and continuously joining them). We call these sets 'cuts'. If we knew geometry, we would say that the all-embracing set comprehends all motions which alter the finger's contacts in space, while the three other sets respectively correspond to a progressive limitation of this alteration, first to a determinate surface, then to a curve belonging to the surface, then to a point belonging to the curve. We thus recognize that if the last set, which leaves the contact invariant, has $n$ dimensions, the space corresponding to the all-embracing set leaving free alteration of contacts – that means, objectively, free mobility of the finger – will have $n + 3$ dimensions.

Now, observation makes us aware that we very often put into the same class all the internal displacements which correspond to sequences of kinesthetic impressions, provided those impressions leave our given point invariant. That is why $n$ becomes equal to zero. Then, the classification of our impressions becomes simpler, and the efficiency of our behavior is strengthened. What then does this new set of observations amount to?

(1) Experience has taught us that two points $M$ and $M'$, here two successive minimal tactile data, are identical when we do not experience any sequence of changes of internal impressions, or more exactly, when we have the peculiar sensation of being tired because we do not make any move. In this case, two tactile impressions of my finger's point are located at the same space's point, because tact being a contact, our internal impressions tell us that the finger's point did not move, so that its two successive contacts, which are referred to the external object $X$, are *ipso facto* located at the same point. When those contacts are enough alike, we say that there was no external displacement; when there is a difference, we interpret it by an external displacement of the object, which at two successive times, is said to occupy the same point.

(2) Conventional identifications are also explicitly used. Suppose my finger's point felt an impression, which I interpreted as resulting from its contact with the external object at space's point $M$. Next, a sequence of kinesthetic impressions, which I have learned to identify with the internal displacement $S$, takes place and is accompanied by a change of tactile impressions. Then I carry out a move, which, previous experiences have taught me, corresponds with the internal displacement $S^{-1}$, inverse to $S$. It then very frequently happens that my former tactile impression is restored. I say my finger's point is again at space's point $M$. If my expectations are wrong, I say that an uncontrolled external displacement of the object took place during my two internal and compensated displacements. So, if $\Sigma$ is the displacement which corresponds to a given sequence of kinesthestic impressions which preserve a given tactile impression, the sequence $\Sigma S S^{-1}$ will preserve it too.

(3) We also observe that there are kinesthetic impressions, which leave our finger's point invariant, although we are unable to put their corresponding displacements either into the first class of natural resemblance (derived from our impressions of staying still) or into the second class of conventional identification obtained by adding compensating internal displacements. We thus give a new enlargement to our conventional identification. To $\Sigma$ and to $\Sigma S S^{-1}$ we add $\Sigma S S^{-1} \sigma$, where we call $\sigma$ a rotation. This enlargement shows how the second law of conventional identification works. Internal displacements $\sigma$ produce

the same effect as $\Sigma$ or $SS^{-1}$: they preserve a given tactile impression, the permanence or the reviving of which makes us aware of our internal displacements as compensating processes.

We are then left with a great number of the skin's perceiving 'points', each of which is endowed with a tridimensional continuous manifold of internal displacements. If space is to have only three dimensions, all those kinesthetic spaces have to be identified. Here the second sense of conventionality will come to light. We shall argue as if all our skin were reduced to two points respectively belonging to the finger $F_1$ and to the finger $F_2$.

Suppose now the following fourfold sequence. At time one, the finger $F_2$ receives impression $A_2$ from contact with object $A$. At time two, the internal displacement $S$ changes the contact of finger $F_2$ with $A$ into the contact of finger $F_1$ with $A$, which produces tactile impression $A_1$. At time three, the impression $A_1$ of the contact of $F_1$ with $A$ is conserved, although we experience kinesthetic changes of impressions which correpond to the displacement $\sigma_1$. At time four, we carry out the move $S^{-1}$ (inverse to $S$). Experience teaches us that, generally, at the end of this move, finger $F_2$ will again experience $A_2$, that is, it will again be in contact with object $A$.

What is the meaning of our expectation? We have called $\sigma_1$ any move of our body which preserves contact of $F_1$, even if it is different from $\Sigma$ and from $\Sigma S^{-1} S$ which we have already identified. Now we notice that at time four, the primitive situation is restored; that is, contact of $F_2$ with $A$ (which produces impression $A_2$) is preserved by the whole succession of operations we performed. We are then justified in identifying this succession with $\sigma_2$, defined as preserving contact of $F_2$. So:

$$\sigma_2 = S\sigma_1 S^{-1}.$$

By similar reasoning, we would have

$$\sigma_1 = S^{-1}\sigma_2 S.$$

But these identities allow us to draw important consequences. The first one is that there exists an isomorphism between the two continua of $F_1$ and $F_2$, and each of them has three dimensions. Let us call the two continua the two spaces $E_1$ and $E_2$. Indeed, if to $\Sigma$ in $E_1$ there corresponds a point $M$, there will be a biunivocal transformation $S\Sigma S^{-1}$ which will make $N \in E_2$ correspond to $M \in E_1$. Moreover if two points of $E_1$ are identical, one easily demonstrates that their images in $E_2$ by this transformation will also be identical[3]. But this is not enough. We indeed identify these isomorphic spaces, because it very often happens that when $\sigma_1$ preserves the impression $A_1$, $\sigma_2$ preserves the impression $A_2$, while if $A_1$ is changed, $A_2$ is also changed. The object $A$, which we posit as the cause of $A_1$, is then said to be identical with the object we posit as the cause of $A_2$, since the impressions we refer to them as their causes are together preserved or changed.

Were we geometers, we would interpret our situation as follows. The groups $G_1$ and $G_2$, which are tied to our two epidermal points, are considered similar, because one results from the other by a change of the coordinate system. Moreover, our coordinate systems are equivalent. We say that two figures are equivalent when they are transformed into one another by a group transformation. Let us then consider a figure, say $A_1$, which is invariant only under the identity transformation of $G_1$, which identity transformation $I_1$ is now conceived of as all transformations of the kind $\Sigma S S^{-1} \sigma_1$. Every group transformation $S$ will transform $A_1$ into an equivalent figure $A_2$ in such a way that $S$ is uniquely determined. If now the figure formed by $A_2$ and a point $A'_2$ is equivalent to the figure formed by $A_1$ and a point $A'_1$, we shall say that $A'_2$ has the same relative position to $A_2$ as $A'_1$ to $A_1$. A primitive coordinate system will then result in another equivalent system, if we choose for $A'_2$ the primitive coordinates of the point $A_1$. The primitive system determines the law which refers a point to the primitive figure $A_1$: the new system defines by the same law the position of the point in relation to the figure $A_2$. To pass from $A_1$ to $A_2$, we use the group transformation $S^{-1}$; to pass from $\sigma_1$ to $\sigma_2$, we use the inverse transformation $S$ since $S^{-1}\sigma_2 \in I_2$, and $S\sigma_1 = S^{-1}\sigma_2$.

A set of figures is said to belong to a field if the different figures which are equivalent to any figure of the set belong to the set. Now, all our tactile points form a field. This field is finite, because its elements only depend on three arbitrary parameters, which then can be regarded as the coordinates of these elements.

Three-dimensionality of space thus results from a new level of abstraction and of convention. The first convention corresponded to the restoration or preservation of *one* impression by compensating two changes. To show that $G_1$ had only three dimensions, we could have chosen the following situations. An impression $A_1$ is preserved by $\sigma_1$; then $A_1$ is changed into $A'_1$ (always at the same finger's point $F_1$); then a new rotation is seen to preserve $A'_1$ too. We would have shown, as we do in Euclidean geometry, that a scale of angles can always be transported from one point into another one. The rotation $\Omega_1$ at the second point would have been seen to result from the rotation $\Omega$ at the first point by the sequence

$$\Omega_1 = T^{-1}\Omega T$$

in such a way that the composition of rotations at the first point is preserved when transported into the second one

$$\Omega_1 \Omega'_1 = T^{-1}\Omega T T^{-1}\Omega' T = T^{-1}\Omega \Omega' T.$$

You may think that the case is different when $\sigma_2$ preserves a contact other than $\sigma_1$. But those contacts belong to the same chain of natural resemblance, since they successively take place at the same epidermal point. We say that we move inside the same space, and we need no new convention to identify the two scales of angles at the two points.

On the other hand, we shall now identify two sequences of impressions which are qualitatively different, since they belong to two different epidermal points. Our identification therefore requires a new convention.

Two rotations $\sigma_1$ and $\sigma_2$ were previously classed into the same equivalence class of the displacements leaving a point invariant, either because they preserved the same external impression (a contact) or because they preserved in the same way two external impressions, $A_1$ and $A'_1$, of the same kind. Now, two rotations $\sigma_1$ and $\sigma_2$ are identified, because they preserve the same relation between different pairs of two external impressions, respectively, $(A_1, A'_1)$ and $(A_2, A'_2)$. Identification of rotations $\sigma_1$ and $\sigma_2$ implies that our knowledge is by structure, not by intuition. Space has three dimensions, because among our kinesthetic changes to which displacements correspond, we distinguish these which are correlators between any external changes, as referred to any point of the skin.

While natural discernibility deals with absolute data, which are called by Wittgenstein *das Mystische*, science deals with structural relations between idealized impressions. Our new convention by compensation thus gives a good illustration of a psychological source for the concept of ordinal similarity. To define the space as the form of phenomena, we must widen our classes of equivalence, and we put into the same one many sequences of internal displacements which should previously have formed different classes as preserving qualitatively different sequences of external changes.

## IV. HOMOGENEITY AND ISOTROPY

Poincaré is careful to remind us that representative spaces are neither homogeneous nor isotropic. As Epicurus noticed, verticality is a distinguished representative direction; as Kant remarked against Leibniz, our intuition allows us not only to distinguish between left and right, but also to determine what in the world is left, and what is right. So, only geometrical space will be homogeneous and isotropic, and only the group of displacements will be able to endow space with such qualities.

Geometers will say that rigid bodies, as invariant configurations of points, are permitted an idealized 'free mobility' and they correlatively speak of the 'free mobility in infinitesimal', which a real continuous group of three-dimensional space has in a real point $P$.[4]

Homogeneity entails that by suitable congruent mappings we can carry any point into any other. Isotropy entails that if we keep a point fixed we can carry any line direction at that point into any other at that same point. Finally, if a point and a line direction are kept fixed, one can by congruent mapping carry any surface direction through them into another direction. Reciprocally, if a point, a line direction through the point, and a surface direction through the point and the line are given, then there exists no congruent mapping besides the identity under which this system of incident elements remains fixed. Such are the two last celebrated axioms of Sophus Lie for geometry, which with a slight inaccuracy (Vuillemin, 1962 p. 42) are called the axioms for the free mobility of rigid bodies.

Now, our axioms do not completely determine space. The description of space as being illimited, continuous, three dimensional, homogeneous and isotropic indeed fits all spaces with constant curvatures, that is, it fits as well as the Euclidian space with curvature zero the Bolyai-Lobatschewskyan space with constant negative curvature or the Riemannian space with constant positive curvature[5]. Experience, by itself, is not able to choose among these three kinds of geometries. What recommends the Euclidian one is merely its intrinsic simplicity. If astronomical observations should one day apparently distinguish one of the non-Euclidian geometries, they necessarily would result from testing together geometrical and optical hypotheses, and simplicity would force us to keep Euclidian geometry and to change our physical axioms.

From a true premise, Poincaré (1902, p. 96) drew a false prophecy. The premise was that experience simply decides about the whole incorporated theory of geometry and physics. The prophecy is that geometry has nothing to fear from such joint confirmations, since were we obliged to change our mixed hypotheses, it would always be more economical to keep Euclideanism while changing our ideas about light rays than to keep optics while changing geometry. Despite some interpreters, this sense of the word 'convention' does not differ from the sense it has in the Duhem-Quine thesis. As Nicod noticed, Poincaré's peculiarity is merely to have believed that however complicated the joint hypothesis would become through new experiences the intrinsic simplicity of Euclidianism would always commit us to keep it unchanged, although extrinsic complications could result from combined examinations of physics and geometry.

V. CONCLUSION

It is ironic that the group concept, which seemed *a priori* so well suited to the relativism of space, was temporarily discredited by the general theory of relativity. Spaces of variable curvature were here required, which Poincaré had excluded from his examination, by alledging that they merely had analytical importance.

It seems, however, that two relevant points can be made. First, except for Poincaré's pretension to save Euclidianism in all circumstances, none of his three concepts of conventionality has to be abandoned. It is not necessary to speak for the Duhem-Quine thesis; Hegel's modern revival rather invites us to reexamine logical atomism with some sympathy and sounder arguments. Concerning the second meaning of conventionality, no defense is needed, since everyone, I think, will agree with the structural character of our knowledge. As for the first sense of group compensation, physics requires revision. Let me simply quote here Cartan (1952–55, p. 1737) according to whom the group point of view showed the true nature of the differential geometry itself as developed by Riemann: "This geometry is not opposed to projective or to algebraic geometry. It is only opposed to the geometry of complete space." Poincaré's prejudice was to extend to this complete space the idealized laws we constructed on the occasion of an experience taking place and making sense only in the neighborhood of the 'point' our body and this earth occupy. Now, Cartan showed how the group concept is amenable to express compensation no longer between tactile and kinesthetic impressions of a same point of view, but between two different views taken from arbitrarily moving coordinate systems.

By the way, we should notice that Cartan, and before him Lie and Poincaré used the word 'group' where we would speak of pseudo-groups.

I touched the second point when I spoke of the difference between internal changes and internal displacements. I insisted on a certain ambiguity in Poincaré's theory. Does Poincaré intend to analyse some mechanism of animal behavior or the origin of geometrical thinking? There are, it seems, two different problems that we should not mix. The first one belongs to psychology, namely, compensation in itself as a means of classifying impressions. The second one belongs to history, that is, the problem of displacements, inasmuch as it involves an 'active' representation of coordinate changes. Although he always pointed to the idealizations which are necessary to pass from representation to geometrical 'spaces', Poincaré was never clear about this last distinction.

*Collège de France*

## NOTES

[1] I take the word in the sense of Speiser and Brandt.
[2] This correction is made in order to keep the Riemannian geometry possible.
[3] $M = M' \in E_1$, if $\Sigma'_1 = \Sigma_1 \sigma_1$ ($\Sigma'_1$ being the sequence which in $E_1$ corresponds to $M'$). The image of $M$ will be $N$, to which $S\Sigma_1 S^{-1}$ corresponds in $E_2$; to the image $N'$ of $M'$ will correspond in $E_2 S\Sigma'_1 S^{-1} = S\Sigma_1 S^{-1} S\sigma_1 S^{-1} = S\Sigma_1 S^{-1} \sigma_2$, which means that $N' = N$.
[4] Lie (1888–93, p. 477) so introduces the following definition: "A real continuous group of the third dimensional space in the real point $P$ has free mobility in infinitesimal,

provided the following conditions are satisfied: If a point $A$ and any real line element going through it are kept fixed, their continuous motions must still be possible; on the contrary, if, apart from $P$ and this line element, any real surface element, which goes through them is kept fixed, then no continuous motion can be more possible".

[5] Klein (1925, II, p. 199) reminds us that in every non-Euclidean geometry (of constant curvature) there is a six-dimensional group of 'motions', which leave angles and distances invariant. We come to the Euclidean system when we distinguish, among the motions, the subgroup of parallel displacements; such subgroups do not exist in non-Euclidean geometries.

## BIBLIOGRAPHY

Cartan, E., *Oeuvres complètes*, Gauthier-Villars, Paris, 1952–55.
Klein, F., *Elementarmathematik vom höheren Standpunkte aus*, 3rd. ed., Springer, Berlin, 1925.
Lie, S., *Theorie der Transformationsgruppen*, III, Teubner, Leipzig, 1888–93.
Poincaré, J. H., *La science et l'hypothèse*, Flammarion, Paris, 1902.
Poincaré, J. H., *La valeur de la science*, Flammarion, Paris, 1916.
Vuillemin, J., *Philosophie de l'Algèbre*, Presses Universitaires de France, Paris, 1962.

CRAIG HARRISON

# ON THE STRUCTURE OF SPACE-TIME

## I. RELATIVISTIC VS. NEWTONIAN SPACE-TIME

What is absolute space and absolute time? We have often heard that the theory of relativity has banished them, and yet Newton's (1946) famous characterization of these ideas leaves us in the dark as to exactly what relativity has discredited:

> Absolute, true and mathematical time of itself and from its own nature flows equibly without regard to anything external. Absolute space in its own nature without regard to anything external, remains always similar and immovable.

Despite the fact that Newton assures us that the concepts of absolute space and time are not being defined, we have the same feeling of uneasiness that accompanies Euclid's (1956) equally famous characterization of straight lines:

> A line is breadthless length.
> A straight line is a line which lies evenly with the points on itself.

In each case, we are informed in almost sepulchral tones of the existence of something perfect against which all relative, operational measures are to be compared, and yet we lack the standard by which such perfection is to be understood. Thus Smart's (1963, p. 136) obvious question (reminiscent of the third man argument) about how time flow is to be measured – in seconds per what? The answer of course requires an absolute super time, and then the same question can be asked about it!

Objections to Newton's absolutism, which makes space a substance, are as old as Leibniz, and the credibility of the relational theory seemed to increase with doubts about Newton's mechanics. Yet it is just as useful to assume the existence of space and time in relativity as it is in classical mechanics. And contrary to what has sometimes been claimed, neither the special nor the general theory of relativity *requires* the elimination of space and time as such.

Reichenbach (1956, p. 25), for example, thinks the special theory does so, because the time order of a pair of events is invariant if and only if they are causally connectable. And, he might have added, the spatial order of two events (e.g., which is west of which) is determinate if and only if they are *not* causally connectable. But under what conditions are two events causally connectable, as opposed to causally connected? According to the theory of relativity, the answer is that the distance divided by the time interval between them must not exceed the speed of light. We might, perhaps, find conditions equivalent to this which do not presuppose space and time, on the basis of which a relational theory of space and time could be devised, but the point is that the theory of relativity does not *force* us to do so.

Others have held that the general theory of relativity has abolished space by showing that its very existence depends upon the presence of matter. Now, it is indeed true that in the general theory of relativity there is a certain relation between the geometry of space and acceleration, or, equivalently, the distribution of matter. It has sometimes been held that the nature of this relation is one of dependence: that the distribution of matter entirely determines the geometry of space to the extent that without matter there would be no space. If this is true, then general relativity has vindicated Mach and Leibniz against Newton, in the sense of rendering superfluous and unnecessary the premise that space has its existence and its nature 'without regard to anything external'.

It has not proved possible, however, to obtain definite solutions to the relativistic field equations without making assumptions (in the boundary conditions) about the geometry of space without matter, or with matter 'at infinity'.[1] Even if it had been possible, this would have established only that space without matter lacks a metric, not that it does not exist. Indeed, to say that space exists only when matter is present still concedes that it does exist when there is matter.

Of course, if we found a way of avoiding the need to make geometrical assumptions in the absence of matter (or with it 'at infinity'), we could, with Mach, view distance as a relation between particles only. But such a view is far from being forced on us by the theory of relativity. Instead, the assumption of space and time has proved to be a convenient basis for expressing theoretical ideas, and not a hindrance, like ether.

What, then, remains of the claim that relativity has brought about the elimination of absolute space and time from physics? For if what we have said is true, this cannot mean the elimination of space and time. The essential difference, then, between classical and modern physics must be found elsewhere. This difference can be stated most precisely in terms of ideas which the theory of relativity has made commonplace, in particular in terms of the space-time continuum. Thus, with the hindsight afforded by a well worked out alternative to classical theory, we can say that classical and modern views do not differ essentially as to the *existence* of a continuum of space-time points, but rather as to the relation between various *structures* imposed upon it.

More specifically, it is necessary to define certain relations or functions on the set of space-time points that endow it with a metric and determine its geometry. This structure must be provided *in addition* to the set of points of space-time, for assuming space-time to be infinitely divisible, there is no property that a point set has by itself, such as cardinality, which could uniquely determine a distance function or a metric without one having to be provided.

The points of space-time do not in fact constitute a metric space in the theory of relativity with respect to distance or duration, nor need we suppose that they do in classical theory either. In both theories, sets of *simultaneous* space-time points determine a metric space with respect to distance. And in both theories, the points in space form a metric space with respect to duration when points in space are thought of as collections of space-time points. For this purpose, one first needs to determine which points in space-time belong to the same point – that is, one needs to provide a relation of *genidentity* on the set of space-time points. In classical theory, the relation of simultaneity and the temporal duration function are the same for all systems regardless of the relation of genidentity; in relativity, once genidentity and simultaneity have been determined for one system, a choice of a different genidentity relation for another system will determine what the simultaneity relation is for it, as we will see.

All this, as we have said, is from the modern point of view, informed by hindsight. Neither Galileo nor Newton ever thought of spelling things out in this way (to say nothing of explicitly asserting the Galilean transformations on the space-time continuum), because they did not have the relativistic viewpoint with which to contrast it. Moreover, they had no motive for considering a four-dimensional space-time continuum as anything more than a formal construction. In relativity, the orientation of both the time axis and the space axes are to a certain extent arbitrary, for the relativity of simultaneity precludes the selection of a unique spatial slice through the continuum, and the relativity of motion precludes a unique selection of points at different times to coincide spatially with given ones.

Relativity of motion is of course inherent in classical theory, too, insofar as no physical measurement can detect uniform motion through space; but Newton and his followers were more inclined to think of this as a puzzle than to regard space and time as being to some extent interchangeable. Thus, instead of a four-dimensional space-time continuum, we have a three-dimensional spatial continuum which persists through time. In this view, only distances and duration are meaningful, whereas space-time distances would be 'bastardizations' devoid of physical content.

## II. NEWTONIAN SPACE-TIME: AN AXIOM SYSTEM

I have attempted a formal treatment of the classical ideas, which I hope will capture what is physically meaningful in them, and which can be extended to account for the relativistic outlook as well. In comparing these two views, I found it convenient to adopt the modern outlook utilizing the notion of a space-time continuum. This approach is equally useful in the formal development. In fact, as we shall see, with a space-time continuum, it is possible to capture certain features of Newton's notion of absolute space and time which from the classical point of view would seem to require a 'super space' or a 'super time' (for example, we can postulate that points keep the same distance from each other at all times). So we start with a space-time continuum with a certain structure, out of which space and time can be constructed. While this does make space-time ontically prior to space and time, we do *not* introduce any such magnitudes as the space-time interval. Later, it will be easy to generalize our ideas to bring in special and general relativity, and we can then relax our stricture against the space-time interval.

But in any case, since we are committed to the existence of space-time points, we consider classical and modern ideas about space and time to reduce ultimately to assertions about relationships between various structures on the space-time continuum. Statements about various coordinate *representations* of this structure I consider useful modes of expression, but not basic; for though coordinates are convenient ways of *naming* space-time points, they are not the space-time points themselves.

We now turn to the formal development of the ideas we have been discussing. We begin by defining a *Newtonian space-time system* $\mathcal{N}$ to be a sextuple which consists of the following items:

(i) A set $\mathcal{S}$ of *space-time points* having the power of the continuum. (The elements of $\mathcal{S}$ will be denoted by $s_1, s_2, \ldots$, or by $p, q, r, s, \ldots$.)

(ii) A real closed field $\mathcal{F}$. (The set of nonnegative elements of this field will be denoted by $R$.)[2]

(iii) A binary relation $G$ on $\mathcal{S}$ called *genidentity*.

(iv) A binary relation $S$ on $\mathcal{S}$ called *simultaneity*.

(v) A function $\sigma$ from $\mathcal{S} \times \mathcal{S}$ onto $R$ called the *distance function*.

(vi) A function $\delta$ from $\mathcal{S} \times \mathcal{S}$ onto $R$ called the *duration function*.

We first assume:

Axiom 1. – $G$ is an equivalence relation.
Axiom 2. – $S$ is an equivalence relation.

There are two ternary relations and one quaternary relation which we shall be using, and which can be depicted as follows:

DEFINITION 1:

(a) $\begin{bmatrix} & s_3 \\ & G \\ s_1 S & s_2 \end{bmatrix}$ if and only if $s_1 S s_2$ and $s_2 G s_3$

(b) $\begin{bmatrix} s_2 S & s_3 \\ G & \\ s_1 & \end{bmatrix}$ if and only if $s_1 G s_2 S s_3$

(c) $\begin{bmatrix} s_4 S & s_3 \\ G & G \\ s_1 S & s_2 \end{bmatrix}$ if and only if $s_1 S s_2 G s_3 S s_4 G s_1$.

Points and instants are regarded as *G*-equivalence classes and *S*-equivalence classes, respectively. The next axiom says that every position and time determines a unique space-time point, in the sense that every instant and point have in common exactly one space-time point.

Axiom 3. – For any $s_1$, $s_2$ in $\mathscr{S}$, there is exactly one space-time point $s_3$ such that
$$\begin{bmatrix} & s_2 & \\ & G & \\ s_1 S & s_3 & \end{bmatrix}.$$

We next define two relations on $\mathscr{S}$ which will be used to determine a spatial geometry after Tarski.

DEFINITION 2:

$$Bs(s_1, s_2, s_3) \quad \text{if and only if} \quad \sigma(s_1, s_2) + \sigma(s_2, s_3) = \sigma(s_1, s_3).$$

'$Bs(s_1, s_2, s_3)$' is to read '$s_2$ is spatially between $s_1$ and $s_3$'.

DEFINITION 3:

$$(s_1, s_2) \, Es(s_3, s_4) \quad \text{if and only if} \quad \sigma(s_1, s_2) = \sigma(s_3, s_4).$$

'$(s_1, s_2) \, Es(s_3, s_4)$' is to read 'the pairs $(s_1, s_2)$ and $(s_3, s_4)$ are equidistant'. The next axiom says that space is Euclidean.

Axiom 4. – Each instant is a Euclidean metric space with respect to the function $\sigma$ and the relations $Bs$ and $Es$ restricted to it. That is, Tarski's system $\mathscr{E}_3$ of axioms for three-dimensional Euclidean geometry is satisfied, relativized to any given instant, and, moreover, each instant is a metric space with respect to the function $\sigma$ restricted to it.[3]

The next axiom may be said to express part of what is meant by saying absolute space 'remains always similar and immovable', since it says that any two points keep the same distance from each other.

Axiom 5. – If
$$\begin{bmatrix} s_4 S & s_3 \\ G & G \\ s_1 S & s_2 \end{bmatrix},$$
then

$$\sigma(s_1, s_2) = \sigma(s_3, s_4) = \sigma(s_1, s_3) = \sigma(s_4, s_2).$$

We now define two additional relations on $\mathscr{S}$ which will be used to characterize the 'geometry' of space points.

DEFINITION 4:

$$Bt\,(s_1, s_2, s_3) \quad \text{if and only if} \quad \delta(s_1, s_2) + \delta(s_2, s_3) = \delta(s_1, s_3).$$

'$Bt(s_1, s_2, s_3)$' is to be read '$s_2$ is between $s_1$ and $s_3$'.

DEFINITION 5:

$$(s_1, s_2)\,Et\,(s_3, s_4) \quad \text{if and only if} \quad \delta(s_1, s_2) = \delta(s_3, s_4).$$

'$(s_1, s_2)\,Et\,(s_3, s_4)$' is to be read 'the time elapsed between $s_1$ and $s_2$ is the same as the time elapsed between $s_3$ and $s_4$'. The next axiom says that each point is a one-dimensional continuum.

Axiom 6. – Each point is a linear continuum with respect to $Bt$ and $Et$. That is, Tarski's system $\mathscr{E}_1$ of axioms for one-dimensional space is satisfied when relativized to any point. Moreover, each point is a metric space with respect to the duration function $\delta$ restricted to it.[4]

In analogy to Axiom 5, which says that points keep the same distance from each other, we shall assume that the time interval between any two instants is everywhere the same, which is to say that time 'flows at the same rate' everywhere:

Axiom 7. – If

$$\begin{bmatrix} s_4 S & s_3 \\ G & G \\ s_1 S & s_2 \end{bmatrix},$$

then

$$\delta(s_1, s_4) = \delta(s_2, s_3) = \delta(s_1, s_3) = \delta(s_4, s_2).$$

And with this we conclude our axiomatic development. What we have done is to impose a geometry on each instant and each point, and also certain relationships between points and instants. The points themselves form a three-dimensional Euclidean geometry. We simply define a distance function $\rho$ on them such that $\rho([s_1]_G, [s_2]_G) = \sigma(s_1, s_2)$, and define betweenness and equidistance in analogy to Definitions 2 and 3. Similarly, a duration function $\tau$ can be defined on instants as follows: $\tau([s_1]_S, [s_2]_S) = \delta(s_1, s_2)$. Thus, the classical assumptions about points and instants can be obtained from ours about space-time points each of which can be thought of as a 'point at an instant'.

### III. RELATIVISTIC AND NEWTONIAN IDEAS RECONSIDERED

We have already observed that Newton considered space a substance, as something physically real and eternal.[5] So, in reality, there can be only one physical space: the actual one. There can of course be many *relative* spaces which are in motion with respect to each other (we would say that they correspond to Newtonian space-time systems having different genidentity relations). But there is only one system which remains always 'similar and immovable'. For Newton seems to have thought that the structure of space is somehow inherent in the continuum of its points, so that its very existence guarantees its uniqueness. While we cannot accept this as an argument for the uniqueness of space, it is true that any two Newtonian space-time systems on a set $\mathscr{S}$ which have the same genidentity and simultaneity relations have essentially the same structure.

Tarski has shown that a *representation theorem* can be proved for $\mathscr{E}_1$ and $\mathscr{E}_3$. For $\mathscr{E}_3$, it says that every model of $\mathscr{E}_3$ is isomorphic to the three-dimensional Cartesian space $\mathscr{C}_3(\mathscr{F})$ over a real closed field $\mathscr{F}$, which consists of a betweenness relation $B$ on $\mathscr{F}^3$ and an equidistance relation $E$ on $\mathscr{F}^3$, $B$ and $E$ having been defined with respect to a distance function $\phi$ exactly as we defined $Bs$ and $Es$. ($\phi$ is defined by the condition $\phi(\langle x_1, y_1, z_1 \rangle, \langle x_2, y_2, z_2 \rangle) = ((x_1 - x_2)^2 + (y_1 - y_2)^2 + (z_1 - z_2)^2)^{1/2}$, which gives us the *metric of* $\mathscr{C}_3(\mathscr{F})$, Tarski, 1969, pp. 168–170.) Now each instant of $\mathscr{S}$ is a model of $\mathscr{E}_3$ with respect to $Bs$ and $Es$, and thus is isomorphic to $\mathscr{C}_3(\mathscr{F})$. Of course, there are infinitely many such models for $\mathscr{E}_3$ obtainable from Newtonian space-time systems with different distance functions, but all of them have the same structure. Similar remarks apply to $\mathscr{E}_1$: any point is a linear continuum with respect to $Bt$ and $Et$. So, if one assumes the genidentity and simultaneity relations on a given set $\mathscr{S}$ to be fixed, then there is only one space-time system up to isomorphism. In this sense, Newton was right.[6]

Newton of course thought that there is a uniquely correct choice of $G$ and $S$, though he admitted that there is no way to find out what $G$ is. Thus, two systems with differing $G$'s represented for him different 'relative' spaces. We can say the same for systems having different $S$. We are thus led to consider certain canonical classes of space-time systems, no two of which have both the same $S$ and the same $G$, and where the choice of $\sigma$ and $\delta$ for one determines them for the others in a relatively simple way.

In particular, if any two systems in it have the same genidentity relation, then they share the same set of points. As we want only one duration function defined on a given point, since two of them on the same point would yield the same structure, we shall require that the duration functions of two such systems restricted to a given point be the same for both. Moreover, since the common set of points forms a Euclidean space, we require that the distance function $\rho$ on them and, thus, the distance function $\sigma$ on $\mathscr{S}$ be common to both systems. Similarly, if two systems have the same simultaneity relation, they share the same set of instants, each of which has a unique structure. So the distance function restricted to any given instant should be the same for both systems. Moreover, since the common set of instants forms a linear continuum, the two systems should have a common duration function $\tau$ defined on them, and thus have the same duration function $\delta$ on $\mathscr{S}$. We summarize these requirements in the following definition:

Let $\mathscr{N} = \langle \mathscr{S}, \mathscr{F}, G, S, \sigma, \delta \rangle$ and $\mathscr{N}' = \langle \mathscr{S}, \mathscr{F}, G', S', \sigma', \delta' \rangle$ be any systems in $\Gamma$. Then $\Gamma$ is a *canonical class* of space-time systems if and only if

(i) Every system in $\Gamma$ has the same set $\mathscr{S}$ of space-time points.

(ii) If $G = G'$, then $\sigma = \sigma'$, and for any $s$ in $\mathscr{S}$, $\delta \mid [s]_G = \delta' \mid [s]_{G'}$.

(iii) If $S = S'$, then $\delta = \delta'$, and for any $s$ in $\mathscr{S}$, $\sigma \mid [s]_S = \sigma' \mid [s]_{S'}$.

Let us call a canonical class *Newtonian* if each system in it is a Newtonian system, each one has the same simultaneity relation, and any two members of it are at rest with respect to one another; that is, the relation of genidentity is the same for all of them. By definition, any Newtonian class is a unit class.

Not all canonical classes are unit classes, however. When they are not, we need some way of classifying them. Let $\mathcal{N} = \langle \mathcal{S}, \mathcal{F}, G, S, \sigma, \delta \rangle$ and $\mathcal{N}' = \langle \mathcal{S}, \mathcal{F}, G', S', \sigma', \delta' \rangle$ be two Newtonian systems defined on the same set $\mathcal{S}$ of space-time points. We say that $\mathcal{N}$ and $\mathcal{N}'$ are in *uniform motion* with respect to each other if and only if for any $s_1, s_2, s_3, s_4,$ and $s_5$ in $\mathcal{S}$ where $s_1, s_2,$ and $s_3$ are distinct, $s_1 G' s_2 G' s_3$,

$$\begin{bmatrix} & s_2 & \\ & G & \\ s_1 S & & s_4 \end{bmatrix} \text{ and } \begin{bmatrix} & s_3 & \\ & G & \\ s_1 S & & s_5 \end{bmatrix}, \text{ we have } Bt(s_1, s_2, s_3) \text{ if and only if}$$

$Bs(s_1, s_2, s_3)$, and $\sigma(s_1, s_4)/\delta(s_4, s_2) = \sigma(s_1, s_5)/\delta(s_5, s_3)$. (See Figure 1.) This ratio is the *relative speed v of* $\mathcal{N}$ and $\mathcal{N}'$.

We define a *Galilean class* to be a canonical class of Newtonian space-time systems in uniform motion with respect to one another, and all of them having the same simultaneity relation. If $\mathcal{N}$ and $\mathcal{N}'$ are two

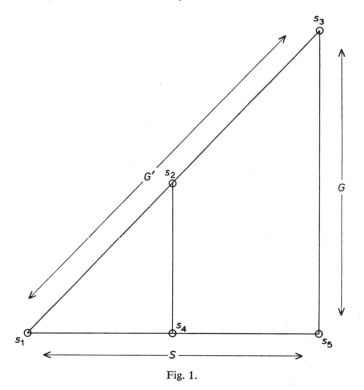

Fig. 1.

Newtonian systems in a Galilean class, then by definition $\delta' = \delta$, and $\sigma' \mid [s]_{S'} = \sigma \mid [s]_{S'}$ for $s$ in $\mathscr{S}$. Since for any $s_1, s_2$ in $\mathscr{S}$ there is exactly one space-time point $s_3$ such that

$$\begin{bmatrix} & s_2 & \\ & G' & \\ s_1 S & s_3 & \end{bmatrix}, \text{ and exactly one } s_4 \text{ such that } \begin{bmatrix} & s_2 & \\ & G & \\ s_1 S & s_4 & \end{bmatrix},$$

and $\sigma'(s_1, s_2) = \sigma'(s_1, s_3)$ and $\sigma(s_1, s_2) = \sigma(s_1, s_4)$, it follows, in view of the fact that $s_1, s_3$, and $s_4$ are all simultaneous, that $\sigma'(s_1, s_2) = \sigma(s_1, s_3)$, and $\sigma(s_1, s_2) = \sigma'(s_1, s_4)$.

We could also supply a coordinate system for each Newtonian system in a Galilean class (or any canonical class) in the following way: Let an isomorphism[7] $f$ between a Newtonian system $\mathscr{N} = \langle \mathscr{S}, \mathscr{F}, G, S, \sigma, \delta \rangle$ and a Newtonian system $\mathscr{C} = \langle \mathscr{S}', \mathscr{F}, G', S', \sigma', \delta' \rangle$ be called a *reference frame* if

(i)  $\quad \mathscr{S}' = \mathscr{F}^4$.

Let $f(s_i) = \langle x_i, y_i, z_i, t_i \rangle$ for $s_i$ in $\mathscr{S}$. Then

(ii) $\quad f(s_i) \, G' \, f(s_j) \quad$ if and only if $\quad \langle x_i, y_i, z_i \rangle = \langle x_j, y_j, z_j \rangle$
(iii) $\quad f(s_i) \, S' \, f(s_j) \quad$ if and only if $\quad t_i = t_j$.
(iv) $\quad \sigma'(f(s_i), f(s_j)) = ((x_i - x_j)^2 + (y_i - y_j)^2 + (z_i - z_j)^2)^{1/2} = \sigma(s_i, s_j)$.
(v) $\quad \delta'(f(s_i), f(s_j)) = |t_i - t_j| = \delta(s_i, s_j)$.

$\mathscr{C}$ is called a *coordinate system* by virtue of (i), and is *Cartesian* by virtue of the *metric* imposed on it in (iv).

Now, if $\mathscr{N}$ and $\mathscr{N}'$ are two systems in a canonical class, and $\mathscr{C}$ and $\mathscr{C}'$ their respective coordinate systems under the reference frames $f$ and $f'$, we shall require the origins to coincide and the axes to have the same orientation; that is, we require that $f(w) = f'(w) = \langle 0, 0, 0, 0 \rangle$ for some $w$ in $\mathscr{S}$, and if $f(s) = \langle x, 0, 0, 0 \rangle$, then $f'(s) = \langle x', 0, 0, t' \rangle$ and, moreover, $x \cdot x'$ should be greater than or equal to 0. If $f(s) = \langle 0, y, 0, 0 \rangle$ then $f'(s) = \langle 0, y', 0, t' \rangle$, and $y \cdot y' \geq 0$. And, finally, if $f(s) = \langle 0, 0, z, 0 \rangle$, then $f'(s) = \langle 0, 0, z', t' \rangle$ and $z \cdot z' \geq 0$. The transformation $\mathscr{T}$ between $\mathscr{C}$ and $\mathscr{C}'$ is then defined by the condition $\mathscr{T}(f(s)) = f'(s)$ for $s$ in $\mathscr{S}$, or equivalently, $\mathscr{T} = f^{-1} \circ f'$. It should be noted that although $\mathscr{C}$ and $\mathscr{C}'$ are isomorphic (as are $\mathcal{N}$ and $\mathcal{N}'$), $\mathcal{T}$ is not an isomorphism between $\mathscr{C}$ and $\mathscr{C}'$.

It can be shown that the transformation between the coordinates of any two space-time systems in a Galilean class belongs to the Galilean group.

The theory of relativity obliterates an absolute distinction between past and future. Simultaneity and genidentity are of course well defined for any given system, but they are no longer the same for all systems – indeed, any two relativistic systems with differing genidentity relations also have different simultaneity relations.

We define a *Lorentz class* to be a canonical class of Newtonian space-time systems which are in uniform motion with respect to each other, such that the function $\gamma$ defined by the condition $\gamma(s_1, s_2) = (\sigma^2(s_1, s_2) - c^2 \delta^2(s_1, s_2))^{1/2}$, is the same for all systems in it. Suppes has shown that the transformation between the coordinates of any two space-time systems in a Lorentz class is a Lorentz transformation.[8]

For general relativity, the systems need not be Newtonian in that we no longer require Playfair's axiom or Axiom 5 to hold. Further, we drop the requirement that any two systems be in uniform motion with respect to each other. An 'Einstein class' is the widest of all.

IV. SOME PHILOSOPHICAL CONCLUSIONS

What I have been trying to do is to provide a conceptual framework in terms of which the differing structure of the space-time continuum in various theories can be exhibited, in the hope that various senses of the notion of 'absolute' space and time can be clarified. In terms of this framework, space-time is absolute in the sense of being a 'given', having thus an ontic priority to other concepts. While such differences have usually been represented in terms of transformations between coordinate systems, I do not see any particular advantage to supposing that coordinates are simply numbers, and not the coordinates of a point.

If not, the coordinate systems we have been considering are after all the images of mappings from *something*, i.e., the set of space-time points, onto the set of quadruples of real numbers.

That something does not possess eternal duration, like Newton's points. It simply is. And yet it is necessary to assume that it has a structure. Newton would surely want to say that if so, this structure also is, despite appearances. Science depends on observation to correct mistaken suppositions. Thus, the discovery of the distinction between appearance and reality presupposes the possibility of a more knowledgeable observer who possesses data lacking to those who have only the appearance. Or at least the appearance is not a real attribute of what gives rise to it, but only a secondary quality. Surely the structure of space must be more than an appearance, or a disposition. But if it is, what observer can distinguish between the appearance and what the structure really is? Newton left that to God, but we have discovered all too often that what seemed to be primary, more than appearance, was not. General relativity has even taught us that talk about a non-Euclidean space is intelligible (contrary to Kant) and true, and quantum theory has even cast doubt on the supposed primacy of position and speed, solidity and endurance, and undulation. All that remains are statements *that* certain appearances can be expected under certain circumstances, and these in turn are deductions made from mathematical equations. What do we usually assume about space and time when doing physics? A continuum of points, with a structure that appears differently to different observers, with no one to mediate.

Yet many feel uncomfortable about supposing that the space-time continuum is physically real, since it cannot be observed. The degree of discomfort is perhaps related to how seriously ontic commitments are taken. Scientific discourse does, I think, make ontic commitments unavoidable, but no particular one is any more irrevocable than the theory which justified it. And perhaps less so. For even without any changes in physical theory, we do not exclude the possibility of some more extended theoretical basis for physics in which space-time points are constructions introduced by definition. Or again, some other relational structures (e.g., between events) might be found isomorphic to (or which can be imbedded into) our space-time systems or something similar. In this case, the coordinates could be correlated to the corresponding relations directly, with space-time points playing at most an intermediate role. There are, of course, many philosophical motives for finding such structures, and with these I have considerable sympathy. Indeed, some authors have claimed to have already discovered them (Grünbaum, 1963; Reichenbach, 1956; for example), but, unfortunately, I do not find their claims convincing.

Yet, contrary to those who hold that the theory of relativity makes a relational theory of time necessary, we have seen that relativity and classical theory alike can very well be expressed within a framework in which spatiotemporal relations are primitive and not defined, for example, in causal terms. But while we have found no physically compelling reasons for either accepting or rejecting a space-time continuum as such, we *have* found a theoretical basis for claims about its structure. And even if we eventually find a way of dispensing with the continuum itself, it helps, I think, to be clear about what is being discarded.

## NOTES

[1] This is fairly well known. For a discussion of these issues and additional references, see Grünbaum (1963).

[2] We shall make use of certain geometrical representations involving constructions on real closed fields first proved by Tarski (1969). A real closed field $\mathscr{F}$ is an ordered field, all of whose nonnegative elements are squares, such that every polynomial of odd degree has a zero in $\mathscr{F}$. This is the definition used in Tarski (1969).

[3] See Tarski, (1969, pp. 166–167). The axioms are for two dimensional space $\mathscr{E}_2$. But to obtain $\mathscr{E}_3$, we make the negation of the upper dimensional axiom into a new lower dimensional axiom. This says essentially that if three points are equidistant from two given distinct points, they are collinear. For an upper dimensional axiom, we might assume that if four points are equidistant from two distinct points, they are co-planar. This last can be stated as a variant of Pasch's axiom. In any case, essentially the same metamathematical results hold for $\mathscr{E}_3$ (or $\mathscr{E}_1$) as for $\mathscr{E}_2$. See Tarski (1969, p. 169, n. 5).

[4] Again, we need to obtain $\mathscr{E}_1$ from $\mathscr{E}_2$. Thus the upper dimension axiom (Tarski, 1969, p. 167) becomes the negation of the lower dimension axiom of $\mathscr{E}_2$, and we could assume for the lower dimensional axiom (in appropriate language) that any two points can be separated into disjoint classes by a point between them.

[5] One consequence of our axioms is that points are uncreated and indestructible, i.e.,

$$\text{if} \begin{bmatrix} & s_3 & \\ & & G \\ s_1\, S & & s_2 \end{bmatrix} \quad \text{then} \quad (\exists\, s_4) \begin{bmatrix} s_4\, S & s_3 \\ G & \\ s_1 & \end{bmatrix}.$$

[6] This is not *quite* true. For the representation theorem did not say that the axioms for $\mathscr{E}_n$ are categorical, since of course no first-order theory is. Thus, though every $\mathscr{E}_n$ is isomorphic to some $\mathscr{C}_n(\mathscr{F})$, there are nonstandard models for real closed fields. See, e.g., Robinson (1969). This means that some Newtonian systems may have distance functions with a nonstandard range. However, this does not really vitiate our decision to allow one distance function to represent every system with a given genidentity relation in the *canonical classes* defined below. Some canonical classes would simply have the property that space and time for the systems in them would be non-Archimedean. We could, of course, have adopted a second-order axiomatization of geometry with the distance function primitive. Such a system can be found, for example, in Bluementhal (1961, p. 156, and modifications for *n*-dimensional space, p. 175). The axioms for New-

tonian space-time systems would then be categorical, though we could no longer prove completeness or decidability (cf. Tarski, 1969, p. 170).
[7] There are various ways of constructing such an isomorphism. One chooses an origin and erects three mutually perpendicular axes. The choice of the orientation of the axes is arbitrary, both as to their direction and whether they form a left-handed or right-handed system.
[8] In fact, he also proved this for the weaker condition that $\gamma(s_1, s_2)$ be invariant whenever $c\delta(s_1, s_2) > \sigma(s_1, s_2)$. See Suppes (1959, pp. 291–307).

## BIBLIOGRAPHY

Bluementhal, L. M., *A Modern View of Geometry*, Freeman, San Francisco, 1961.
Euclid, *The Thirteen Books of Euclid's Elements*, 3 vols, Dover, New York, 1956 (translated by T. L. Heath).
Grünbaum, A., *Philosophical Problems of Space and Time*, Knopf, New York, 1963.
Newton, I., *Principia*, Univ. of California Press, Berkeley, Calif., 1946 (Cajori translation).
Reichenbach, H., *The Direction of Time*, Univ. of California Press, Berkeley, 1956.
Robinson, A., 'The Metaphysics of the Calculus', in J. Hintikka (ed.), *The Philosophy of Mathematics*, Oxford University Press, London, 1969, pp. 153–63.
Smart, J. J. C., *Philosophy and Scientific Realism*, Humanities Press, New York, 1963.
Suppes, P., 'Axioms for Relativistic Kinematics With or Without Parity', in L. Henkin, P. Suppes, and A. Tarski (eds.), *The Axiomatic Method, with Special Reference to Geometry and Physics*, North-Holland, Amsterdam, 1959, pp. 291–307.
Tarski, A., 'What is Elementary Geometry?', in L. Henkin, P. Suppes, and A. Tarski (eds.), *The Axiomatic Method, with Special Reference to Geometry and Physics*, North-Holland, Amsterdam, 1959, pp. 16–29. (Republished: in J. Hintikka (ed.), *The Philosophy of Mathematics*, Oxford University Press, London, 1969, pp. 164–75.)

CLARK GLYMOUR*

# TOPOLOGY, COSMOLOGY AND CONVENTION

### I. INTRODUCTION

Aus der Annahme, dass der uns umgebende Raum eine euklidische oder hyperbolische Struktur aufweist, lässt sich keinswegs folgern, dass dieser Raum eine unendliche Ausdenung besitzt; denn die eukldische Geometrie ist z.B. durchaus mit der Annahme einer endlichen Raumausdehnung verträglich, eine Tatsache, die man früher übersehen hat. Diese Möglichkeit, dem Weltall auch bei beliebiger Struktur einen endlichen Inhalt zuzuschrieben, ist besonders wertvoll, weil die Vorstellung einer unendlichen Ausdehnung, die zunächst als wesentlicher Fortschritt des menschlichen Geistes betrachet wurde, mannigfache Schwierigkeiten, z.B. bei dem Problem der Massenverteilung, mit sich bringt. Man sieht hieraus, wie tief alle diese Überlegungen in die kosmologischen Probleme eingreifen.

F. Klein (1928)

It has been said that it was Reichenbach's merit to have realized that the conventional aspects of geometry are precisely the metrical aspects of geometry. I disagree. The most suggestive parts of Reichenbach's work on space and time (1958) are his remarks on the possibility of alternative topologies for physical space. For, if Reichenbach is correct, the topology of space has an intimate connection with both the causal structure of the universe and with the identity of objects in time. We have alternatives for the first just because we have alternatives for the latter two. My intent is to apply Reichenbach's insights on the relativity of topology, and some of Felix Klein's work on the connections between local geometry and global geometry, to the subject of relativistic cosmology.

My contention is that within the context of relativistic cosmology, the topology of physical space is conventional in two very different ways. In the first place, there are cosmological models in which the topology of physical space may be taken to be either open or closed (more precisely, not compact or compact), as one pleases. In the second place, for each of a class of fashionable cosmological models there is another (unfashionable) model different from the first in the topology it ascribes to space-time, and there are good reasons to think that any two such cosmological models are, both in fact and in principle, experimentally indistinguishable. Any bit of evidence which we can account for with one model, we can account for with another, and conversely.

I must make clear at the outset why I regard these two kinds of cases as being different, and importantly different. In the first case the possibility of choosing space to be either open or closed derives simply from the fact that in relativistic cosmology the notion of space is not invariant. Merely by changing coordinates we can change which events are co-temporal, but we do not thereby change the cosmological model or the characteristics we ascribe to space-time. A relativistic cosmology does not in general determine a unique space at all, only a unique space-time. The license afforded us in our specification of the topology of space does not reflect any indeterminacy in our knowledge. It is not as though we had two theories of space that could not both be true but between which we could not decide by any empirical means. In the context of a particular cosmological model, the claims that space is open and that space is closed may be perfectly reconcilable, and, indeed, may both be true,[1] for when they use the term 'space' the two claims do not refer to the same thing.

The second case is very different, for what is claimed is that for each of a broad class of cosmological theories we can find another, different, cosmological theory empirically equivalent to the first. The difference between the two does not, as in the first case, reduce to a pun; they will differ as to how things move, what worlds collide, and even as to what there is. They may both be false, but they cannot both be true. No coordinate transformation, no remetrization, no finagling which preserves relativistic structure, will permit us to translate one into the other. The existence of such pairs of cosmological theories signifies, if one of them is true, the impossibility of obtaining inductive evidence to show that that one, and no other is indeed true. When, in this second sense, I say that topology is conventional I mean that if general relativity is true, then there are cases in which no possible evidence can alone decide between pairs of cosmological theories which differ as to the topology of space-time. I do not say that there is no truth to the matter, but only that we cannot find it out. In the same way, I shall argue that, outside the context of relativity, the topology of space is conventional, and this conventionality has an importance which is lacking in the conventionality of metric properties. Since the latter subjects are simpler than the relativistic issues and have already been discussed by Reichenbach, I shall begin with them.

## II. THE RELATIVITY OF GEOMETRY AND THE RELATIVITY OF TOPOLOGY

Reichenbach maintained that with suitable additional hypotheses any pair of physical geometries positing the same topology can be made empirically equivalent. More specifically, of any two geometries having the same topology, one can be obtained from the other by positing suitable 'universal forces'. I do not think either Reichenbach or his philosophical successors have given very clear accounts of what sort of mathematical object a universal force might be and how it relates to the notion of force used in classical physics. Reichenbach did say that universal forces have the 'same effect' everywhere and that gravitation is a universal force. The best way to begin the comparison of geometric and topological conventions is to try to square all of these claims with a mathematically clear notion of universal force on the one hand, and with the classical notion of force on the other.

In an inertial coordinate system, $x^i, i = 1, 2, 3$, the acceleration vector of a point particle with position coordinates $x^i(t)$ has the components

(1) $$\frac{d^2 x^i}{dt^2}.$$

If, however, we consider some arbitrary coordinate system $y^k = y^k(x^1, x^2, x^3)$ then we have for the acceleration (1)

$$\frac{d^2 y^k}{dt^2} + \sum_{l,m} \left( \sum_i \frac{2y^k}{2x^i} \frac{2^2 x^i}{2y^l 2y^m} \right) \frac{dy^l}{dt} \frac{dy^m}{dt}.$$

The terms in parentheses are just an affine connection expressed in the $y$ coordinates.[2] Expressing such terms by $\Gamma^k_{lm}$ we have for the general form of the acceleration

$$\frac{d^2 y^k}{dt^2} + \Gamma^k_{lm} \frac{dy^l}{dt} \frac{dy^m}{dt},$$

where repeated indices are understood to be summed over. In general form then, the first law requires that particles subject to no forces follow paths which are solutions to the equations

$$\text{(2)} \qquad \frac{d^2 y^k}{dt^2} + \Gamma^k_{lm} \frac{dy^l}{dt} \frac{dy^m}{dt} = 0,$$

and we know from differential geometry that these are the equations of the geodesics of any geometry compatible with the connection. Suppose now that a class of particles, the free ones, follow Equation (2) and that their motion is taken to characterize the geodesics of the space. Then to obtain the equations of motion of any system subject to forces, we must multiply the left-hand side of Equation (2) by the mass of the system and subtract a force expression. The general form of the second law is thus

$$\text{(3)} \qquad m\left(\frac{d^2 y^i}{dt^2} + \Gamma^i_{jk} \frac{dy^j}{dt} \frac{dy^k}{dt}\right) - F^i = 0.$$

Now Reichenbach's idea, I suggest, is that if we use the proper kind of force, Equation (3) will entail that the motions of particles subject only to this force follow a different set of geodesics than those determined by $\Gamma^i_{jk}$. Thus for a force to be universal it is necessary that it act on all objects and that for a body of mass $m$ the components of the force vector have the value

$$F^i = m\phi^i,$$

where $\phi^i$ is solely a function of position, time and time derivatives of position. Then for *every* system subject only to this universal force, Equation (3) will reduce to

$$\frac{d^2 y^i}{dt^2} + \Gamma^i_{jk} \frac{dy^j}{dt} \frac{dy^k}{dt} - \phi^i = 0,$$

which may be a set of geodesic equations for some other geometry.

Now it is easy to show that given any two Riemannian geometries on the same differentiable manifold, the geodesics of one can be obtained from those of the other by subtracting a suitable universal force. Suppose that $g$ is the metric tensor field of some geometry compatible with the connection $\Gamma^i_{jk}$ and that $h$ is the metric tensor field of any other geometry on the same manifold. Let

$$^0\Gamma^s_{ik} = \tfrac{1}{2}h^{sl}\left(\frac{2h_{kl}}{2y^i} + \frac{2h_{li}}{2y^k} - \frac{2h_{ik}}{2y^l}\right)$$

be the affine connection associated with $h$. The version of the first law which accords with this geometry is that particles subject to no forces follow

(4) $$\frac{d^2 y^k}{dt^2} + {}^0\Gamma^k_{lm}\frac{dy^l}{dt}\frac{dy^m}{dt}.$$

Assume that no class of particles actually follow Equations (4), but particles subject to no identifiable forces do follow Equations (2). We may reconcile this observation with the assumption that the geometry of space is given by $h$ by further assuming that a universal force acts on all bodies. The force is given by

(5) $$F^k = m({}^0\Gamma^k_{lr} - \Gamma^k_{lr})\frac{dy^l}{dt}\frac{dy^r}{dt}.$$

Multiplying the acceleration on the left-hand side of Equation (4) by $m$ and subtracting the universal force Equation (5), we obtain the equations of motion (2). It should be noted that the quantity given by Equation (5) is in fact a vector since the difference of two connections is a tensor.

Note that according to Newtonian theory gravitation cannot be a universal force of this kind since the gravitational force does not depend on the square of the velocity.

I do not wish to claim that considerations such as these put the relativity of affine geometry beyond question. But supposing that they do, the relativity of geometry does not appear to present any large philosophical puzzle. One could, perhaps, maintain the position that these alternative physical theories are contradictory and that there is, therefore, a great mystery as to which of them is true, but I see no compelling reason to take this view. As long as we view the matter through differential geometry we know how to build models of some theories from models of other theories; we have seen, for example, how to characterize the geodesics of one geometry in terms of curves (which are not geodesics) in another geometry. There is no reason why a given differentiable manifold need have only a single metric or a single connection, and no reason why it cannot have relations enough to satisfy a plethora of alternative geometries. Two such alternative theories may, perhaps, not say the same thing, but there is no obvious reason why both cannot be true, granted that we understand them to use 'geodesic', 'congruence', etc., in different ways.

According to Reichenbach the topology we ascribe to physical space may depend on how we reidentify objects. If we change our reidentification habits we will also have to chance the topology we ascribe to space, and if we wish to cling to a certain topology, come what may, events may force us to change our reidentification procedures. Consider, for example, a two-dimensional being living on the surface of a sphere. Following a great circle, he returns to where he began and thus determines that his space is closed. Following two great circles, one after the other, he discovers that they intersect twice, and so determines that his space is spherical, not elliptic. Assuming that he wishes to maintain that his space is open, not closed, can he reconcile his theory with the evidence? When, after following a great circle, he returns to the place where he began, he may refuse to allow that it is the *same* place. It may look very similar to the place from which he began, and may even stand in very peculiar causal relations with the place from which he began, but it is not, he maintains, the same place. Each circumnavigation of the sphere takes him, according to his lights, farther from where he began, and with each tour he comes into another place very much like the one which he left. The universe, on his account, is redundant; indeed, it even reveals a kind of pre-established harmony.

About the formal connection between alternative topologies Reichenbach was even less explicit than he was about the connection between alternative geometries. When expressing the relation between two geometries having the same topology he wrote

$$g_{ik} = h_{ik} + U_{ik},$$

where $g_{ik}$ and $h_{ik}$ are the coefficients (in some common coordinate system) of the metric tensor fields of the two geometries, at some point, and $U_{ik}$ is the universal force acting at that point. While this is an entirely trivial decomposition using objects, the $U_{ik}$, having no intuitive connection with the classical notion of force, it does at least make mathematical sense since the two tensor fields are defined on the same manifold, and the collection of all second-order tensors at a point forms a real vector space.[3] The analogous formula which Reichenbach wrote for the case of geometries having different topologies

$$g_{ik} = h_{ik} + U_{ik} + A$$

(where $A$ is the 'causal anomaly') is, however, nonsense.[4] There is an elementary formal connection between the topologies of manifolds obtainable from each other by changing reidentification procedures, one which Reichenbach used but did not describe. Consider for example the case of the torus and the real plane. On the real plane we define the following equivalence relation

$$(x_1, y_1) \approx (x_2, y_2) \quad \text{if and only if} \quad \begin{cases} (x_2 - x_1) \text{ is integral} \\ \text{and} \\ (y_2 - y_1) \text{ is integral}. \end{cases}$$

Then every point is equivalent to some point on the unit square, $0 \leqslant x \leqslant 1, 0 \leqslant y \leqslant 1$, and points on the boundary of the square which can be connected by lines parallel to the coordinate axes are also equivalent.

If now, we take the set $E$ of equivalence classes under the above relation and consider the mapping

$$h : R^2 \to E$$

given by

$$h((x, y)) = e \in E \quad \text{iff} \quad (x, y) \in e,$$

we can define a topology, the identification topology, on $E$ as follows: A subset $B$ of $E$ is open iff

$$\{(x, y) \mid h((x, y)) \in B\}$$

is open in $R$. It is well known that the topology thus defined on $E$ is that of the torus. Reichenbach's example (1958, p. 58 f.) of a torus topology interpreted as Euclidean does not use exactly this construction, but an analogous one. The covering manifold Reichenbach uses is Euclidean space with one point removed.

The connection between identification topologies and the relativity of topology is obvious. The equivalent points on the plane are just those which the Euclidean theorist takes to be distinct, but the torus theorist holds identical. The device is not limited to two dimensions, nor to the torus and the Euclidean plane. The topology of the Klein bottle can be obtained from a spherical topology, and by a sufficiently queer equivalence relation, elliptic space can be obtained from Euclidean space,[5] and so on. In each of these cases if under 'normal' reidentification procedures the manifold has topology $A$, and $A$ is homeomorphic to some identification topology generated from $B$, then by changing reidentification one can account for his experiences and maintain that the manifold has topology $B$.

Considering these examples of alternative topologies, I find myself reluctant to grant that both of a pair of alternatives might in fact be true. The alternatives seem to be most naturally understood as contradictory and irreconcilable, and while I have no conclusive argument for this view, there are considerations which support it. In the case of alternative geometries with the same topology there seemed to be no fundamental disagreement as to the basic individuals in the universe, but in the topological cases there appeared to be just such disagreements. We could try to construe the advocate of an open topology as quantifying over, not objects themselves, but ordered pairs of objects and natural numbers.[6] So understood, many of his claims, even those about identity, would be true in a closed universe. But we would then have difficulty interpreting his views about the relations among sets and physical objects, assuming his views to be (on the interpretation I prefer) rather ordinary ones. For instance, if '$\phi$' is an ordinary physical predicate (e.g., 'yellow') we may imagine the open-topology theorist to be committed to all of the following:

(i) $\quad \exists x \, \phi(x)$

(ii) $\quad \sim \exists x \, [\phi(x) \, \& \, \{0\} \in x]$

(iii) $\quad \sim \exists x \, [\phi(x) \, \& \, \{1\} \in x]$

and so on. But on the proposed translation, (i) would be translated by

$$\exists x \, \exists y \, \exists z \, [x = \{y, \{z\}\} \, \& \, z \text{ is a natural number} \, \& \, \phi(y)].$$

Claim (ii) would be translated by

$$\sim \exists x \, \exists y \, \exists z \, [x = \{y, \{z\}\} \, \& \, z \text{ is a natural number} \, \& \, \phi(y) \, \& \, \{0\} \in x]$$

and for each natural number we would obtain a similar sentence. The collection of translations clearly smacks of $\omega$-inconsistency. No doubt there are devices which would avoid such problems, but I suspect they would, unlike the present proposal, be rather contrived. In the geometric cases it is plausible to think there is an ambiguity, that common predicates, e.g., 'geodesic', are being used differently in two or more alternative theories. But in the topological cases the only predicate that seems likely to suffer such ambiguity is the identity predicate. The identity predicate, however, is a logically very special one, rather unlike 'geodesic'. It seems likely that if the open-topology theory is formulated within a second-order logic which includes the identity of indiscernibles, then to maintain the truth of the theory in a closed universe we would be forced to contend either that the identity predicate occurs ambiguously within the open theory alone or else that the logic of the open theory is nonstandard.

I do not want to rest the case that there is an epistemic difference between the relativity of geometry and the relativity of topology on the preceding examples. Admittedly, they arise by the use of unusual criteria for identity and that alone may render them suspect. Alternative topologies can, however, arise in ways that do not require any systematic differences in usage. All that is required is that it be possible for there to be regions of space, or even points, which are causally isolated from the rest of space.

A simple example is suggested by the stereographic projection of the sphere on the plane. Let $S$ be an ordinary two sphere with the ordinary spherical metric induced by embedding $S$ in a three-dimensional Euclidean space. Let $S$ be tangent to a plane $E$ (topologically $R^2$) at point $p$. Let $q$ be antipodal to $p$ on S. We can give the plane $E$ a spherical metric as follows: For each point $x \in E$, let $Sx$ be the point on $S$ determined by the line segment connecting $x$ and $q$. Then for $x, y \in E$ we define the distance between $x, y$ to be the distance on $S$ between $Sx$ and $Sy$. $E$ is then open but finite, since every geodesic has length $2\pi r$, where $r$ is the radius of $S$. One might then assume that ordinary Euclidean rods contract as they move away from the point of tangency $p$. So metrized, the geometry of the plane is exactly like that of a sphere except that there is a point (intuitively, $q$) which one can never reach. Suppose, then that someone lives on the plane $E$, and his rods do expand and contract so as to give the spherical geometry: How is he to decide whether space is Euclidean but finite, or spherical but contains an unreachable point? It seems clear that he cannot.

It might be objected that there are obvious Occamite grounds for rejecting any theory which posits a region of space which cannot causally interact with the rest of space. Just such situations arise quite naturally, however, in general relativity. We cannot dismiss the possibility of such situations without dismissing a good deal of relativistic cosmology; we will discuss some examples subsequently.

## III. RELATIVISTIC COSMOLOGY

Einstein says that the space-time of the universe forms a differentiable semi-Riemannian manifold on which there is everywhere defined two tensor fields, the metric and the energy-momentum tensors.[7] The field equations of general relativity specify a *local* relation between the components $g_{uv}$ and $T_{uv}$ of the metric and energy-momentum tensors in any coordinate system. They are

$$R_{uv} - \tfrac{1}{2} g_{uv} R + \Lambda g_{uv} = k T_{uv},$$

where $R_{uv}$ is the Ricci tensor, which at any point $p$ is completely determined by the value of $g_{uv}$ throughout any neighborhood of $p$, $R$ is a scalar determined locally by $R_{uv}$, and $k$ and $\Lambda$ are universal constants. The energy momentum tensor represents all energy and momentum sources at the point $p$, save for gravitational energy.[8]

It is important to realize that the field equations are indeed local conditions. The two tensor fields, $g_{uv}(p)$ and $T_{uv}(p)$ each determine, at each point, $p$ a specific tensor which we denote by $g_{uv}$ and $T_{uv}$, respectively; the field equations relate the latter. Besides the field equations there are two principles which Reichenbach would no doubt call coordinating definitions: light rays follow null geodesics and free particles – that is, particles subject to no forces other than gravity – follow timelike geodesics.

A relativistic cosmological model is, then, a theory of the universe specifying a differentiable manifold and two tensor fields satisfying Einstein's equations. To test such a model, the mathematical coordinate systems must be related to physical systems; moreover, to reduce the plethora of possible models, certain very strong assumptions are made:

(1)  There exists a coordinate system $(t, x^1, x^2, x^3)$ such that the galaxies have constant $x^1, x^2, x^3$.

(2)  The space sections given by $t =$ constant are isotropic, i.e., all

of the sectional curvatures at any point of the hypersurface are equal to one another.

(3) In the coordinate system of 1, the square of the interval, $ds^2$, equals $dt^2$ for constant $x^1, x^2, x^3$.

In such a coordinate system $t$ is called the cosmic time.[9] It can be shown (Noonan and Robertson, 1968, p. 336f.) that in a coordinate system and manifold satisfying these conditions, the line element (or, more properly, the metric tensor) must have the form

$$ds^2 = dt^2 - R^2(t)\left[\sum_i \frac{dx^i\,dx^i}{\left[1 + \frac{\bar{k}}{4}\sum_l x^l x^l\right]^2}\right],$$

where $R(t)$ is an as yet unspecified function of time, and the expression in parentheses is the familiar form of the line element for a space of constant curvature.

The geometry of a cosmological model satisfying the above assumptions is, then, determined *locally* by a specification of the function $R(t)$ and the value of $\bar{k}$. These quantities determine the geometry in a neighborhood of each point. The physical distances[10] between coordinate space points will vary with the function $R(t)$. If $R(t)$ is constant the distances will be static. If $R(t)$ increases or decreases monotonically, then the universe will expand or contract in the same way. If $R(t)$ is periodic, so will be the expansion and contraction of the universe. The observational tests of relativistic cosmological models – galaxy counts, radio source counts, red-shift and magnitude relations, background radiation measurements, mass density measurements, etc. – are all attempts to determine something about $\bar{k}$ and about $R(t)$ and its derivatives.[11] These quantities completely determine the local geometry and the expansion and contraction of space.

The important point is that $\bar{k}$ and $R(t)$ do not determine a unique cosmological model, even if $\Lambda$ and $k$ are specified. They determine the local geometry but not the global. Indeed for each specification of $\bar{k}$ and $R(t)$ there are differentiable manifolds such that the hypersurfaces $t = $ constant have different topologies. Both manifolds will satisfy the cosmological assumptions and both will of course satisfy the field equations for suitable choices of $\Lambda$ and $T_{uv}$, and since they agree on $\bar{k}$ and

$R(t)$, their local geometries will be exactly alike. They will agree on $T_{uv}$ for the observable region of the universe and on the values (and methods of measurement) of all measurable quantities which can be derived from the line element.

Except for changes in distances which depend on $R(t)$, the geometry of space sections $t =$ constant is only restricted by $\bar{t}$: the curvature $\bar{t}$ is the only topological restraint on such space sections. But for each of the cases $\bar{t} = 0$, $\bar{t} = -1$ there are both closed and open geometries; indeed, in general there are a great many geometries having the same constant curvature but different topologies. Looking first at the two-dimensional case, for $\bar{t} = 0$ we have the Euclidean plane, the torus, the Klein bottle, the cylinder, and all other surfaces applicable on the plane. All of these structures, so different topologically, admit a metric geometry which is everywhere locally Euclidean. For $\bar{t} = 1$ both the spherical and elliptical planes are models, but they have different topologies. For the hyperbolic case, $\bar{t} = -1$ it is known that there are an infinite number of compact, topologically distinct surfaces, all having connectivity greater than three (Klein, 1928, p. 262f.). In the three-dimensional case there are an infinite number of topologically different spaces of constant positive and of constant negative curvature; there are eighteen topologically different zero curvature cases. Wolf (1967) has given a complete classification of the three-dimensional Euclidean and spherical space forms.

Let us take a closer look at one example, the torus with a Euclidean geometry. If this seems unlikely, it is no doubt because one thinks of the torus as embedded in a three-dimensional space, and a Euclidean geometry cannot be induced from a Euclidean three space on a surface homeomorphic to a torus. But in a four-dimensional Euclidean space we can imbed a surface homeomorphic to a torus and having the geometry of the Euclidean plane.[12] The surface is defined by the parametric equations

$$x_1 = \cos u \quad x_3 = \cos v$$
$$x_2 = \sin u \quad x_4 = \sin v$$

and the line element is

$$ds^2 = dx_1^2 + dx_2^2 + dx_3^2 + dx_4^2$$
$$= \cos^2 u\, du^2 + \sin^2 u\, du^2 + \cos^2 v\, dv^2 + \sin^2 v\, dv^2$$
$$= du^2 + dv^2.$$

Hence the surface is locally isometric with the $u$, $v$ Euclidean plane. Tori of arbitrary dimensions having a Euclidean geometry can also be characterized intrinsically, that is, without reference to an embedding space.

What is the upshot? What is the difference between space-times having space with the same local geometry but different spatial topologies? Consider only those regions in which there is a timelike geodesic, parametrized, say, by $t$, such that the union of all spacelike hypersurfaces projected out at each value of $t$ will cover the space-time region. If these hypersurfaces all have topology $T$, then the space-time region will have topology $T \times I$, where $I$ is the interval of the real numbers over which $t$ ranges. Thus, in general, two space-times of the kind considered having space sections with different topologies will, as space-times, have different topologies. The static Minkowski cosmology has topology $R^3 \times R$, for example, but its torus analogue has topology $S \times S \times S \times R$.

## IV. INDISTINGUISHABLE SPACE-TIMES AND EXPANDING UNIVERSES

Our observations may always be topologically local, but that does not mean that locally isometric but topologically different space-times cannot be distinguished experimentally. Marder (1962) has argued by example that in some cases space-times which differ only globally are distinguishable. Consider the following metric for the exterior field of a static infinite cylinder:

$$ds^2 = dt^2 - r^2 d\varphi^2 - A^2(dr^2 + dz^2),$$

where $A$ is constant. For all $A \neq 0$, this metric is locally flat and hence the metrics with any two nonzero values of $A$ are locally isometric. Only for $A = 1$, however, is the space-time homeomorphic to Minkowski space (with a cylinder removed). For if we substitute

$$r' = Ar, \quad z' = Az, \quad \varphi' = \varphi/A$$

we have

$$ds^2 = dt^2 - r'^2 d\varphi'^2 - dr'^2 - dz'^2$$

with $\varphi'$ ranging only from 0 to $2\pi/A$. Now we consider $z =$ constant and release two light beams which pass symmetrically on either side of the cylinder. For the case $A > 1$ the light rays will intersect at $B(=C)$.

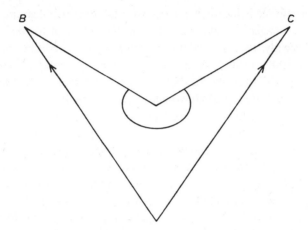

For $A \leqslant 1$ the rays will not intersect. Hence by determining the intersection of nonintersection of two light rays we can distinguish between locally isometric but topologically distinct space-times.

Marder's experiment is not really local, since it involves regions so large that in competing models they are not isometric. But it does suggest a general procedure for discriminating among locally isometric space-times. We can send out light waves in different directions from a point-event and count how many times they intersect or how many times they reach how many of the same spatial points. Since the light waves travel faster than we do, such experiments might not actually be performable in certain space-times, roughly because if we send an observer to a place to tell us if light pulses intersect there he may never be able to return the information to us. But evidently in some cases we actually could perform such experiments. For example, such an experiment could be used to distinguish between the spherical and elliptic versions of the static Einstein cosmology.

The intersection properties of light rays emanating from a place at a time provide us some access to the global topology of space-time. Those aspects of the topology which are discriminable by light intersections may change with time; two light rays sent in specific directions may intersect if sent earlier than a special time, but fail to intersect if sent thereafter. As we will see, this is just what happens in many expanding universes. First we must give a more exact characterization of what it is for two space-times to be indistinguishable by means of Marder's kind of experiment.

Let $M$ and $N$ be time-orientable (isochronous) space-times. We will say $M$ and $N$ are *indistinguishable* just if for every point $p \in M$ there are neighborhoods $U \subseteq M$, $V \subseteq N$ and an isometry $b: U \to V$ such that if $l_p$ and $k_p$ are any pair of timelike or null vectors in the future cone of the tangent space $T_p$, and $l$, $k$ are the corresponding geodesic segments in $M$, then $b_*(l_p)$ and $b_*(k_p)$ lie in the future cone of $T_{b(p)}$ (where $b_*$ is the linear mapping of the tangent spaces which is induced by $b$). Further, if $b(l)$ and $b(k)$ are the geodesic segments in $N$ determined by $b_*(l_p)$ and $b_*(k_p)$, respectively, then $l$ and $k$ must intersect $x$ times in $M$ if and only if $b(l)$ and $b(k)$ intersect $x$ times in $N$. Finally, we require that the union of the ranges of such isometries cover $N$, so that the definition is symmetric in $M$ and $N$.

The definition given is inadequate. It would, for example, entail that the static Einstein model and the spatially elliptic model it covers are indistinguishable, which is clearly wrong. Let $x \in M$ and denote by $P(x)$ the set of all points $y$ in the past of $x$ which are connected to $x$ by an everywhere time-like curve. Then the correct definition is that $M$ and $N$ are indistinguishable if and only if for every maximal time-like curve $\sigma$ on $M$ ($\mathfrak{J}$ on $N$) there is a maximal time-like curve $\mathfrak{J}$ on $N$ ($\sigma$ on $M$) and $\bigcup_{x \in \sigma} P(x)$ is isometric to $\bigcup_{y \in \mathfrak{J}} P(y)$. Subsequent definitions still make sense, and subsequent arguments showing that examples are indistinguishable still hold. More details, and topological conditions sufficient for a space-time to admit a non-homeomorphic, indistinguishable counterpart, are given in 'Indistinguishable Space-Times and the Fundamental Group', in preparation.

We need also to define a notion of indistinguishability after a certain time. Two time-orientable space-times, $M$ and $N$, will be called *t-t' indistinguishable* just if there are cosmic time functions $t_M$ and $t_N$[13] such that the manifold obtained by restricting $M$ and $N$ to $t_M > t$ and $t_N > t'$, respectively, are indistinguishable in the sense of the preceding definition.

These definitions could be generalized in any of several interesting ways, but they are complicated enough and will do for the purpose at hand. Note that the relation of being $t$-$t'$ indistinguishable for particular values of $t$ and $t'$ is a coordinate dependent relation, but the relation of being $t$-$t'$ indistinguishable for some pair $(t, t')$ is invariant. There is evidently an interesting general classification problem here, but our concern at this point is simply whether there are space-times which are topologically different but $t$-$t'$ indistinguishable for some or all $t, t'$. In fact, cosmological models for an expanding universe provide many examples; one of the simplest and most interesting is the De Sitter model.[14] The De Sitter universe can be viewed as a hyperboloid

$$x^2 + y^2 + u^2 + v^2 - z^2 = R^2$$

imbedded in five-dimensional pseudo-Euclidean space with line element

$$ds^2 = dz^2 - dx^2 - dy^2 - du^2 - dv^2.$$

The $z$ coordinate represents time. Just as on the hypersphere, the geodesics are the intersections of the hyperboloid with hyperplanes through the origin. The spacelike geodesics are hyperellipses, the timelike geodesics are hyperbola branches, and null geodesics are the straight lines which are the generators of the hyperboloid. The relations are most clearly seen by taking a cross section $u = 0$, $v = 0$. One then gets a reduced model, a hyperboloid in three space, which exhibits the same structural relations as the full model.

In the reduced model a plane through the origin determines a spacelike geodesic if its angle with the $x$, $y$ plane is less than 45°, timelike hyperbola branches if it is more than 45°, and two parallel straight lines if the angle is precisely 45°. Each null geodesic of such a pair contains all of the antipodal points of the other. The collection of null geodesics, generators of the hyperboloid, from two families of straight lines. Two straight lines are in the same family if they are skew, in opposite families if they are parallel or intersect. Through any point-event on the hyperboloid there are exactly two straight lines and these determine the past and future light cones of the point-event.

Through any point $p$ on the hyperboloid we may draw a straight line (not a line on the hyperboloid) through the origin of the embedding space. Extending this line until it again cuts the hyperboloid we obtain the antipodal point, $-p$, of the point $p$. Consider the null geodesics, $L1p$, $L2p$, intersecting at $p$ and the null geodesics, $L1-p$ and $L2-p$ intersecting at $-p$. Since $p$ and $-p$ are antipodal, each of the first pair of geodesics will be parallel to one of the second pair. Let us suppose that $L1p$ is parallel to $L1-p$. What then is the relation between $L1p$ and

$L2 - p$ and between $L2p$ and $L1 - p$? The parallel pairs belong to different families, hence $L1p$ and $L2 - p$ must belong to the same family, and they are therefore skew to each other. Similarly for $L2p$ and $L1 - p$. We have, then, that the light cones of $p$ and $-p$ do not intersect either in the past or in the future. There can be no causal relation between a point-event and its antipode. There is no way in which an observer at $p$ can learn about what happens at $-p$, no matter where he goes from $p$, and similarly, there is no way an observer at $-p$ can learn about what happens at $p$. An observer wandering about through time can only learn about or influence half of the universe; any half he chooses, but only half.

The De Sitter metric can be expressed in Robertson-Walker form

$$ds^2 = R^2 dt^2 - R^2 \cos h^2 t [d\chi^2 + \sin^2 \chi (\varphi\theta^2 + \sin^2 \theta d\phi^2)]$$

by means of transformations which are not singular anywhere. Any hypersurface $t =$ constant is a three-dimensional sphere. If we identify antipodal points on each of the $t =$ constant hyperspheres we obtain a model which is very much like the De Sitter space except that the constant-time hypersurfaces are elliptic spaces.[15] The two models are locally isometric, but the elliptic spaces have only half the radius of the corresponding spherical spaces. The elliptic and spherical De Sitter spaces are not $t$-$t'$-indistinguishable for all $t$, $t'$. If we send out two light rays in opposite directions at $t = -\infty$ in the spherical De Sitter space they will just meet at $t = \infty$; in the elliptic De Sitter space they will have intersected $t = 0$ and just returned to their place of origin at $t = \infty$. Once the De Sitter space begins expansion, however, things are different: the elliptic and spherical De Sitter space-times are indistinguishable for all $t > 0$. Light rays leaving in opposite directions at $t > 0$ in spherical De Sitter space can, in the rest of time, move less than 180° apart (where the angle is the azimuthal angle around the time axis). Hence in the elliptic interpretation the corresponding light rays will not, in the rest of time, reach a point-event where they intersect.

There are interesting flat-space examples of $t$-$t'$ indistinguishable space-times. Consider a space-time $M$ which is topologically $R^4$ and has the metric

$$ds^2 = dt^2 - e^{Kt}(dr^2 + r^2 d\theta^2 + r^2 \sin^2 \theta d\phi^2),$$

where $r$, $\theta$, $\phi$ are ordinary spherical coordinates for Euclidean space and $K$ is a positive constant. A light ray moving in the radial direction has coordinate velocity (Tolman, 1934)

$$\frac{dr}{dt} = \pm e^{-1/2Kt}.$$

The distance, $r$, that a radial light ray leaving the origin at $t_2$ travels in time $\Delta t = (t_2 - t_1)$ is

$$r = \int_{t_1}^{t_2} e^{-1/2Kt} \, dt$$

and hence the maximum $r$ coordinate, $r_1$, which can be reached by a light ray leaving the origin at $t_1$ is

$$r_1 = \int_{t_1}^{\infty} e^{-1/2Kt} \, dt = \frac{2}{K} e^{-1/2Kt_1}.$$

Thus for every time $t_1$ there is a coordinate two sphere, $S$, in three space such that a light ray leaving the origin at $t_1$ never reaches this sphere. Consider a cube in which $S$ is inscribed. If on opposing faces of the cube we identify points which can be connected by a straight line orthogonal to both of the opposing faces, we obtain a three-dimensional torus. Now in our original space-time $M$ we choose a time $t_1$ and a cube containing the two sphere of coordinate radius $r_1$; such a cube exists in every $t =$ constant hypersurface of $M$ (remember, the coordinate radius $r_1$ of the cube is fixed, it is not a function of time). For each time slice construct the torus determined by the cube; the construction induces a natural Euclidean metric everywhere on the torus. The result is a space-time which is topologically $S \times S \times S \times R$ and which is $t_1 - t_1$ indistinguishable from $M$. Hence the following result: For every time $t$ there is a space-time $t$-$t$ indistinguishable from $M$ and not homeomorphic to $M$.

I do not mean my definitions to be persuasive, and I do not claim that the discussion has firmly established that $t$-$t'$ indistinguishable spacetimes really are empirically indistinguishable after a certain time. I do think that there is a strong *prima facie* case, one that could be demolished or supported by analysis of further proposals for crucial experiments to decide between alternative space-times. For example, in the elliptic analogue of De Sitter space-time discussed above, observers after the crucial time *can* receive light rays sent in different directions from a point-event occurring *before* the crucial time, but that is not possible in the De Sitter universe itself. So when astronomers in the elliptic universe look in opposite spatial directions after the crucial time they will observe a complete symmetry. Everything seen to be going on in one part of the heavens at a given time will also be seen to be going on, at that same time, in the opposite portion of the heavens. In general, we should expect any quotient space-time to show special symmetries not necessarily present in the space-times which cover it. Could we not use such properties to distinguish empirically any pair of *t-t'* indistinguishable space-times? Perhaps, but the matter is not obvious. In the first place, such symmetries are possible in the De Sitter space-time and in other universal covering manifolds; it is only that they are not required. There may be a compelling theory of inductive inference which would discriminate between a quotient space-time with 'built-in' symmetries and its covering space-times with symmetries determined by the distribution of matter, but if there is I am unfamiliar with it. Second, if we take the usual cosmological assumptions seriously, all of the models under discussion require a completely uniform distribution of matter and radiation in space, so that the special observable symmetries of the multiply connected space-times would not be special at all. It can be objected that the uniformity assumptions are no more than idealizations which are not to be taken literally in the small. But if strict uniformity of matter distribution is given up, then the strict symmetries of quotient space-times will have to be given up as well. Third, we have so far considered very few examples. The complete, isochronous space-times are classified (up to global similarity) by the quotient manifolds $D/G$, where $D$ is the De Sitter space-time and $G$ is any group of rotations about the time axis in the ambient five-dimensional Minkowski space which is isomorphic to the fundamental group of a three-dimensional Riemannian spherical space form.[16] The elliptic model discussed previously is only the simplist of these quotient space-times. I have not investigated the indistinguishability, in the technical sense, of these quotient structures, but it seems likely that any pair of them will be *t-t'* indistinguishable for some *t, t'*. More to the point, and more problematic, is whether some pair of such space-times will be indistinguishable for all *t, t'*, that is, just indistinguishable. For clearly, if there is a pair of nonhomeomorphic indistinguishable space-times then they will not be empirically distinguishable by any of the observable symmetries of the kind under discussion. Finally, it should be noted that if we consider geodesically incomplete space-times then there obviously are pairs which are indistinguishable (in the technical sense) for all times, and thus between which we could not decide by examining the symmetries of the heavens. The portions of the De Sitter space-time and its elliptic analogue which occur after the crucial time provide an example.

I claimed at the beginning of this paper that topology in relativistic cosmology can be conventional in both an important and a trivial way. Indistinguishable space-times exemplify the important way, and we can use the De Sitter model again to exemplify the trivial conventionalism. We note that in the coordinate system given, the spaces $t = $ constant are closed. This representation is not, however the usual one; a far more common representation of the De Sitter metric is given by the transformations

$$\bar{x} = \frac{Rx}{y+z} \quad \bar{u} = \frac{Ru}{y+z} \quad \bar{v} = \frac{Rv}{y+z} \quad \bar{t} = \log \frac{y+z}{R}$$

and the line element takes the form

$$ds^2 = R^2 d\bar{t}^2 - e^{2\bar{t}}(d\bar{x}^2 + d\bar{u}^2 + d\bar{v}^2).$$

The space sections $\bar{t} = $ constant have a Euclidean geometry, and in fact they are open. In the reduced model, they are a family of parabolae determined by parallel planes passing through the hyperboloid at an angle of 45° with the $x, y$ plane. Thus simply by changing coordinate systems one alters not only the geometry of space sections, but also the topology.[17] This is no deep liberty, for all that has been done is to change what counts as space.

## V. CONCLUDING REMARKS

There are three philosophical morals which I should like to draw from this discussion. Reichenbach (1958) himself was equivocal as to the empirical determinateness of the topology of space and space-time.[18] But on one thing he was clear: "Topology is an empirical matter as soon as we introduce the requirement that no causal relations must be violated [p. 80]." By an "empirical matter" Reichenbach meant a question which experience and inductive inference could decide. But whether we understand 'topology' in this passage to mean spatial topology or the topology of space-time, it is wrong, for the De Sitter universe is a counterexample to both interpretations. The really interesting and unexamined question is this: What is a causal anomaly?

Mach's principle, whatever it was,[19] has been interpreted in innumerable ways by both philosophers and physicists. One interpretation, suggested by Einstein (1923) himself, is that the spatial universe must be closed; more recently this view of Mach's principle has been adopted by Wheeler (1964) with the intent of making the principle testable. Our discussion suggests that the assumption of a closed universe may turn out instead to be entirely an article of philosophical faith.

According to Reichenbach the importance of conventional alternatives in physics is as an antidote to Kantian philosophy. Conventional alternatives say the same thing in different words, and therefore our inability to decide among them experimentally does not indicate any noninductive limitations on our knowledge. Reichenbach attempted to justify this view by appeal of his own version of the verifiability theory of meaning. Putnam (1963) has attempted to justify the same view of conventional alternatives by arguing that alternative, empirically equivalent theories are intertranslatable.

I see no reason to think that empirically equivalent theories must say the same thing, or that alternative geometries – especially not those having different topologies – are intertranslatable, and I have elsewhere given reasons for these opinions (Glymour, forthcoming). Nor do I think, like Professor Grünbaum, that the existence of conventional alternatives has some ontological significance. Some conventions in physics arise just because more than one theory is in fact true, and in such cases any appearance of contradiction is illusory. Thus there is a strong similarity between the conventionality of geometry which results from different choices of metric standards and the conventionality of the topology of space which results from different coordinate systems in space-time. Just as we can change what we are talking about when we talk of space, so we can change what we are talking about when we talk about congruence.

Cases do arise, however, where it is unreasonable to think that two theories can both be true, yet we cannot decide between them on empirical grounds. It is with such cases that conventions in physics have a real importance, and that importance is epistemic. The significant cases of conventional topologies are those in which the world is one way or the other, not both, but things are so arranged that we cannot discover which way the world is.

*Princeton University*

## NOTES

\* This paper has benefited from conversations and correspondence with John Earman, Richard Grandy, Gilbert Harman, Hartry Field, and Michael Friedman. I am indebted to Bas van Fraassen for reading preliminary versions and encouraging their refinement.

[1] Those who like Minkowski think that what is, is what is invariant, may well dispute this on the grounds that there is no such thing as space. Nonetheless it seems correct to me.

[2] The notion of an affine connection is explained in almost every text on differential geometry. See, for example, Trautman (1965).

[3] See Laugwitz (1965, p. 93). Although neither Reichenbach nor his philosophical successors have suggested the use of such decompositions outside of classical contexts, very similar devices have been used extensively in general relativity, e.g., by Gupta (1957) and by Thirring (1961).

[4] Nonsense, because a Riemannian tensor field is defined on a differentiable manifold which is, among other things, a topological space. The metric topology on the manifold always agrees with its topology *qua* differentiable manifold (cf. Sternberg, 1964, p. 203). Hence metric tensor fields associated with different topologies cannot be defined on the same manifold and their addition, subtraction, equality, etc., is not well defined.

[5] The last claim is justified as follows. Take two points in the real plane to be equivalent if they are an integral distance apart and lie on a line passing through the origin (0, 0), or if they are both an integral distance from the origin. Then every point is equivalent to some point in the circle of unit diameter and center (0, 0). But points on the boundary of this circle and connected by straight lines through the origin are also equivalent. Hence, under the identification topology, this circle is homeomorphic to the hemisphere with antipodal equatorial points identified, and thus to elliptic space.

[6] A maneuver suggested to me by Richard Grandy.

[7] For accounts of these notions, see Bishop and Goldberg (1968), or Sternberg (1964) and Trautman (1965).

[8] Actually, authorities differ as to whether or not the gravitational field can act as its own source. Compare the remarks of Wheeler (1962) with those of Moller (1962), and similarly, Noonan and Robertson (1968, pp. 373–374) with Adler *et al.* (1965, p. 275).

[9] Most discussions of cosmology add a further principle, namely, the cosmological principle or the homogeneity of space. This is put in various ways, e.g., "Any two points in a three space belonging to a given fixed time are equivalent" (Adler *et al.*,

1965, p. 339) or "The view of the universe in its entirety, obtained by any one observer in natural motion, is identical with the view obtained by any other contemporary observer in natural motion" (Noonan and Robertson, 1968, p. 336). If these conditions are meant only to require that spatial curvature be constant, they are redundant, since, by an old theorem of Schur, isotropy entails constant curvature (see Laugwitz, 1965, p. 136). If they are meant to require that space be homogenous in the technical sense – i.e., that the group of global isometries act transitively – then they exceed the empirical evidence but do not affect our examples.

[10] As determined by any of the standard procedures for measuring astronomical distances, e.g., the bolometric distance.

[11] For a recent discussion and review, see Davidson and Narlikar (1969).

[12] The following example is taken from Hilbert and Cohn-Vossen (1952, p. 342).

[13] That is, scalar functions which increase along every future directed timelike or null curve and whose surfaces of constant $t_M$ are spacelike three-dimensional submanifolds without boundary. See Hawking (1968).

[14] The following discussion of De Sitter space-time is taken largely from Schrodinger (1956).

[15] The space-time described here should not be confused with that called "Elliptic De Sitter Space-Time" by Schrodinger. He identifies antipodal points on the hyperboloid. We identify antipodal points on the constant-time hypersurfaces. Schrodinger's model is not time orientable; ours is. His sections $t=$ constant are elliptic only at $t=0$ ours are elliptic everywhere.

[16] See Calabi and Markus (1962). The fundamental group is defined in every text on homotopy. The space-times classified are those of constant positive curvature. $G$ must be finite.

[17] This point has been made by Earman (1970).

[19] Compare his remarks on p. 266–268 with those on p. 66.

[19] For a competent review, see Grünbaum (1963, Chapter 14).

## BIBLIOGRAPHY

Adler, R., Bazin, M., and Schiffer, M., *Introduction to General Relativity*, McGraw-Hill, New York, 1965.

Bishop, R. and Goldberg, S., *Tensor Analysis on Manifolds*, MacMillan, New York, 1968.

Calabi, E. and Markus, L., 'Relativistic Space Forms', *Annals of Mathematics* **75** (1962), 63–76.

Davidson, W. and Narlikar, J., 'Cosmological Models and Their Observational Validation', in R. Taylor, W. Davidson, J. Narlikar, and M. Ruderman (eds.), *Astrophysics*, Benjamin, New York, 1969, pp. 57–149.

Earman, J. 'Space-Time, or How to Solve Philosophical Problems and Dissolve Philosophical Muddles Without Really Trying', *Journal of Philosophy* **67** (1970), 259–277.

Einstein, A., 'Cosmological Considerations on the General Theory of Relativity' in H. A. Lorentz, A. Einstein, H. Minkowski, and H. Weyl (eds.), *The Principle of Relativity*, Methuen, London, 1923, pp. 175–188.

Glymour, C., 'Theoretical Realism and Theoretical Equivalence', in R. Buck and R. Cohen (eds.), *Boston Studies in Philosophy of Science*, Vol. VIII, D. Reidel, Dordrecht, 1971.

Grünbaum, A., *Philosophical Problems of Space and Time*, Knopf, New York, 1963.
Gupta, S., 'Einstein's and Other Theories of Gravitation', *Reviews of Modern Physics* **29** (1957), 334.
Hawking, S. W., 'The Existence of Cosmic Time Functions', *Proceedings of the Royal Society* **308** (1968), 433–435.
Hilbert, D. and Cohn-Vossen, S., *Geometry and the Imagination*, Chelsea, New York, 1952.
Klein, F., *Nicht-Euklidische Geometrie*, Springer, Berlin, 1928.
Laugwitz, D., *Differential and Riemannian Geometry*, Academic Press, New York, 1965.
Marder, L., 'Locally Isometric Space-Times', in *Recent Developments in Relativity*, Polish Scientific Publishers, Warsaw, 1962, pp. 333–338.
Moller, C., 'Discussion of McVittie, G. C., Cosmology and the Interpretation of Astronomical Data', in M. A. Lichnerowicz and M. A. Tonnelat (organizers), *Les théories relativisties de la gravitation*, Centre National de la Recherche Scientifique, Paris, 1962, pp. 273.
Noonan, T. and Robertson, H., *Relativity and Cosmology*, Saunders, Philadelphia, 1968.
Putnam, H., 'An Examination of Grünbaum's Philosophy of Geometry', in B. Baumrin (ed.), *Delaware Seminar in Philosophy of Science*, vol. **2**, Interscience, New York, 1963, pp. 205–255.
Reichenbach, H., *The Philosophy of Space and Time*, Dover, New York, 1958.
Schrodinger, E., *Expanding Universes*, Cambridge University Press, Cambridge, 1956.
Sternberg, S., *Lectures on Differential Geometry*, Prentice-Hall, Englewood Cliffs, N.J., 1964.
Thirring, W., 'An Alternative Approach to the Theory of Gravitation', *Annals of Physics* **16** (1961), 96–117.
Tolman, R., *Relativity, Thermodynamics, and Cosmology*, Clarendon Press, Oxford, 1934, pp. 387–389.
Trautman, A., 'Foundations and Current Problems of General Relativity', in S. Deser and K. Ford (eds.), *Brandeis Summer Institute in Theoretical Physics*, Vol. **1**, Prentice-Hall, Englewood Cliffs, N. J., 1965, pp. 1–248.
Wheeler, J. A., 'Geometrodynamics and Issue of the Final State', in B. Dewitt and C. Dewitt (eds.), *Relativity, Groups and Topology*, Gordon and Breach, New York, 1964, pp. 317–520.
Wheeler, J. A., 'Discussion of McVittie, G. C., Cosmology and the Interpretation of Astronomical Data', in M. A. Lichnerowicz and M. A. Tonnelat (organizers), *Les théories relativisties de la gravitation*, Centre National de la Recherche Scientifique, Paris, 1962, pp. 269–273.
Wolf, J., *Spaces of Constant Curvature*, McGraw-Hill, New York, 1967.

MICHAEL FRIEDMAN*

# GRÜNBAUM ON THE CONVENTIONALITY OF GEOMETRY

Physical geometry is an especially good case study for one interested in the problem of conventionalism in science for at least two reasons. First, there exist a large number of *alternative* geometrical theories which have all been extensively investigated. This contrasts sharply with most other areas of science in which 'alternatives' are largely conjectural. Thus, in the case of geometry, claims that our theories are in some sense conventional or arbitrary, in that we could just as well have had others, at least have more substance. Second, the issue of the conventionality of geometry appears to depend on one particular problem: the status of our methods of measuring length or our 'congruence definitions'. Whether physical geometry is in some sense arbitrary or conventional depends on whether there is a sense in which our methods of measurement or our 'congruence definitions' are arbitrary or conventional. This, of course, simply reflects the mathematical fact that all the properties of a geometry are determined once a metric is determined.

Adolf Grünbaum argues that there is an especially significant sense in which our geometrical theories are conventional – especially significant because conventionalism in this sense is not true of our scientific theories in general. That is, Grünbaum wishes to distinguish a sense in which geometry is conventional from the sense of 'conventional' on which *all* our theories are conventional. To do this he tries to establish that the metric of physical space is conventional or arbitrary in a more important way than other physical concepts (e.g., mass) are, because our freedom to use alternative metrics is based on certain alleged *facts* about physical space which Grünbaum labels its "metric amorphousness." It is because of this 'structural property' of physical space that we are free to use different methods of measurement and to thereby obtain different physical geometries as descriptions of the same empirical facts.

The purpose of this paper is to dispute this central contention of Grünbaum's. I will argue that he has failed to give any real support to the claim that the conventionality of geometry is based on certain 'structural properties' of physical space; and that consequently, geometry, if it is conventional or arbitrary at all, is so only in the trivial sense in which *all* our theories and concepts are conventional or arbitrary. My strategy will be the following: First, I will illustrate the way in which the geometry we ascribe to physical space is dependent on our methods of measurement. Then I will endeavor to show that Grünbaum's attempt to find a basis for the conventionality of our measuring procedures (and thus of our geometry) in certain *facts* about physical space is unsuccessful. Finally, I will argue that the conventionality of geometry is *at best* a special case of a trivial kind of conventionality that holds for all scientific theories.

I. THE RELATIVITY OF GEOMETRY

To illustrate the way in which the geometry we ascribe to physical space depends on our methods of measurement I will consider an example used by Reichenbach (1958, p. 11) to make just this point. He asks us to consider two surfaces: a hemisphere on top and a plane below as in the figure.

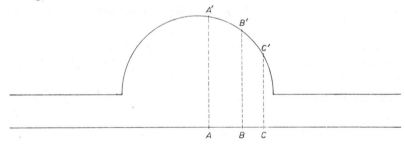

Creatures on the hemisphere measure length in the normal way, finding, e.g., $A'B'$ to be congruent to $B'C'$, and also discovering thereby that the geometry of their surface is non-Euclidean, i.e., that they do indeed live on a hemisphere. Creatures on the plane also measure length in the normal way and find that their geometry is Euclidean, i.e., that they do indeed occupy a plane.

Now, Reichenbach continues, suppose that the hemisphere is transparent and that light from above strikes the plane, casting shadows of the measuring rods on the hemisphere onto the plane. Suppose also that the creatures on the plane *change their methods of measurement* so as to use as standards not rigid rods in the plane, but the shadows that rigid rods on the hemisphere cast onto the plane. For example, people on the plane now judge that $AB$ is congruent to $BC$ because they are projections of congruent segments on the hemisphere. In this case, Reichenbach observes, the plane-people would judge that their geometry was non-Euclidean and would suppose that they really lived on a hemisphere.

What is the 'normal' method of measurement, and what does 'changing our methods of measurement' mean? Reichenbach looks at the matter this way: Normally we think that the length of a solid rod does not depend on its position or orientation in space. If we wished, however, we could suppose that the length of a rod *was* a function of its position and/or orientation by positing certain 'universal forces' which cause uniform shrinkages and expansions in all materials. We could suppose, for example, that a 'universal force' causes measuring rods on the plane to shrink as they move from $A$ to $C$, thus yielding a longer length than normal for the interval $BC$. It seems clear, however, that Reichenbach's use of the notion of 'universal forces' only clouds the issue by introducing extraneous questions about our right to posit such ad-hoc physical forces, questions which are quite different from questions about our right to change our measuring procedures. In contrast to Reichenbach, therefore, I would describe the situation thus: Our normal method of measuring length consists in taking the length of an interval to be (roughly) the number of times a standard meter stick can be laid off from one endpoint to the other. Other methods of measurement are also conceivable, however. We could take the length of an interval to be a certain *function* of the number of times a standard meter stick can be laid off along it, a function that may depend on the position and/or orientation of the interval. And, as in Reichenbach's example, different methods of measurement result in different geometries for the space in question.

To show more precisely how this dependence of geometry on measuring procedures comes about, I will consider Reichenbach's example in a bit more detail. To this end we can represent both the hemisphere and the plane parametrically (cf. Kreysig, 1968, Sec. 16, pp. 55ff.) by vector-valued functions $X(u^1, u^2)$ (I will use capitals to represent vectors) which map the point $(u^1, u^2)$ of the $u^1$-$u^2$ plane onto the point with position vector $X(u^1, u^2)$ in Euclidean three space. In this way we obtain representations of surfaces embedded in Euclidean space with coordinates

($u^1$, $u^2$) induced by the mapping. Denoting $X_i \cdot X_k$ (where $X_i$ represents the partial derivative of $X(u^1, u^2)$ with respect to $u^i$) by $g_{ik}$, we find as an expression for the element of arc on a surface:

$$ds^2 = g_{ik} du^i du^k.$$

This expression is called the first fundamental form of the surface; $g_{ik}$ is called the metric tensor of the surface. (For details, see Kreysig, 1968, Sec. 20.)

In Reichenbach's example we can represent the hemisphere with radius $r$ by the function

$$X(u^1, u^2) = (r \cos u^2 \cos u^1, r \cos u^2 \sin u^1, r \sin u^2)$$
$$(0 \leq u^1 < 2\pi, 0 \leq u^2 \leq \pi/2),$$

and the portion of the plane immediately below the hemisphere by the function

$$Y(v^1, v^2) = (v^1, v^2, 0) \quad (0 \leq (v^1)^2 + (v^2)^2 \leq r^2).$$

The first fundamental form of the hemisphere is

$$ds^2 = r^2 \cos^2 u^2 (du^1)^2 + r^2 (du^2)^2.$$

For the plane the first fundamental form is of course

$$ds^2 = (dv^1)^2 + (dv^2)^2.$$

Calculation yields for the Gaussian curvature of the hemisphere (a quantity which depends only on the metric tensor; cf. Kreysig, 1968, Sec. 27) the value $1/r$, and for the plane the value 0. Thus the geometry of the hemisphere is elliptic, that of the plane Euclidean.

We can now represent Reichenbach's projection of shadows from the hemisphere onto the plane by a *mapping* from the hemisphere onto the plane. This mapping clearly takes the point ($u^1$, $u^2$) on the hemisphere onto the point ($r\cos u^2 \cos u^1$, $r\cos u^2 \sin u^1$) on the plane; i.e.,

$$v^1 = r \cos u^2 \cos u^1$$
$$v^2 = r \cos u^2 \sin u^1.$$

To simplify matters we introduce new coordinates ($v^{1*}$, $v^{2*}$) on the plane related to the old coordinates by

$$v^1 = r \cos v^{2*} \cos v^{1*}$$
$$v^2 = r \cos v^{2*} \sin v^{1*},$$

so that the coordinates on the plane are the same as those on the hemisphere. The plane is now represented by

$$Y^*(v^{1*}, v^{2*}) = (r\cos v^{2*} \cos v^{1*}, r\cos v^* \sin v^{1*}, 0)$$
$$(0 \leqslant v^{1*} < 2\pi, 0 \leqslant v^{2*} \leqslant \pi/2).$$

Its first fundamental form is therefore

$$ds^2 = r^2 \cos^2 v^{2*} (dv^{1*})^2 + r^2 \sin^2 v^{2*} (dv^{2*})^2.$$

Its Gaussian curvature is still 0, however, since changing the coordinates on the plane cannot change its geometry.

How, then, can occupants of the plane come to attribute an elliptic geometry to their 'world'? Obviously, they must alter the expression for the first fundamental form to

$$ds^2 = r^2 \cos^2 v^{2*} (dv^{1*})^2 + r^2 (dv^{2*})^2,$$

a form identical to that of the hemisphere. They can do this by making the distance between two points depend on the new coordinates $(v^{1*}, v^{2*})$ of the points in such a way that the length of an interval $I = [a, b]$ is equal to

$$\int_{a_I}^{b} ds,$$

where $ds$ is given by the above fundamental form. This would involve the adoption of a nonstandard measuring procedure in the sense discussed above. Using such a nonstandard measuring procedure to instantiate the hemispherical fundamental form, the occupants of the plane would find a value of $1/r$ for the Gaussian curvature, and would conclude that the geometry of their 'world' was elliptic.

Whether the geometry of physical space is Euclidean or non-Euclidean depends on our measuring procedures. If standard measuring procedures are adopted – i.e., the length of an interval is not a function of its position or orientation – one geometry will result; if nonstandard measuring procedures are adopted – i.e., the length of an interval *is* a function of its position and /or orientation – a different geometry will result.Changing our methods of measurement also changes our 'congruence definition' in the sense that two intervals having equal length on one system of measurement will in general have unequal lengths on a different system. Thus Reichenbach (1958) writes:

> The determination of the geometry of a certain structure depends on the definition of congruence.... *The geometrical form of a body is no absolute datum of experience, but depends on a preceding coordinative definition*; depending on the definition, the same structure may be called a plane, or a sphere, or a curved surface (p. 18).

This fact Reichenbach calls the "Relativity of Geometry."

The question of whether our theories of physical geometry are conventional therefore reduces to the question of whether our measuring procedures or 'congruence definitions' are conventional. Grünbaum's answer to this question is that our measuring procedures are indeed conventional or arbitrary; we are free to choose whichever method of measurement we like because physical space itself has no metric – physical space is 'metrically amorphous'. Metrical properties can only be imposed on physical space 'from outside' via our measuring procedures, which are therefore perfectly arbitrary. I will now turn to a critical discussion of Grünbaum's argument.

## II. GRÜNBAUM'S ARGUMENT

Grünbaum argues for the conventionality of geometry by arguing for the conventionality of our 'congruence definitions'. A 'congruence definition' is a rule which determines for any two intervals whether they are congruent or not. As noted above, 'congruence definitions' and methods of measurement come to roughly the same thing: a method of measurement automatically yields a 'congruence definition', and a 'congruence definition', together with the choice of a unit interval, yields a natural measuring procedure. Grünbaum is especially concerned to argue that the conventionality of congruence is not simply an instance of the trivial fact that the word 'congruent' is conventional. His method consists in trying to base the conventionality of congruence on certain facts, on certain 'structural properties' of physical space. The 'structural property' in question is the *lack of metrical properties* in physical space; metrical properties are not 'intrinsic' to physical space.

Grünbaum reaches this surprising conclusion in the following way: If physical space were discrete or granular the distance between two points could be defined in terms of the number of elements or space-atoms between them. There would then be a sense to saying that two intervals contained the same amount of space: i.e., that they contained the same number of space-atoms. Congruence would be an 'intrinsic' property of pairs of intervals. But since physical space is in fact *continuous*, there is no 'intrinsic' property of spatial intervals in virtue of which they can be said to be equal or unequal to each other. In a continuous space the distance between two points *cannot* be defined in terms of the number of points between them, for the cardinality of the set of points between *any* two points in a continuous space is $2_{\aleph_0}$ – all such sets have the same cardinality. Thus, metric properties are not 'intrinsic' to a continuous space, it is 'metrically amorphous':

... in the case of a discretely ordered set, the "distance" between two elements can be defined *intrinsically* in a rather natural way by the cardinality of the smallest number of intervening elements. On the other hand, upon confronting the extended continuous manifolds of physical space and time, we see that neither the cardinality of intervals nor any of their other topological properties provide a basis for an *intrinsically* defined metric. The first part of this conclusion was tellingly emphasized by Cantor's proof of the equicardinality of all positive intervals independently of their length. Thus, there is no *intrinsic* attribute of the space between the end points of a line-segment *AB*, or any relation between these two points themselves, in virtue of which the interval *AB* could be said to contain the same amount of space as the space between the termini of another interval *CD* not coinciding with *AB*. Corresponding remarks apply to the time continuum. Accordingly, the continuity we postulate for physical space and time furnishes a *sufficient* condition for their *intrinsic metrical amorphousness* (Grünbaum, 1968, pp. 12–13).

Since intervals of physical space do not stand to one another in relations of equality or inequality *prior* to the establishment of measuring procedures or 'congruence definitions', such procedures do not *ascertain* relations of equality and inequality among spatial intervals but serve to *define* them. Our measuring procedures are not constrained by a previously existing objective metric and are therefore arbitrary or conventional:

Only the choice of a particular *extrinsic* congruence standard can determine a unique congruence class, the *rigidity* or self-congruence of that standard being *decreed by convention*... the role of the spatial or temporal congruence standard *cannot* be construed with Newton or Russell to be the mere ascertainment of an otherwise intrinsic equality obtaining between the intervals belonging to the congruence class defined by it... the congruence of two segments is a matter of convention, stipulation, or definition and not a factual matter concerning which empirical findings could show one to have been mistaken. And hence there can be no question of an *empirically* or factually determinate metric geometry or chronometry until *after* a physical stipulation of congruence (Grünbaum, 1968, p. 14).

This, then, is the core of Grünbaum's argument. How should we evaluate it? An initial difficulty is understanding precisely what Grünbaum means by an 'intrinsic' property. It should be pointed out that there is a familiar sense of 'intrinsic property' as applied to geometrical spaces according to which the intrinsic properties of a space are those properties which are invariant under 'bending' transformations, transformations which deform a space without stretching or tearing – e.g., the transformation which takes the plane onto a cylinder. These intrinsic properties turn out to be precisely those properties of a space which depend on its first fundamental form alone (cf. Kreysig, 1968, Sec. 53, pp. 165–167). Since distance obviously depends only on the first fundamental form it is an intrinsic property of space in this sense of 'intrinsic property'. Therefore, Grünbaum is certainly not using 'intrinsic property' in this familiar sense.

To see what Grünbaum means we must examine more closely his central example of an 'intrinsic' property, the metric of a discrete space. Grünbaum nowhere gives a very precise characterization of a discrete space, but I think it is fair to say that it is intuitively any set isomorphic in a certain way to a subset of $I^n$ – the $n$th Cartesian product of the set of integers with itself. One way of defining a discrete space is as follows:

(1) An *integral space* is a couple $\langle I^n, A \rangle$, where $A$ is a two-place relation defined on $I^n$ such that $A(\langle a_1, ..., a_n \rangle, \langle b_1, ..., b_n \rangle)$ iff for all $i$, $a_i = b_i$ or $|a_i - b_i| = 1$.

(Intuitively, the relation $A$ represents the relation of being immediately adjacent to.)

(2) A *discrete space* is a couple $\langle X, R \rangle$, where $X$ is a set, $R$ a two-place relation defined on $X$; and there is a 1-1 function $f$ from $X$ into $I^n$ such that for all $a, b$ in $X$, $R(a, b)$ iff $A(f(a), f(b))$; and for all $a, b$ in $X$, there exists a finite set $\{c_1, ..., c_k\} \subset X$ such that $R(a, c_1), R(c_1, c_2), ..., R(c_{k-1}, c_k)$, $R(c_k, b)$.

(This last condition insures that a discrete space is 'connected'.)

Given this notion of a discrete space we can construct a metric for $\langle X, R \rangle$ in the way suggested by Grünbaum:

(a) A set $\{a, c_1, ..., c_k, b\} \subset X$ which satisfies the conditions in (2) is called a *curve* with endpoints $a, b$.

(b)  The *interval* [a, b] is any curve with endpoints a, b of smallest cardinality.
(c)  $d_G(a, b) = \text{card } [a, b] - 1$.

Clearly $d_G(a, b)$ satisfies the metric space axioms and is therefore a metric for $\langle X, R \rangle$. I will call it the *Grünbaum metric*.

What is the force of the claim that the Grünbaum metric is 'intrinsic' to a discrete space? It might seem that Grünbaum means *topological property* by 'intrinsic property'; e.g., in the passage quoted above he says "... neither the cardinality of intervals nor any of their other topological properties provide a basis for an *intrinsically* defined metric" when referring to continuous manifolds. If this is what he means, then he is clearly wrong in calling the discrete metric an intrinsic property, as is easily shown: Let $X = \langle 1, 2, 3, 4, 5 \rangle$ be one discrete space and let $Y = \langle 6, 7, 8, 9, 10 \rangle$ be another. Let the topology for $X$ be $P(X)$ (the power set of $X$); that for $Y$, $P(Y)$. Define a mapping $f$ from $X$ onto $Y$ as follows: $f(1) = 6, f(2) = 8, f(3) = 9, f(4) = 10, f(5) = 7$; $f$ is trivially a homeomorphism. However, the Grünbaum metric is not invariant under $f$: $d_G(1, 5) = 4$, while $d_G(f(1), f(5)) = 1$. Therefore, the Grünbaum metric is not a topological property. This is because while cardinality is trivially preserved under topological mappings, order relations are not. And order relations are obviously essential to the definition of the Grünbaum metric. Thus, it seems that Grünbaum must be using 'intrinsic property' to mean any property definable in terms of topological properties and/or *order relations*. In this sense of 'intrinsic property' the metric of a continuous space does not seem to be an 'intrinsic' property.

But the philosophically relevant question is not whether there is *some* sense of 'intrinsic' on which the metric of a continuous space is *not* 'intrinsic'; rather, it is whether the sense in which the metric of a continuous space is *not* 'intrinsic' gives support to the claim that metrical properties are not *really objective* properties of a continuous space, that a continuous space objectively *lacks* metrical properties. And it seems to me that Grünbaum's discussion of discrete and continuous spaces gives no support at all to this latter claim. What has Grünbaum shown in demonstrating that the metrical properties of a continuous space cannot be defined in terms of its topological properties and order relations? He has shown just that: that metrical properties cannot be so defined. In our combined theory of the topology and geometry of physical space its metric is a *primitive notion*. Does it follow from this that metrical properties are not objective properties of physical space? Not at all – both 'mass' in classical mechanics and 'charge' in classical electrostatics are primitive terms, not definable by means of other terms of their respective theories. Would Grünbaum wish to claim that neither mass nor charge is an objective property of bodies? I think not, especially in view of his specific assertion (Grünbaum, 1968, pp. 34–35) that mass is an 'intrinsic' property. And, of course, in our combined theory of physical topology and geometry, both cardinality and order relations are primitive notions, not definable in terms of more basic concepts. But Grünbaum clearly wishes to claim that *they* are objective properties of physical space.

Thus, it appears that there is a further claim implicit in Grünbaum's argument, namely, that topological properties and order relations *are* objective properties of physical space and that no properties not definable in terms of them are objective or *really there* ('built in' is a favorite term of Grünbaum's). Is there any reason to think that this further claim is true? It *cannot* be supported by arguing that topological properties are objective in a way in which metrical properties are not *because* the latter are conventional while the former are not. Grünbaum thinks that a property cannot be truly (nontrivially) conventional unless it is *not* objective or 'built in' to the domain in question. We therefore cannot know whether topological properties are truly conventional unless we *first* know whether or not they are objective. Nevertheless, in the only place in which Grünbaum (1968) comes close to arguing for the above crucial claim, he argues that topological properties, specifically nondenumerability, are not *conventional* in the sense of possessing feasible alternatives:

> [claims that continuity is a conventional property] lose much of their force, it would seem, as soon as one applies the *acid test* of a convention to the conventionalist conception of continuity in physical geometry: the feasibility of one or more *alternative* formulations dispensing with the particular alleged convention and yet permitting the successful rendition of the same total body of experiential findings.... Upon applying this test, what do we find? Attempts to dispense with the continuum of classical mathematics (geometry and analysis) by providing adequate substitutes for the mathematics used by the total body of advanced modern physical theory have been programmatic rather than successful.... Thus, for example, the neointuitionistic endeavors to base mathematics on more restrictive foundations involve mutilations of mathematical physics (p. 29).

Two comments: First, Grünbaum's reference to intuitionist versions of the continuum only obscures the issue. An alternative theory of physical space on which it was a discrete space would not need *intuitionist* versions of the *classical* continuum but a *classical* theory of the discrete; e.g., the calculus of finite differences. Second, this argument can only show something about the *objectivity* of topological properties if there is a close connection between the objectivity of a certain property or properties and the existence of alternative theories about the property or properties. But (a) Grünbaum never argues for the existence of this connection, and (b) it is quite implausible to suppose that such a connection should exist. The existence of alternative theories is a matter of our ingenuity and creativity; whether a property is objective or not is presumably a matter of ontological fact. And on Grünbaum's own account there is no necessary connection between the two. For example, according to Grünbaum, physical time objectively lacks metrical properties, even though it is clearly an open question whether systems of mechanics embodying arbitrary alternative time metrizations are feasible. It is a contingent fact that we were able to develop non-Euclidean geometries. Even if we had not had this ability, however, the metric of physical space would have had the same ontological status. Therefore, it seems to me, Grünbaum cannot use the availability or nonavailability of feasible alternatives to support the claim that metrical properties are not objective properties of physical space (and time) while topological properties like cardinality *are* objective.

Grünbaum's whole case for the special (nontrivial) status of the conventionality of geometry rests on the claim that metrical properties are not objective unless definable in terms of topological properties and/or order relation, which *are* objective. But that there is a distinction between metrical and topological properties of physical space with respect to their *objectivity* has not been established. Further, I can think of no grounds on which such a distinction could be made. Metrical properties are just as much objective properties of physical space as topological properties are. Thus, Grünbaum's attempt to justify our right to alternative 'congruence definitions' in terms of space's lack of metrical properties is a failure. It remains to be seen whether any other justification for the conventionality of geometry can be given.

III. CONVENTIONALISM TRIVIALIZED

An alternative method of trying to justify the conventionality of geometry is what Grünbaum calls the "model theoretic trivialization." He calls it a "trivialization" because, in contrast to his own view, the conventionality it establishes for physical geometry and congruence is extended to all physical concepts and theories. According to this view, a scientific theory is best seen as an uninterpreted formal calculus whose terms receive semantic interpretation by being correlated with certain empirical properties via 'correspondence rules'. In the case of most physical properties these 'correspondence rules' take the form of measuring procedures. For example, in thermodynamics the theoretical term 'having a temperature of $r$ degrees centigrade' is given an interpretation in the empirical realm by being correlated with the readings of a measuring instrument like a mercury thermometer. Furthermore, if the theory (as uninterpreted calculus) allows for more than one interpretation of its terms via such measuring procedures, the choice among them is arbitrary or conventional. No factual considerations constrain us to choose one rather than another; the choice among them can be based only on considerations of simplicity, tractability, and elegance.

This view applies neatly to the case of geometry. In this case, the term 'congruent' – as imbedded in Hilbert's axioms for congruence for example – is capable of being interpreted in many, mutually contradictory, ways. If standard measuring procedures are adopted one family of congruence classes of intervals will result, if nonstandard measuring procedures are adopted a different family of congruence classes will result. Depending on which measuring procedures are adopted different fundamental forms will be realized, and, in general, different geometries will result. But there is no way of choosing among these various alternatives except on grounds of convenience and economy. Thus, congruence is in this sense conventional.

Although Grünbaum often speaks as though he considers Reichenbach an exponent of *his* view, it seems fairly clear that Reichenbach's arguments for conventionalism are closer to this 'model theoretic trivialization' than they are to Grünbaum. For example, Reichenbach (1958) writes:

Physical knowledge is characterized by the fact that concepts are not only defined by other concepts, but are also coordinated to real objects. This coordination cannot be replaced by an explanation of meaning but simply states that this *concept* is co-ordinated to *this particular thing*... certain preliminary coordinations must be determined before the method of coordination can be carried through any further; these first coordinations are therefore definitions which we shall call *coordinative definitions*. They are *arbitrary* like all definitions; on their choice depends the conceptual system which develops with the progress of science (p. 14).

There is a close parallel between Reichenbach's 'coordinative definitions' and our model-theoretic rules of interpretation or 'correspondence rules', as has been noted frequently in the literature. Indeed Reichenbach's treatment of geometry is one of the historical sources for the model theoretic view of scientific theories. In regard to physical geometry Reichenbach (1958) says:

Whenever metrical relations are to be established, the use of coordinative definitions is conspicuous.... The problem does not concern a matter of *cognition* but of *definition*. There is no way of knowing whether a measuring rod retains its length when it is transported to another place; a statement of this kind can only be introduced by definition. For this purpose a coordinative definition is to be used, because two physical objects distant from each other are *defined* as equal in length. It is not the *concept* equality of length which is to be defined, but a *real object* corresponding to it is to be pointed out. A physical structure is coordinated to the concept equality of length, just as the standard meter is coordinated to the concept unit of length (pp. 14, 16).

Aside from Reichenbach's appeal to testability or knowability, this is a typical statement of the model-theoretic view. And, in accordance with that view, Reichenbach does not limit his ascription of conventionality to physical geometry, but extends it to all physical theories. Contrary to Grünbaum, he even considers the *topology* of space to be conventional, requiring a coordinative definition which rules out 'causal anomalies' (Reichenbach, 1958, p, 80).

Sometimes Grünbaum (1968) seems to almost endorse this model-theoretic view himself, as when he approvingly describes Poincaré's views:

... suitable alternative semantical interpretations of the term 'congruent' (for line segments and angles), and correlatively of 'straight line', etc., can readily demonstrate that, subject to restrictions imposed by the existing topology, it is always a live option to give *either* a Euclidean *or* a *non*-Euclidean description of any set of *physico*-geometric facts (p. 108).

On the whole, however, Grünbaum vehemently denounces this view, as he must if the conventionality of geometry is to be nontrivial. He has basically three objections.

First, he argues that the model theoretic defense of conventionalism for 'congruence' fails to do justice to the fact that our freedom to use 'congruent' in different ways, corresponding to different procedures for measuring length, is only a *reflection* on the semantic level of a fundamental *structural property* of physical space. It is therefore not to be confused with the trivial conventionality that applies to every concept of science, and which is not in general based on any structural properties of the domain in question (cf., e.g., Grünbaum, 1968, pp. 99–100). The 'structural property' referred to is of course the alleged lack of an objective metric in physical space, its 'metric amorphousness'. Since Grünbaum has failed to establish that physical space *has* any such structural property, this argument counts for very little.

Second, Grünbaum states that the conventionality of congruence does not simply amount to our freedom to use 'congruent' as we like, because even *after* it is already preempted to be used as a *spatial equality predicate* the class of intervals congruent to a given interval is still indeterminate.

> These axioms [of congruence] preempt 'congruent' (for intervals) to be a *spatial equality predicate* by assuring the reflexivity, symmetry, and transitivity of the congruence relation in the class of spatial intervals. But although having thus preempted the use of 'congruent', the congruence axioms still allow an *infinitude* of *mutually exclusive* congruence classes of intervals (Grünbaum, 1968, p. 13; this point is specifically made as an objection to the model-theoretic view on p. 26).

All this is prefectly correct, of course, but why should Grünbaum think it constitutes an *objection* to the model-theoretic view? The point of that view is not that we are free to use 'congruent' in *any* way whatever, but that there are many different ways of realizing spatial equality predicates – all equally compatible with the congruence axioms. And since they are all so compatible, we are free to choose among them arbitrarily.

Third, Grünbaum (1968) points out that not all terms of geometry can be reinterpreted in such a way as to yield a choice between Euclidean and non-Euclidean descriptions of the same facts. Therefore, he claims, it is incorrect to equate the possibility of reinterpretation which holds for 'congruence' with the possibility of reinterpretation which holds for *every* term of an uninterpreted calculus:

> ... suitable alternative spatial interpretations of the term 'congruent'... show that it is always a live option... to give *either* a Euclidean of a *non*-Euclidean description of the same... facts; and by contrast, the possibility of alternative spatial interpretations of

such *other* primitives of rival geometrical calculi as 'point' does *not* generally issue in this option (pp. 26–27).

Again, this seems perfectly correct, but what does it show? It shows only that the possibility of reinterpreting 'congruent' is much more *interesting* than the possibility of reinterpreting (say) 'point'. But the model-theoretic view is simply an attempt *to justify* our right to reinterpret certain terms by altering our measuring procedures; it is not a theory of which reinterpretations are *interesting*. As far as our right of reinterpretation goes, 'point' is on a par with 'congruent'.

Thus, it seems to me that Grünbaum has failed to offer any convincing objections to the model-theoretic view. *If* geometry is conventional, it is so only in the same sense on which all our scientific theories and concepts are conventional. However, I am dubious about the justification for conventionalism that the model-theoretic view provides. According to this view, the only constraints on the measuring procedures we use in interpreting theoretical terms are the axioms of the theory in question, considered as an uninterpreted formal calculus. Within the range of measuring procedures which lead to the satisfaction of the theoretical axioms, our choice is perfectly arbitrary. This appears to be an oversimplified account of scientific practice. A scientific theory is not an uninterpreted calculus in search of an interpretation; rather, it is supposed to describe and explain a *pretheoretical* domain of objects, properties, and relations. Its application is further constrained by our pretheoretical conception of this domain.

For example, thermodynamics is intended to describe and explain temperature phenomena. The interpretation we give to the term 'temperature' (as it occurs in this theory) via measuring procedures is constrained by our ordinary understanding of temperature phenomena. We cannot set up our measuring procedures in such a way that an ice cube has the same 'temperature' as a blast furnace. If we had a theory containing the term 'temperature', which could be successfully interpreted using such a weird measuring procedure, it seems to me that it would not be a theory of *temperature*. A property which is shared to an equal degree by an ice cube and a blast furnace just is not temperature. Similarly, if there were a method of measuring 'mass' on which the earth had the same 'mass' as a billiard ball, this would not be a method of measuring *mass*. A property which is shared to an equal degree by the earth and a billiard ball just is not mass. A theory containing the term 'mass' which could be successfully interpreted by such a measuring procedure would not be a theory of *mass*.

Of course, it does not follow that an interpretation of a theory must agree with *all* our pretheoretical views about the domain in question. These views can be wrong. Sophisticated methods of temperature measurement, for example, can detect differences of temperature which are indistinguishable to the senses. Many of our ordinary notions about temperature can be in error or only approximately true, and an interpreted scientific theory can lead us to correct these views. Nevertheless, there comes a point when a sufficient divergence from our ordinary understanding of temperature phenomena makes it reasonable to say that we are no longer theorizing about *temperature*.

These remarks also apply to 'congruence' and 'distance' as these terms function in physical geometry. Our measuring procedures are not constrained solely by the axioms for congruence. We could not adopt a method of measurement on which these two segments

| |

had the same 'length', although such a method could very well satisfy the congruence axioms and might even satisfy a particular (no doubt non-Euclidean!) geometry. But what reason is there for supposing that such a theory is a theory of *congruence*? A property which is shared to an equal degree by the height of Mt. Everest and my mechanical pencil just is not *length*. It is also true that some of our pretheoretic ideas about length may be mistaken. Indeed, it seems that our ordinary notions of 'congruence' and 'distance' are inconsistent when applied to physical space, since it is in reality non-Euclidean. We must alter either our natural ideas about spatial measurement or our natural ideas about geometry. From this it does not follow that we can use *just any* methods of measuring 'spatial distance' and still be measuring *spatial distance*. Both our measuring procedures and our theories can be revised, but there comes a point when we are no longer theorizing about the same properties.

In conclusion, it seems to me that whatever truth there is in the thesis of the conventionality of geometry lies along the lines of the model-theoretic view, and therefore applies equally to all scientific theories.

Considerations like the above, however, suggest that the term 'conventional' is highly inappropriate. To be sure, the world does not constrain our theorizing so tightly that there are no alternatives; but there is no reasonable sense in which our theories, geometrical or otherwise, are arbitrary.[1]

*Princeton University*

### NOTES

\* I am indebted to Clark Glymour for correcting several mistakes in my argument.
[1] Since this paper was written (April, 1970) Professor Grünbaum has significantly reformulated and expanded his position (cf. Grünbaum, 1970). This paper is directed only towards his earlier formulation (Grünbaum, 1968), although I believe that the paper's main criticisms apply to his later position as well.

### BIBLIOGRAPHY

Grünbaum, A., *Geometry and Chronometry in Philosophical Perspective*, University of Minnesota Press, Minneapolis, Minn., 1968.
Grünbaum, A., 'Space, Time, and Falsifiability', *Philosophy of Science* **37** (1970), 469–588.
Kreysig, E., *Introduction to Differential Geometry and Riemannian Geometry*, University of Toronto Press, Toronto, 1968.
Reichenbach, H., *The Philosophy of Space and Time*, Dover, New York, 1958.

ARTHUR FINE

# REFLECTIONS ON A RELATIONAL THEORY OF SPACE*

## I. A HISTORICAL PERSPECTIVE AND INTRODUCTION

The modern discussion of the philosophy of space and time continues in the pattern set by the controversy between Leibniz and Newton. The positions in that controversy are usually identified by the labels 'relational' vs. 'absolute' and the opposition thus marked is generally taken to be between the conceptions of space as "a lattice of quantitative relations" and as "a unity which precedes and makes possible all relations that can be discovered in it."[1] Both positions, so described, are difficult to understand. There is, moreover, associated with each of them a special difficulty that I want to focus on, for these special difficulties will suggest a point of view that may aid our understanding of the controversy and its contemporary continuations.

The relational view takes for granted the existence of bodies that bear to one another certain qualitative 'spatial' relations. Some bodies are shorter or longer than others, some are nearer or more distant from others, some are straighter or more bent than others. The difficulty for the relational view is to quantify this network of qualitative relations, for as Newton (=Clarke) puts it to Leibniz, "space and time are quantities; which situation and order are not" (Alexander, 1956, pp. 32, 49, 105). Leibniz's response is to say that order also has its quantity (Alexander, 1956, p. 75) and his effort to implement this remark constitutes a first rudimentary theory of measurement, of the sort associated with Helmholtz, Campbell and their more recent descendents.[2] The basic idea is to take some physical body as a standard for the relation in question and to specify a regimen for deploying this standard so as to construct a quantitative scale the ordering of which coincides with that of the given qualitative relation. In the case of spatial relations the various scales are forged together to form a metric geometry. Questions about space, then, become questions about this geometry and these questions ultimately turn on the deployment of physical standards.

It is more difficult to get a grip on the absolute view and that is precisely the difficulty I wanted to focus on. A whole empiricist tradition, from Berkeley through to the Logical Empiricists, has found Newton's writings on space unintelligible. What is this unity, this container which is absolute space? The gist of Newton's response, however, is I think quite clear. In the famous Scholium on space and time, Newton discusses absolute space together with absolute motion. He defines absolute (= true) motion as change of location in absolute space. To determine the true motions of bodies, then, is to determine how they move in absolute space. The Scholium ends with the words, "But how we are to obtain the true motions... shall be explained more at large in the following treatise. For to this end it was that I composed it." And surely the implication is that space is just what physics says it is. The idea, I take it, is that a comprehensive physics is written in the language of space and time. There is a metric space-time geometry that underwrites physics in a way that inseparably entwines the two. Questions about space, then, become questions about this geometry and these questions ultimately turn on what we take to be the true physical theory.[3]

The opposition that I have been suggesting is certainly not an exclusive one. Clearly if a relational view makes questions of space turn on the deployment of physical bodies, then insofar as physics is concerned with the motion and interactions of bodies, these questions will involve physics. Conversely, if the absolute view makes questions of space turn on questions about physics, then insofar as physics has to do with relations among physical bodies, the absolute view will likewise have to do with these relations. Furthermore the opposition that I have sketched does not seem to cut deeply enough. In particular it does not seem to have the ontological commitments required for a true account of relational or absolute views of space. Those views are usually taken to be views about the nature of space itself, whereas my opposition is at an epistemological level. It is an opposition between different settings for questions about space. There are, to be sure, certain standard routes from epistemology to ontology. In the case of the relational view, from Leibniz on, some sort of verifiability principle has commonly been used. In the case of the absolute view, starting with Newton, some way of implementing a realist (as opposed to instrumentalist, idealist, etc.) account of scientific theories has been employed. But neither verificationism, on the one hand, nor realism on the other is a necessary part of a relational or an absolute view of space. In examining some of Adolf Grünbaum's views below, we shall be seeing a relational view that is not built on any sort of verificationism. Thus in drawing the opposition between relational and absolute views of space, I have not included paths that lead to positions concerning the nature of space itself for, historically, there have been many different paths and, conceptually, it is at least possible that no paths are required at all.

The contrast between relational and absolute that I have been pointing to is a contrast between programs for approaching questions of space (and time). The Leibnizian program suggests that within the range of legitimate questions about the physical universe, there is an autonomous sphere of inquiry that has to do with questions about space and time. These questions assume a quantitative form and the research required to answer these questions concerns the possible deployment of physical bodies as standards of measurement. To carry on this research one may certainly employ auxiliary physical laws, but the import of the investigation is geometric (i.e., has to do with space and time). Ideally, the geometric results should be cleanly separable from the auxiliary physical research aids. By contrast, the Newtonian program is founded on a more holistic conception of the realm of legitimate geometric inquiry. In the Newtonian program that realm is relative to and completely circumscribed by physics itself. The research required to answer questions about space and time is research in physics. It concerns either the further development of physical theory or the investigation of the geometric commitments of current theory. The results of this research are part and parcel of physics. There is no clean separation between physics and geometry.

Most notable among those who have carried on the program set by Newton are surely Poincaré and Einstein.[4] Indeed the general theory of relativity certainly marks a high point in the historical development of the Newtonian scheme.[5] Those who have followed the program set by Leibniz include important members of the school of Logical Positivism; notably, Schlick, Reichenbach and Carnap. In recent years a thorough and creative exploration of the relational program has been carried out in a number of important works by Adolf Grünbaum. One of the early essays has recently been re-issued in a separate book [GCPP] that includes emendations, implications and a response to Hilary Putnam's criticism of the original. I should like to discuss the philosophy of space (I shall not discuss time) set out in this book concentrating on a few selected topics. At the conclusion I shall give a brief summary of some of the interesting topics that I will not have touched on, so that the reader may get a better idea of the content of the book as a whole.

## II. GRÜNBAUM'S CONVENTIONALISM

In some sense Grünbaum's position on space and time is a conventionalist one and it is a major task that continues throughout the chapters of GCPP to make that sense clear. Grünbaum approaches this task by considering a large array of conventionalist theses, tracing them to important authors, and distinguishing them from his own brand. (The reader can see most of them discussed together in Chapter III, Section 3, pp. 243–258.) The most important distinction is between what Grünbaum calls trivial semantical conventionalism (TSC) and geochronometric conventionalism (GC).[6] The thesis of TSC is a general one. It holds that any semantically uncommitted expression may be assigned a conventional use (i.e., intension and/or extension). In particular, if we consider 'congruent' to be an uncommitted expression, then its use may be conventionally determined. Insofar as this use fixes a metrical spatial geometry, one might hold that geometry is conventional (in the sense of TSC). This, however, is not the special sense that Grünbaum espouses. Rather Grünbaum holds to GC which maintains that even after the term 'congruent' has been pre-empted so as to denote a spatial equality relation, no unique such relation is thereby determined. There are indeed many different congruence relations. The choice of one among them on which to found a metrical geometry can only be made on the basis of conventions. Once the conventions are fixed, however, what geometry actually obtains is then a matter of empirical fact. This is the sense, according to GC, in which geometry is conventional.

Let us take a closer look at the items in GC. It is based on the view that questions about space can be given a quantitative form as questions about the metrical geometry of space. The metrical geometry is determined, once the class of points is given, by the properties of the metric, the function that assigns to each pair of points a nonnegative real number that is the distance between the points. All the other geometrical notions can be defined in terms of the metric, and so it is appropriate that discussions of geometry focus on the metric. This is a relatively modern view of geometry, one that is due to Fréchet.[7] Grünbaum, however, generally writes from the older standpoint of differential geometry where the fundamental relation is that of congruence as determined by the differential **ds** of arclength. This is somewhat unfortunate for the framework of differentiable manifolds that Grünbaum supposes takes for granted a set of assumptions stronger than those of the framework of metric spaces, and in this sort of epistemological investigation one would have thought that the weaker the starting assumptions the better. I propose, therefore, to operate from the point of view of metric spaces. I think this will enable us to get a bit clearer about certain aspects of Grünbaum's views.

### III. CONGRUENCE, METRIC AND CONTINUITY

A first question to ask is what are the connections among the congruence relation, the metric function and the geometry of a space. In criticizing Carnap and Reichenbach, Grünbaum points out (Chapter 1, Section 3, Part (iii)) that one and the same geometry – say a Euclidean one of zero curvature – is compatible with different metrics. Thus the geometry does not determine the metric. Metric properties, however, do specify the geometry, so it is true as stated previously that the metric determines the geometry. How then is congruence related to the metric? Grünbaum writes as though it were the case that, up to a scale factor, congruence determines the metric. Is this correct?

Suppose in a metric space $\langle M, \mathbf{d} \rangle$ where $M$ is the point set and $\mathbf{d}$ the metric function, we define point pairs $(x, y)$ and $(x', y')$ to be congruent just in case $\mathbf{d}(x\ y) = \mathbf{d}(x'\ y')$. This defines an equivalence relation, the equivalence classes of which are called congruence classes. The question is whether these congruence classes fix the metric. That is, suppose we start with the set $\mathbf{C}$ of congruence classes. If under two metrics $\mathbf{d}, \mathbf{d}'$ the same classes $\mathbf{C}$ are obtained by using the above definition of congruence, is it the case that for some constant $k$

$$\mathbf{d}' = k \cdot \mathbf{d}?$$

The answer is no and one can show this by means of a standard example.

Consider the function $\mathbf{d}'$ defined by

$$\mathbf{d}' = \frac{\mathbf{d}}{1+\mathbf{d}}.$$

One can easily verify that $\langle M, \mathbf{d}' \rangle$ is a metric space whose congruence classes constitute the same set $\mathbf{C}$ as those under the metric $\mathbf{d}$. But clearly $\mathbf{d}'$ is no multiple of $\mathbf{d}$. (One might suppose that if congruence does not fix the metric it might still determine the geometry. But if in the above example $\langle M, \mathbf{d} \rangle$ is the plane under a Euclidean metric $\mathbf{d}$, then $\langle M, \mathbf{d}' \rangle$ fails to have standard Euclidean features. The law of cosines, for example, fails in $\langle M, \mathbf{d}' \rangle$. Thus congruence, as above, will not determine the geometry either.)

It might be thought that the move from $\mathbf{d}$ to $\mathbf{d}'$ is too bizarre and could be ruled out on topological grounds. That is, it might be thought that congruence together with topology determines the metric. But the metric topology of $\langle M, \mathbf{d} \rangle$ is the same as the metric topology of $\langle M, \mathbf{d}' \rangle$. So moving from $\mathbf{d}$ to $\mathbf{d}'$ does not alter the topology. That move does, however, alter a very basic relation. In any metric space $\langle M, \mathbf{d} \rangle$ the relation of betweenness can be defined by

*q is between p and r*    iff    $\mathbf{d}(p\ q) + \mathbf{d}(q\ r) = \mathbf{d}(p\ r)$.

Then one can say that three points are collinear just in case one of them is between the other two. And one can define the interval between two points as the set of all points between them. What the move from $\mathbf{d}$ to $\mathbf{d}'$ does is to destroy betweenness and, along with it, collinearity. Suppose, for example, that $q$ is between $p$ and $r$, and at unit distance from each of them. Thus $\mathbf{d}(p\ q) = \mathbf{d}(q\ r) = 1$ and $\mathbf{d}(p\ r) = 1 + 1 = 2$. Under $\mathbf{d}'$ one would have

$\mathbf{d}'(p\ q) = \mathbf{d}'(q\ r) = \frac{1}{2}$,    but    $\mathbf{d}'(p\ r) = \frac{2}{3} \neq \mathbf{d}'(p\ q) + \mathbf{d}'(q\ r)$.

So under $\mathbf{d}'$, $q$ is no longer between $p$ and $r$. One might now suggest that if the betweenness relation is fixed, then congruence determines the metric. This requirement certainly does rule out the preceding example of $\mathbf{d}'$.

Let me set this latest suggestion somewhat more precisely. I have been speaking of congruence as a relation holding between pairs of point pairs. But it is more customary to treat congruence as a relation between line segments or spatial intervals and Grünbaum, in particular, treats it in this way. Suppose, therefore, we take as our basic framework a point set $S$ together with a ternary betweenness relation **B**. (The exact postulates for **B** are not important here. Intuition is a safe enough guide for now.) Betweenness gives us the family **J** of intervals $I_{xy}$, where $I_{xy}$ is the set of points of $S$ that are between the point $x$ and the point $y$ of $S$. Once the family of intervals is specified, any equivalence relation on this family looks like a candidate for the congruence relation. (Recall that TSC and CG differ here in that according to GC, but not TSC, 'congruent' must denote such an equivalence relation.) But clearly some constraints must be imposed. For example, the universal relation on **J**, the one that makes every two intervals equivalent, is surely ill-suited for congruence. The general idea is that a pair of intervals should be congruent just in case they have the same length; that is, just in case the distance between the endpoints of one is the same as the distance between the endpoints of the other. So we want to consider only those equivalence relations on **J** which, considered as congruences, are related in the stated way to a metric on the space $S$. Once we are considering metrics on $S$, however, we must take care that – as in the move from **d** to **d**′ – we do not destroy the family **J** of intervals. These considerations suggest the following definitions.

(1) The interval frame $\langle S, \mathbf{J} \rangle$ *is metrized by* **d** just in case $\langle S, \mathbf{d} \rangle$ is a metric space whose metric betweenness yields the same family **J** of intervals.

(2) If $\langle S, \mathbf{J} \rangle$ is metrized by **d**, then we shall say that $I_{xy}$ is **d**-*congruent* to $I_{x'y'}$, for $I_{xy}$ and $I_{x'y'}$, intervals in **J**, provided $\mathbf{d}(xy) = \mathbf{d}(x'y')$.

Finally we can state the suggestions with which we started.

(3) If $\langle S, \mathbf{J} \rangle$ is metrized by functions **d** and **d**′ in such a way that intervals of **J** are **d**-congruent iff they are **d**′-congruent, then there is a constant $k$ such that $\mathbf{d}' = k \cdot \mathbf{d}$.

Speaking roughly, (3) says that once the betweenness relation is fixed, any two metrics that give rise to the same class of congruent intervals are merely constant multiples one of the other. But, unfortunately, even in this sense congruence does not determine the metric. Consider the following example.

$$S = \{a, b, c, d\}.$$

**J** can be read off from the line segments of Figure 1. (Depending on how one understands the betweenness relation singletons may or may not be intervals. If they are intervals, just add them to **J**.)

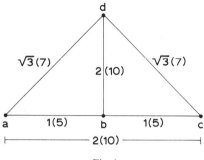

Fig. 1.

One can metrize this interval frame by functions **d** and **d'**, where the **d** distances between points are those not in parentheses in Figure 1 and the **d'** distances are enclosed in parentheses. So,

$$\mathbf{d}(a\ b) = 1 = \mathbf{d}(b\ c) \quad \text{and} \quad \mathbf{d}(a\ c) = 2$$

whereas

$$\mathbf{d'}(a\ b) = 5 = \mathbf{d'}(b\ c) \quad \text{and} \quad \mathbf{d'}(a\ c) = 10;$$

thus both according to **d** and **d'** we have that $b$ is between $a$ and $c$. If **d'** were related to **d** as in (3), then the only possible constant for $k$ would be 5. But $7 \neq 5\sqrt{3}$ and hence (3) fails to hold.

The failure of (3) should not come as a very great surprise, for a little reflection will show that the antecedent of (3) is just tantamount to postulating that there is a function **f** such that $\mathbf{d'} = \mathbf{f}(\mathbf{d})$. If one wants to get that **f** is linear, which is what the conclusion of (3) asserts, then something more must be assumed. The obvious (and as far as I have been able to tell the only general) assumption is one of continuity. More precisely, suppose we define a line in $S$ to be a set $L$ of points such that for any three points of $L$ one of them is between the other two. Then (on the understanding that each pair of points in $S$ determines a line) we want to require that the lines of $S$ are continuous; i.e., that (4) *each line is connected in the topology induced on the line by the betweenness relation.*

It follows from (4) that each interval in **J** is topologically equivalent to an interval on the line of real numbers. This strongly restricts the metrics that can be introduced. In particular (4) implies that the function **f** connecting any two metrics is continuous and that **f** satisfies the Cauchy equation,

$$\mathbf{f}(x+y) = \mathbf{f}(x) + \mathbf{f}(y).$$

It follows that **f** is linear. Thus under the continuity assumption of (4), congruence does determine the metric; i.e., (3) holds. (Actually we need the additional assumption that there are enough intervals, that each interval of length $x$ contains a subinterval of length $y$ for every $0 \leqslant x \leqslant y$.)

We are now in a position to answer the question about how congruence, metric and geometry are related. If we bring to the discussion the background assumption that space is continuous and if we understand this to embrace the continuity of lines, as expressed by (4), then congruence determines the metric up to a scale factor. Since the metric determines the geometry, we can focus our discussion of geometry on the congruence relation. Grünbaum does focus his discussion on congruence and so one might view the preceding technical considerations as vindicating Grünbaum's approach. It does more than that, however, for it brings out the continuity assumption, which is an important element of Grünbaum's view of space. Not only is the assumption of the continuity of space essential to justify the emphasis on congruence, but that assumption is also the central prop to GC.

## IV. CONTINUITY AND INTRINSIC METRICS

You will recall that GC delegates to the congruence relation the role of spatial equality in a framework like the one employed above. The first claim that GC makes is that this role does not determine the actor. Grünbaum sometimes puts this as the claim that space has no intrinsic metric. Sometimes he employs another terminology that is customary in this context and puts it as the claim that space is metrically amorphous. He traces to Riemann (and, in response to Putnam, also to others – p. 215) the view that the postulated continuity of space yields a sufficient condition for space to be metrically amorphous. Grünbaum endorses this view with the qualification that in addition to postulating continuity one must also assume that the points of space are all alike with regard to magnitude (pp. 33–35 and p. 254). Despite Grünbaum's endorsement of this Riemannian line and his reiteration of it in many places throughout the book, I can find no argument for it. That is, there is no argument that begins with the supposed continuity of space (and the likeness in magnitude of the points) and that concludes with the failure of space to have an intrinsic metric. Grünbaum does offer a contrast between discrete (or atomic) space and continuous space by pointing out that in the former the cardinality of intervals offers a measure of their spatial extent, whereas in a continuous space this measure would not be appropriate since all intervals have the same cardinality (that of the continuum). He also translates 'intrinsic' as 'built-in' and claims that no measure of spatial extent is built into a continuous interval. Finally, he contrasts 'intrinsic' with 'extrinsic' or 'external'. No general thread connecting continuity with metrical amorphousness emerges, however. Let us, therefore, examine the issue ourselves.

Begin with continuity. Grünbaum makes here a small but significant error. He appears to conceive of space as continuous in the standard, topological way that, for example, I have expressed in (4). But when he makes an explicit statement, he equates being continuous with being nondenumerable and dense (p. 236). And this lapse seems to account for some of his other remarks (see Note 8, p. 13 and his acceptance of Putnam's usage, pp. 233–234). There are, however, spaces given, for instance, by an order topology that are nondenumerable and dense but that are not order-complete and thus that are not connected or continuous. The significance of this lapse is that in making it Grünbaum has brought to the fore the cardinality of the spatial intervals. This goes along with his contrasting discrete with continuous space in terms of cardinality, and his evident belief that the contrast supports the supposed intrinsicality of the metric in discrete space and the amorphousness in continuous space. It seems to me that all this emphasis on cardinality fits together. To see this, however, we had better look more closely at what it is for space to be metrically amorphous.

## V. CARDINALITY AND METRICAL AMORPHOUSNESS

Consider again an interval frame $\langle S, \mathbf{J} \rangle$. The relation of congruence is an equivalence relation on the family $\mathbf{J}$ of spatial intervals. However, not every equivalence relation will do. As expressed previously by (1), the congruence must be derivable from a metric on $S$ that preserves the given betweenness relation. This requirement certainly restricts the candidates for congruence but in general it cannot be expected to single out a unique equivalence relation as the spatial congruence. Insofar as there is no unique congruence relation, however, then (if I understand how Grünbaum employs his terms) space is metrically amorphous; space has no intrinsic metric. It now appears that we have the desired conclusion but at too cheap a price. For it is easy to see that a discrete no less than a continuous space will, as above, lack an intrinsic metric. The space $S$ and metrics **d**, **d**' of Figure 1 give an example of this. But this is contrary to Grünbaum's claims and I, therefore, conclude that in addition to the above requirement on the candidates for congruence, Grünbaum must have in mind some additional constraints.

The constraints, whatever their form, must have the effect of narrowing down the class of metrics that one can impose on $S$. In the discrete case (this is where the betweenness relation is such that between any two points there are at most finitely many others) the constraints must issue in a metric that is unique (up to a scale factor) and that is, moreover, a one-one function of the cardinality of the spatial intervals. (Only this, I think, would support Grünbaum's repeated claims about the character of the intrinsic metric for discrete spaces. See esp. p. 12 and p. 148.) A very simple sort of requirement might be this. Say that the frame $\langle S, \mathbf{J} \rangle$ has an intrinsic metric just in case $\langle S, \mathbf{J} \rangle$ can be metrized, in the sense of (1) by a metric that is a one-one function of the cardinality of the intervals in **J**. Then a simple cardinality argument of the sort employed by Grünbaum would yield that discrete but not continuous spaces have intrinsic metrics. (Between the discrete and the continuous cases are a whole range of intermediate ones, and not just the dense, denumerable space that is the only other case Grünbaum considers. I shall not discuss these intermediate cases here.) Probably this simple requirement is too simple, or perhaps too transparent. I shall not worry here about more sophisticated versions, for since their output is clear I think I am in a position to challenge their very introduction.

If one looks at the discrete example in Figure 1, then we can see what it is that Grünbaum must exclude. My space included a linear segment $\{a, b, c\}$ with $b$ between $a$ and $c$. The **d** metric gave distances $\mathbf{d}(a\ b) = \mathbf{d}(b\ c) = 1$, $\mathbf{d}(a\ c) = 2$. Clearly these accord with a metric sensitive to cardinality. But in my example there was also an interval $\{d, b\}$ where

$\mathbf{d}(d\ b) = 2$. Thus my **d** metric is not sensitive to cardinality and Grünbaum's requirements must rule it out. In so doing, however, they rule out the very possibility that distance could be inherent in a discrete space in the sense that the distance between certain pairs of space atoms could be inherently greater than the distance between other pairs, even if there were no intervening points.[8] This certainly seems like a possibility and we might try to tell a story to support it. Suppose, for example, that however we try to signal or to move, it just takes longer to get from *d* to *b* than it does to get from *a* to *b*. This would not, of course, prove that the *db*-distance was greater than the *ab*-distance but it certainly might suggest such a hypothesis. To be sure there are difficulties in filling out my story in more detail. These have to do with the nature of motion and of time in discrete space. Such difficulties stem from the extremely puzzling character of discrete space itself. From the point of view of a physically possible world I, for one, do not understand what the discreteness would amount to. Nevertheless this difficulty does not seem to rule out the hypothesis of inherently greater distances and that hypothesis sounds to me at least as sensible as the idea of discreteness itself. Grünbaum, however, must set requirements that eliminate the hypothesis on a priori grounds. Not only is this a burden, but it seems to me a questionable objective.

If one thinks of the metric as yielding a measure of how much spatial stuff there is in an interval, then, provided every space atom has the same amount of this stuff, one might see cardinality as an appropriate basis for the metric. This is apparently a line of thought that has influenced Grünbaum. If, however, one thinks of the metric as providing a measure for the length of an interval or of the distance between the points of space it is difficult to imagine why cardinality should be thought to be relevant at all. In the usual measure theory we are accustomed to the idea that both denumerable and nondenumerable point sets may have measure zero. We know that adding a point or even a denumerable (and sometimes a nondenumerable) set of points to an interval will not change its measure. Why should we think then, in the discrete case, that stuffing an additional point or two into an interval must somehow make it bigger?

We could not only ask, as above, why cardinality should be thought to be relevant, but we can also ask how cardinality could be relevant to an intrinsic metric. Suppose that we have a metric for a discrete space based on cardinality. In the simplest case we should assign as the length of an interval one less than the number of points in the interval. To judge the number of points, however, is to judge that a one-to-one correspondence can be established between the points of the spatial interval and some appropriate counting class. We do this usually by counting or by some other physical matching process. The point I want to make, however, is not about how we determine cardinality, but it is about the concept itself. The very idea of cardinality involves the idea of a one-to-one correspondence between two classes. Even if both range and domain of the correspondence are sets of spatial points (and why not?), the correspondence itself is external, both conceptually and in practice, to the spatial framework. Thus if I must judge on the basis of cardinality, whether two disjoint spatial intervals are congruent I must imagine some process or procedure outside of the intervals themselves by virtue of which I can attempt to make a one-to-one correspondence between the given intervals. There are a variety of ways to try, but they are all external. It is no more a 'built-in' feature of an interval that it contains ten space atoms than it is a built-in feature that it admits of ten applications of my standard rod. Conceptually, I am in both cases dealing with a relation between a given spatial interval *and something else*. For finite cases one can, to be sure, attempt to construe cardinality in the manner of Russell; that is, one can express statements of cardinality by means of purely logical formulas that involve only quantification and identity. It is not clear, however, just what bearing this sort of reduction has on my claim that the concept of cardinality (as well as that of equi-cardinality) is conceptually linked with the concept of a one-to-one correspondence. (Grünbaum, 1970, argues that the dyadic property of equi-cardinality is intrinsic. I intend my claim here to challenge that argument.) If this claim is correct then insofar as Grünbaum constrains the admissible metrics so that for discrete space the metric must depend on the cardinality of intervals, then he will have lost the connection he wanted between *intrinsic* and *built-in* as well as the contrast between *intrinsic* and *external*.

Let me summarize this discussion. In speaking of intrinsic metrics Grünbaum appears to have the following scheme in mind. One lays down requirements to be met by the metrics used to metrize a given spatial frame. If these requirements are uniquely met, then that unique metric is said to be intrinsic to the space in question. Otherwise, we say that the space is metrically amorphous. These requirements connect the metric with the cardinality of the spatial intervals measured. I have suggested that this connection constitutes an arbitrary restriction of an a priori character that seems to conflict with our basic uses of 'length' or 'distance' concepts. I have argued that the metrics so designated as 'intrinsic' would in fact be extrinsic and, thus, that Grünbaum's whole way of approaching the question of intrinsic metrics is self-defeating.

It is not hard to identify the source of the difficulty with Grünbaum's approach. He begins with a spatial framework that has a limited content. It is, broadly speaking, merely topological. He wants to claim, however, that only what can be defined on the basis of this framework is intrinsically spatial. (Again, I am speaking broadly, i.e., loosely). This procedure is bound to appear arbitrary and indeed to conflict with ordinary modes of thought, unless some independent source of spatial knowledge is available to support the restriction to just this topological setting. But because Grünbaum, in the typical relational way, wants to build up such a fund of spatial knowledge just on the basis of the posited frame, he is in no position to give an independent justification. He can only plead consistency and success. That, however, is not a negligible basis for a defense. Let us, therefore, continue to examine his espousal of GC with this defense in mind.

## VI. ALTERNATIVE METRIZABILITY

I had been looking for an argument to support the Riemannian line that the continuity of space implies its metrical amorphousness. Insofar as this amorphousness amounts to the nonuniqueness of the metric as specified by a certain set of requirements, the presumption is that a simple cardinality argument would do the trick. This, of course, would not make full use of the continuity assumption. It would, however, set the stage for an interesting variety of relational program. If continuity provides a sufficient condition for space not to have an intrinsic metric, then to metrize continuous space is to impose on the spatial frame an external metric. This metric determines the geometry of space and thus questions about space must turn on questions about this externally imposed metric. This, however, is precisely the relational program but founded here not on assumptions of parsimony or verifiability, but on the assumed continuity of space. Because space is continuous it has no intrinsic metric. Because of that we must impose the metric from the outside. In this way it seems that we necessarily fall into a relational mold. This is the mold labeled GC.

Let us suppose, then, that continuity does guarantee the lack of an intrinsic metric. Grünbaum concludes from this that space has alternative metrics; i.e., that there are several different ways to metrize the spatial frame. Indeed he sometimes writes as though alternative metrizability were the same as metrical amorphousness (e.g., p. 21). This is a mistake. If there is no unique metric satisfying certain posited requirements, it may be either that there are many metrics or it may be that there are none at all. If we think, for example, of the space-time frame of general relativity, then clearly there is in general no unique way to project from that a purely spatial component. In certain models, however, it may well happen not that there are many ways to project out a 'space', but that there are no ways to do so. If the argument from continuity is to support GC, therefore, it must yield not merely the metrical amorphousness of space but the stronger conclusion of alternative metrizability. In so doing, however, it may well conflict with entrenched physical theory.[9] This is not a point that can be discussed in detail here, for we have no argument to examine. But it is a difficulty with GC that one should keep in mind.

If this difficulty could be overcome, then we would have moved from the continuity of space to its alternative metrizability. At this juncture GC takes two important steps. First, it says that the very existence of metric or congruence relations between spatial intervals depends on the relations that such intervals bear to an extrinsic metric standard which is applied to them. Thus GC steps into the relational parlor. Second, GC asserts that the choice of a particular external standard from among the available alternatives is a matter of convention. Grünbaum's text suggests, moreover, that these two steps are by no means arbitrary. They are supposed to follow from, or at the very weakest, to be supported by the alternative metrizability claim.

### VII. EXTERNAL CONGRUENCE STANDARDS

The step to external congruence standards seems to be based on the following funny line of thought. Since space lacks an intrinsic metric, then if space is to have a metric at all this metric must be extrinsic. If it is extrinsic then it is external and in that case it must be embodied in some physical object, like a rod, that can be moved around and applied from the outside to compare various spatial intervals. But this line begins with a pun on 'intrinsic'. If space lacks an intrinsic metric, then indeed it is correct to say that whatever metric it has is extrinsic. But, as I have pointed out already, there is no reason to think that 'extrinsic', as used by Grünbaum, has anything to do with ordinary uses of the word. To have an extrinsic metric is to have a metric that is one of several that satisfy certain formal, quasi-topological requirements. The only difference between an intrinsic and an extrinsic metric is that there is only one of the former whereas there are always several of the latter. The move that Grünbaum seems to make here is to go from the fact that an extrinsic metric cannot be defined on the basis of a certain posited framework to saying that it must, therefore, be defined operationally. This is not only a non sequitur, it is a variety of operationalism that is inimical to Grünbaum's entire philosophical perspective.

The issue that Grünbaum faces here is precisely the point raised by Newton against Leibniz. If the assumption is that the basic properties of space are embodied in the requirements on the spatial frame posited by Grünbaum, then when that frame fails to yield metrical properties for space one might have expected Grünbaum to see this as a blow to his position. For, 'space and time are quantities'. The alternative metrizability, however, has made Grünbaum aware that the straightforward Leibnizian parry ('order also has its quantity') will not do. So Grünbaum varies the Leibnizian defense and makes the move to external standards that are conventionally imposed. This move is founded on a bold rejection of Newton's point. In affirming his own framework with its alternative metrizability, Grünbaum is saying that space is not a quantity; he is saying that there are no metrical properties of space itself. But in making the step to external standards, Grünbaum is recognizing that Newton's point has some plausibility. It is based on the quantitive language we actually employ for space. Grünbaum's first step, then, can be seen as part of an explanation for how the use of a metrical geometry arises.

I think that this is a reasonable way of viewing Grünbaum's moves. It at least has the merit of showing that Grünbaum is not employing some crudely operationalist procedure. It does not, however, show that Grünbaum's way is well founded. It does not show that continuity compels us to have recourse to an extrinsic standard (p. 149). Clearly Grünbaum might, at this stage, have given in to Newton. He might have said that indeed there are metrical properties of space but that my framework does not enclose them. In that case Grünbaum could simply **have** taken the straightforward Newtonian tack and fallen back on physics, or he might have enlarged his framework to include some specifically metric postulates. Such an enlarged frame might be thought of as a physical hypothesis about space. One could then embed it in a physical theory and test the whole. Such tests would most likely involve the transport of bodies, as do most physical procedures, but the possibility of such transport would not be thought of as constituting the meaning of the geometrical aspects of the theory. To take this alternative, however, is to wed geometry to physics in an essentially Newtonian way. Grünbaum will not do this. He opts instead to work within the confines of a Leibnizian program. So he moves to external, transportable standards. Although this move does not follow from the failure of space to have an intrinsic metric, it is made possible by the alternative metrizability of space. It is perhaps too strong to expect that such options between fundamentally opposed philosophical programs could be strictly rationally founded.

## VIII. THE CONVENTIONAL CHOICE OF STANDARDS

If we take the first step with Grünbaum and say that we shall impose a metric on the amorphous spatial frame by using an external and transportable standard to determine whether two spatial intervals are congruent (or whether a given interval stays congruent with itself over time) must we also take the second step? Must we say that the choice of such a standard is a matter of convention? Grünbaum certainly thinks that this second step is a consequence of the first, together with the existence of alternative metrics (see pp. 20–21, 148–149, 248–249). I have found it very difficult, however, to locate a logical path that leads this way. The following considerations seem to be central. To impose a metric on our amorphous spatial frame, we select a suitable physical object, like a rod, (or perhaps a set of objects) that is transportable and we prescribe a procedure for using this rod to judge whether two spatial intervals are congruent.[10] This is what is meant by an external standard. Following the prescription will determine the class of congruent intervals. If we now select a unit for a scale of measurement, the metric is fixed. Conventions are supposed to enter at two places: where we choose the rod and the prescription for its use and where we choose the unit. If we admit, with Grünbaum, that there are alternative metrics for space, then there will be no spatial features derivable in Grünbaum's frame that could be used to decide between external standards that yield different alternative metrics. It does not follow, however, that there are no features of space which could determine our choice. For, again, we could accept Newton's charge that such a framework will not comprehend all the relevant spatial facts. If, however, we pursue Grünbaum's Leibnizian option then we should have to accept the premise that all the relevant spatial features are derivable in Grünbaum's frame. It does now follow from alternative metrizability that no purely spatial features can fix our choice of external standard. But unless this conclusion is itself taken to be tantamount to our choice being conventional,[11] there is still a gap to be filled.

To say that we make a conventional choice is not necessarily to say that our choice is arbitrary, that no reasons could be offered in its support. It is to say that even after the relevant constraints have been applied there remain options no one of which are we compelled to accept. We may give reasons for making whatever choice we do make, but these reasons will have to be said to support rather than to compel our choice. (This is too fuzzy. But unless we take some sort of soft line, as above, I think any conventionalist thesis could be felled by sustained logic-chopping. I prefer a different method of attack.) In the case of choosing external standards, Grünbaum himself recognizes some appropriate constraints. In particular he recognizes that the determination of congruence should not alter the meaning of the word. In this regard he distinguishes between the intension and the extension of 'congruent' and he holds that only the extension is at issue when we are confronted with a choice of external standards. (See Chapter III, Sections 3.2 and 6.1, where Grünbaum rebuts change of meaning charges by Putnam.) Since I have some special trouble with the notion of intension, I shall substitute the word 'meaning'. Then, in my terminology, Grünbaum holds that the meaning of 'congruent' is spelled out by the formal conditions which require that it designate an equivalence relation on the spatial intervals. I have pointed out, however, that this is not enough. We certainly require that the particular equivalence relation should be derivable from a metric that preserves betweenness. I think that Grünbaum recognizes the propriety of this condition and has merely neglected to state it in the passages cited. Thus certain important connections between distance (or length) and congruence must be preserved if the meaning of 'congruent' is not to alter. But such requirements are of a formal character and it is by no means clear to me that they could exhaust the meaning (i.e., intension) of the term. It seems to me that the meaning of words like 'distance' or 'length' is also connected with their use in a variety of ordinary, daily affairs. And if that is so, then the meaning of 'congruent' must also have its empirical aspect. Specifically I am suggesting that the meaning of the metrical concepts of geometry is constrained by the ways in which we actually go about making the quantitative judgments necessary to use these concepts. Implicit in this suggestion is the possibility that the constraints are strong enough to make the choice between external standards that lead to alternative metrics not conventional at all.

Let me illustrate how this suggestion might work. In our daily affairs we measure, just as we count, by means of routine and standard procedures. We learn these procedures, at home and in our early school years, when we learn – for example – how to use a ruler and a compass. We have these procedures in mind when we understand that a room or a table is of such-and-such dimensions. Given the scale, we understand how to use these procedures in standard ways so as to check the dimensions. In terms of these procedures we easily come to see that one can alter the unit and thereby change, for example, from feet to meters. These commonplace remarks, however, have a bearing on what we understand by the quantitative concepts of geometry. I shall try to bring this out by telling a short story.

Suppose that we go to buy a carpet and that we require a square one, nine feet by nine feet, just to fit snugly into a small den. Suppose we purchase a carpet and bring it home only to discover that it does not fit at all. Indeed, when we check the dimensions with our yardstick we find that each of the sides has a different length. We call the store and they send out a representative. He is certain that the size of the carpet was accurate, so he begins by measuring the size of our room. He lays his yardstick out three times along one wall and says, "Yes, that is nine feet." He lays his stick out three times along another wall, then takes out his slide rule, does some calculations and says, "Well, this side is $10\sqrt{\pi}$ feet long." He repeats the procedure twice more and then, of course, he begins to argue with us. He claims that our den does not have a square floor; the sides are of unequal length. We tell him that of course it is a square; he himself laid out the yardstick three times along each wall. My son, the scientist, intervenes with the suggestion that perhaps the gentleman from the store thinks there is an inhomogeneous field of distorting forces along the floor. But the gentleman from the store does not even understand the remark. "Look," he says, "that's how the carpet guild measures rooms for carpets. I've a lot of experience, and, take it from me, for carpets that is the best and simplest way. And I assure you that the most eminent and respected firms do it that way. You see, your room just won't fit a square carpet." (See pp. 17–19.) Exasperated, I can only reply that I could not care less what is simplest or what the best firms do; "Buddy, what *you* mean by 'square' and what *I* mean by 'square' are two different things!"

I apologize for the silly story. It is a moral fable whose moral, I am sure, is clear. (If not, one might try retelling it in terms of counting; say, with a representative from a marble factory.)

### IX. CUSTOMARY CONGRUENCE

If we choose an external standard by using, say, the meter bar in the way that we have all learned in school then Grünbaum would call this the 'customary congruence'. This involves certifying that two intervals are congruent just in case laying the meter bar out end over end in both cases yields the same result. Of course if there were different distorting influences along the two intervals, we should have to correct for them. But otherwise, we proceed naively as we have been taught to. Grünbaum describes this naive procedure as stipulating that our rod is self-congruent under transport. If we do not so stipulate, then we shall be employing a noncustomary congruence (as the carpet man does in the story) to realize an alternative metric. This stipulation, according to Grünbaum (p. 149), is the locus for the conventionality in the choice of an external standard. But why does Grünbaum think that we are engaged here in making a stipulation? He seems to proceed as follows. Notice that the situation in which we are supposed to stipulate self-congruence occurs before we have established a metric standard. Thus in precisely the situation where

Grünbaum holds that space is metrically amorphous we are supposed to say that the spatial intervals occupied by our rod on two different occasions are congruent to one another. But features of space as embodied in the amorphous spatial frame could not certify this congruence. Indeed nothing in the frame could count either for or against it. So Grünbaum says that we are (*must be*) making a stipulation.

But consider the following comparison. We can think of the choice of an external standard as a sort of recursive process. In the first stage we fix the congruence relations (or length) for spatial intervals in which our rod fits snugly. Then we follow an iterative procedure for judging more and more (as well as less and less) spacious intervals. Suppose we compare Grünbaum's spatial frame with the framework given by the Peano Postulates for the natural numbers. I shall assume that these are given in a standard set-theoretic context. Then I want to treat the addition of numbers the way Grünbaum treats the congruence relation. I thus lay down formal postulates for addition: it is a binary operation that is closed, commutative, associative, etc. In setting these postulates, however, I carefully refrain from saying what will be the result of adding any specific numbers. It now follows (and this is the analogue of the alternative metrizability claim) that no unique operation satisfies all these formal requirements. Suppose, then, that I set out to impose addition externally; that is, I set out to give a recursive definition of addition. At the first stage I must lay down an answer to the question

$$x+0=?$$

I have a choice here. I might, for example, be able to make $x+0=0$ and then perhaps to continue so that this 'addition' turns out to be multiplication. There is nothing in the Peano Postulates, even as augmented by my special requirements, that would support one alternative over another. I might, therefore, say that I just stipulate the answer to be $x$. So I do. But my stipulation here is not a matter of convention. If I want the operation I am defining to be addition, then there is only one alternative for me to choose.

This was my suggestion in Grünbaum's case. If no issue of distorting influences is at hand, then the question as to whether two intervals occupied by the standard rod on different occasions are congruent must be answered in the affirmative. Confronted with the choice of developing either the customary or a noncustomary congruence, if we are going to use this as a basis for distance or length, then the customary congruence must be selected. This choice is, therefore, not conventional except in the sense of TSC. It is, one might say, a matter of convention that we use the word 'distance' for distance. That sort of conventionality, however, is not what GC is all about.

I hasten to acknowledge that in no sense have the conclusions just stated been established. I have, rather, tried to lay the background for a suggestion that would yield a nonconventional approach to the setting up of external standards, on the assumption that such standards must be externally imposed. Given my enviable role as critic, this is perhaps enough. For Grünbaum claims that a conventionalist approach is necessary and it should be sufficient for me to point out that this claim requires for its defense a line of argument that will rule out the preceding suggestion. This line will have to span the troubled waters of meaning change. I think it is reasonable to expect that no argument of this sort is likely to appear conclusive to a very large audience.

In addition to employing constraints with respect to meaning, I should also point out that there is a Newtonian alternative to conventionalism even at this stage. Faced with a choice of procedures for how to use a standard rod, we might very well fall back on physics. We might notice that the ordinary practice of scientific measurement on the basis of which physical theory is tested, developed and understood already employs a fixed procedure. Thus insofar as general relativity, for example, endows physical space with a metric, the manner in which this metric relates to the uses of standard rods is already fixed. Locally at least, it employs the customary congruence. For anyone with Newtonian predilections, this sort of constraint ought to be a sufficient ground for choosing the customary congruence. To urge, in the face of this, that the choice is nevertheless conventional is just to reject the Newtonian program.[12]

## X. THE AUTONOMY OF GEOMETRY

I have broken GC up into separate pieces. There is the spatial framework, the assumed continuity, the metrical amorphousness, the alternative metrizability, the imposition of external metrical standards and the conventional choice of these standards. This array is supposed to hold together internally by means of its logically interlocking parts. Thus, given the continuity of the framework the rest of the items are supposed to follow logically, in the order listed. I have tried to show, however, that none of the pieces is logically linked to its companion. And I have urged that it would be better if some of these pieces were not so linked.[13] At the same time, however, I have tried to suggest that the array does form a coherent whole, for it all fits snugly into a relational bag. Given this relational covering, one can trace out lines of support from piece to piece. I think that the central prop is the autonomy of geometry vis-à-vis physics. It is the belief in this autonomy that makes it plausible initially, I think, to attempt to set up an independent spatial framework and it is the autonomy, again, that supports the move to external standards. I should like to isolate the autonomy for separate scrutiny.

There is no question that Grünbaum is well aware of the intimate bond between physical geometry and physical theory. His writing is replete with physical examples and with interesting references to the physical literature. Moreover, even with regard to the interpretation of the geometrical terms, Grünbaum himself emphasizes the essential role played by physical theory (see pp. 32–33 and p. 125). He recognizes, nevertheless, that his relational account requires the autonomy of geometry relative to physics. He devotes two sections (Sections 6 and 7) of the essay reprinted here as Chapter I to an exegesis and defense of this autonomy. The issue recurs in Chapter II, which is entitled 'Geometry and Physics', and it is an important part of Grünbaum's rebuttal of Putnam in Chapter III.

Grünbaum is concerned to defend two theses with regard to autonomy. The first is the thesis that from an epistemological point of view geometry is separable from physics. That is, it is possible to isolate at least some geometrical hypotheses from the network of collateral physical assumptions so as to subject them to the usual inductive cannons of evidence; in particular, so as to refute them. (See pp. 114–115.) The second thesis concerns the possibility of actually extricating the geometry of space from the network of physical laws that would be involved in any empirical procedure that is designed to determine that geometry. Grünbaum holds that, modulo the uncertainties involved in any inductive procedure, it is possible – at least in theory – actually to isolate the geometry for independent determination. It is clear that a cogent defense of these theses would seriously undermine the Newtonian program. Such a defense would show that the conceptual ties between geometry and physics do not bind geometrical assertions to presuppositions embodied in physical theory. This would strongly suggest that the only value to a Newtonian program is a methodological one: it might just be the case that the best way of finding out about geometrical facts is to approach them via a total physical theory. Grünbaum need not deny this in order to vindicate his own approach. Let us look, then, at Grünbaum's defense.

## XI. SEPARABILITY

The defense of the first thesis is set in the context of discussing the Duhemian view of falsifiability. Grünbaum has written extensively on this topic and there is a growing secondary literature on his writings alone.[14] I do not want to consider all that here. It will be sufficient to examine one of the proffered geometrical counterexamples to the Duhemian thesis and to indicate what I take to be problematic about such examples. The example concerns some spatial region $R$. The object is to show that a certain geometrical hypothesis $H$ can be isolated from collateral physical assumptions so as to allow for the possibility of its refutation. Let $H$ be the hypothesis that the geometry of $R$ is Euclidean. It is to be understood that this hypothesis is founded on the presupposition that the metric of the geometry is based on some set of 'rigid unit rods' to be deployed so as to constitute the customary congruence. This supposition is understood to be involved in the very meaning of $H$. The collateral assumption $A$ is that the region $R$ is free of deforming influences. From $H$ together with $A$ it follows that the results of measurements made with these rigid rods will be Euclidean. Grünbaum now supposes that certain non-Euclidean results are in fact obtained. Then one could conclude either that $H$ is false or that $A$ is false. Were strong independent evidence available for the truth of $A$, however, then in normal inductive fashion we could conclude that the geometrical hypothesis $H$ is false. Grünbaum suggests that strong evidence for $A$ would be available if we could confirm $C$, that independently of their chemical constitution all rods invariably preserve their initial coincidences under transport in $R$. That is, that any pair of chemically different rods that coincide at one place in $R$ continues to coincide regardless of where the rods are moved to in $R$. Grünbaum claims that $A$ implies $C$ and, on this assumption, he argues that $C$ provides strong inductive support for $A$. I shall not enter into the questions of inductive support here, for they seem to me to be entirely peripheral. The central questions concern the relation between coincidence behavior and deforming influences, as well as the nature of the latter. It is here that geometry joins physics.

The question about coincidence behavior and deforming influences is quite simple. Is it true that $A$ implies $C$; if the region is free of deforming influences must the coincidence behavior of rods be preserved under transport? Let us begin with coincidence and then pass to its supposed link with deformations. Notice, first of all, that the region $R$ of concern must be of vast proportions. (Grünbaum sometimes speaks of the surface of tables. These had better be astronomical tables.) For all the various Riemannian geometries (which is the context within which Grünbaum writes) are locally Euclidean. Thus deformations that engender a non-Euclidean structure for $R$ can be expected to show up only on a large scale. So the breaking of coincidences, in particular, are large-scale phenomena. That is to say, we should be looking for changes of coincidence with rods, say, of length no smaller than that of the diameter of the earth. I claim no understanding of the physical properties of such rods, but it is clear that the question as to whether a pair of them do or do not coincide could scarcely be answered by just looking and seeing. For such rods coincidence is not a relation that can be ascertained by 'direct' observation. It seems, rather, that coincidence is like simultaneity. For nearby rods whose end points are within our visual field, coincidence in the small can be certified by looking. Distant coincidence (i.e., for rods that extend beyond our visual field or that are far away from us), however, like distant simultaneity involves the transmission and reception of physical signals. The laws governing such signals are none other than laws of physics. Thus to say that a pair of rods coincides is to make a theoretical, not an observational, statement. Such statements, even if they concern local coincidence, may be subject to revision should the appropriate physical theories be revised. (My language is Quinean; my point is not. See Note 12.) It follows that even if Grünbaum's coincidence test could be said to confirm $A$, it would all by itself involve the supposition of collateral physical assumptions of precisely the sort that he is attempting to rule out.

But what, after all, is the bearing of coincidence on deformations?

Presumably we are supposed to be able to infer the presence of a deforming influence if a pair of chemically different rods that have coincided at one place in $R$ no longer coincide when moved to another place. Clearly such a breaking of coincidence would constitute a change. It is not so clear, however, that this change is to be counted as a deformation. I have two reservations about this. One is that 'deformation' has a very physical ring and I should feel more comfortable viewing the connections between the physical terms 'coincidence' and 'deformation' as part of physics. My other reservation is this. If (but not only if) $R$ is in fact free of deforming influences and if our rods, whose coincidence relations have been altered, are among the unit rods that are ingredient in the external standard of congruence (and this is a harmless assumption I should think), then it follows from the definition of customary congruence that the spatial intervals occupied by the rods on the occasion when they do not coincide are, nevertheless, congruent. But if the intervals occupied by the rods are congruent, then surely we should say that the rods have the same length (or size, or whatever). But then I think that there is no room to hold that the change of coincidence constitutes a deformation. Thus given a change of coincidence we have the option of saying that no deformation has occurred or saying that the break of coincidence is itself a deformation. Putting these reservations aside, however, suppose we do say that a deformation has occurred. What leads us now to conclude that the region contains a deforming influence? Presumably the principle at work is that deformations occur only when deforming influences are present. This is a strong causal principle and I shall assume that for Grünbaum it does not have an a priori character. Rather it corresponds to a large set of physical laws, each of which connects a specific kind of force (magnetic, thermal, etc.) with specific deformations in bodies of a given type. And since the connection is universal, there must be a general assumption that each possible deformation is covered by some combination of these laws. But sensible and plausible as this is, I hope it is equally apparent that it constitutes a set of strongly causal assumptions about physics. It seems reasonable to conclude, therefore, that each stage in the attempt to support the assumption $A$ that $R$ is free of deforming influences involves suppositions of a physical character.

Although this would be enough, I think there is a deeper and more important problem about Grünbaum's example. The problem is with the very introduction of the assumption $A$. If we assert $A$, then we are saying something like this: there are no forces (or what have you) acting in $R$ that could have the effect of changing the sizes or shapes of physical bodies in $R$. But this assertion makes sense only if we suppose that there could be bodies in $R$ that have a size (or shape). In the context set by the geometrical hypothesis $H$, however, to speak of the size of objects is to speak of the possible results of the procedure involved in implementing the customary congruence when that procedure is carried out in a deformation-free environment. The environment is precisely the region $R$ itself. To put this together, in order for the assumption that $R$ is free of deforming influences to make sense, it must be the case that if $R$ were free of deforming influences then the customary congruence procedure would have certain determinate results. But this is a prescription for going around in circles. If we try to follow it, then we shall find ourselves always short of finding out whether it makes sense to say that $R$ is free of deforming influences.

In cases of circularity, I generally feel that I have merely been trapped by some manipulation of language. To see that no such superficial source is at the root of this circularity, we might ask how it is that we could come to know (or to verify) that $R$ is deformation-free. To see that a spatial region is free of deforming influences is to see that changes (or differences) of a certain kind do not occur. One does this by looking for changes of the relevant sort and then failing to find them. In the case of deformations we should be searching for alterations of sizes or shapes, or the like. That is, we should be looking for alterations in some metrically determined features of bodies. Fundamentally, therefore, we should be after the effects of a change in the basic congruence relations among spatial intervals that are given by the unit rods of the customary congruence. Thus we must be able to trace whatever changes are observed to changes in the congruence behavior of our unit rods. The behavior of the rods is only relevant to congruence, however, if either the rods are deployed in a deformation-free environment or if they are corrected for deformations. The latter would involve the prior determination of the presence of deformations, which is precisely the issue at hand. The former would involve being able to certify the absence of deforming influences. It was just this freedom from deforming influences that we set out to uncover. It cannot be done.[15]

Grünbaum offers a second example in defense of the separability of geometry from physics (pp. 342–350 contain its tightest formulation). This is to show the refutability of a hypothesis about the geometry of a certain region of space in the case where there are deforming influences present in the region. The issues here center around the logic of correcting for the deforming influences. I shall not discuss these issues. I want only to suggest that the difficulties that I have found with the deformation-free example above, seem to me present in this second example. The arguments advanced by Grünbaum seem to depend essentially on strong background assumptions of a physical nature. It also seems that one cannot avoid the inherent circularity that I have focused on above.

XII. EXTRICABILITY

This conceptual circularity complements a procedural circularity noticed by Einstein and used by him to argue against the possibility of extricating the underlying geometry of space from the physics that covers it over. This possibility is what makes up Grünbaum's second autonomy thesis. Einstein's objection (see pp. 123–125 but also pp. 336–338 for a qualification) is that in order to carry out an empirical procedure for determining geometry it is necessary to correct our measuring instruments (rods, etc.) for deforming influences. These corrections are made by means of physical laws, like the law of thermal expansion. These laws, however, are stated in quantitative geometric terms. They thus suppose the applicability of a geometry. It follows that the most our empirical procedure could yield would be a conclusion of the form, 'If such-and-such geometry is assumed together with so-and-so physical laws, then such-and-so geometry is found to obtain'. It appears that the geometrical conclusion here could never be detached.

Grünbaum faces this challenge head on. He admits that the correction for perturbational forces is based on physical laws that involve geometrical quantities. He contends, nevertheless, that there is an approximation procedure that would enable one to extricate the geometry. The procedure goes like this. Begin with whatever geometrically based correctional laws $P_0$ one may have and use them, together with whatever measurement routine one has in mind, to determine a geometry $G_1$. Now employ $G_1$ to recalculate the geometrical quantities that enter into the laws $P_0$ so as to get a new set of laws $P_1$ that are based on $G_1$. With $P_1$ use the geometrical routine again so as to get a geometry $G_2$. Use $G_2$ to get laws $P_2$ and keep repeating this whole procedure. It may happen that at some stage the geometry becomes stable; i.e., that using correctional laws based on a certain geometry the empirical procedure yields up that very same geometry. Clearly if we continue nothing new will arise. If such a stage occurs then Grünbaum would say that the approximation procedure has converged. What is the significance of convergence?

Grünbaum lays down three conditions (p. 139) that the underlying geometry $G$ of space must satisfy. One of these conditions is that $G$ must be a geometry to which some approximation procedure converges. He supposes that these three conditions must be logically bound together and that just one geometry could satisfy all of them. Some years ago in a critical note on the approximation procedure, I pointed out (among other criticisms) that these three conditions are not logically connected and that there is no reason to suppose that they determine a unique geometry (Fine, 1964). Thus one cannot rule out the possibility that if approximation routines begin with different sets $P_0$ of laws (based on different geometries) they will converge to different geometries. In Chapter II, Section 2 (ii), Grünbaum acknowledges this criticism. He tries to get around it by suggesting that it *may* just be the case that some geometry satisfies all of his three conditions. If so, then his approximation routine would have succeeded in uncovering the underlying geometry of space (p. 193).

I find this response of Grünbaum's quite out of character. It is a frivolous response, for it concedes that convergence according to his approximation procedure has no logical connection with 'the underlying geometry' of space, and then it tacks on that just maybe, by accident, a convergent sequence might converge to the right thing. But random guessing is not connected with uncovering the true spatial geometry either and yet I might accidentally think of the right one while so engaged. I should not on that account, however, be entitled to claim any special efficacy for random guessing as a defense of the extricability of geometry. Parity of reasons suggests that Grünbaum's approximation procedure should be granted no higher status. It does not show, even in theory, that modulo the uncertainties involved in any inductive procedure one can actually isolate the geometry of space for independent determination.

The failure of Grünbaum's procedure is hardly surprising. It could be successful only if Grünbaum could explicate what is the underlying geometry of space in such a way that we could recognize it when we find it. The three conditions mentioned above try to do this. As expected, however, they make essential reference to how rigid rods would behave in the absence of deforming influences. They thus involve us in the vicious circularity brought out in the discussion of the first separability thesis. One might have looked to the approximation procedure as a way to cut of this circle as well. But it just does not work.

I conclude that the separability and the extricability theses that make up the claim for the autonomy of geometry require a better defense than the one Grünbaum has offered. His arguments do not rattle the foundations of the Newtonian program, nor do they support the relational path he has chosen to follow. One should not conclude, however, that the autonomy claim should be withdrawn. Accepting my criticisms (not a light suggestion) a confirmed relationalist need only explore the available variations that may lead to a more adequate defense. One might even expect that the circularity I have pointed to could be avoided by choosing a somewhat different way of spelling out the prescription for the use of rods in the choice of a congruence standard. For example, that prescription might simply omit any reference to deformations. Then whatever intervals the rods certify as congruent just are congruent, regardless of the environment. This way with congruence clearly has certain practical difficulties to overcome (what happens if the rods melt, will the world keep changing size in not easily codified ways, etc.) but Carnap (1966), for example, considers it at least a theoretical possibility. And it is surely consistent with a relational program. Another possibility might be to stipulate that certain rods are free from deformation in a certain environment, like the sealed Paris meter bar. Then the concept of 'deformation-free' could be relativized by some quasi-recursive scheme according to which environments at any given stage are free of deformations relative to conditions set by environments in the preceding stages.[16] This idea might well involve some essential physical assumptions, but that remains to be seen. Whatever the particular suggestions, I think one must recognize the vitality that resides in the relational legacy of Leibniz.

## XIII. CONCLUSION

The debate between Newton and Leibniz is a debate between two different, empirical approaches to geometry. In its contemporary version both parties to the debate reject the aprioristic features that one finds in Kant or in the early Russell, or even in the more moderate apriorism of Wiener.[17] It is a rivalry, one might say, between siblings in the same family. Paradigmatically, such rivalry is especially bitter and petty. Historically, the contest between relationalists and absolutists has had this character. Each side has accused the other of being incoherent and, therefore, incomprehensible. And these charges have been looked upon by each side as a knock-out blow to the other. From this perspective it is a particular merit of Grünbaum's work that it seeks to render Newton's position intelligible, although false. (If space had an intrinsic metric, which is a possibility but one that just does not obtain, then Newton would have been correct about absolute space.) This is not an accidental merit. It flows, on the contrary, from the seriousness, and the deep intellectual concern that is clearly evident throughout Grünbaum's work.

The depth of Grünbaum's work is indicated by the range of philosophical topics that it covers. Many of the topics of recent philosophical interest are there: the analytic/synthetic distinction, the observational/theoretical dichotomy, the questions of meaning, criteria and meaning change and, of course, the role of conceptual systems. The background for the work is empiricist, but it is remarkably free of doctrinaire lines.

I have tried to analyze the logical structure of Grünbaum's view of space and of his defense of the claim to autonomy. I have found some incoherence in this defense and, all along the way, I have uncovered what looks to me like important logical gaps. Judged by the criteria of consistency and success, however, there is much to be said in Grünbaum's favor. For all the pieces of the analysis do fit together into a relational whole, even if they are not bound by logical ties. And I should think that there is enough leeway in the relational scheme, to provide filler for at least some of the gaps and to leave room for straightening out the circularity. I find, that is, that Grünbaum does offer the outline for a possibly successful relational view.

## XIV. ADDENDUM

I shall try here to recount, as promised, some of my omissions. The book consists of three chapters. Chapter I is a reprinting of Grünbaum (1962). Chapter II consists of two sections. The first section gives an extended response, properly set, to Schlesinger's (1964) lively and provocative defense of the 'hypothesis' that overnight everything has doubled in size. The second section is concerned with a defense of Grünbaum's anti-Duhemian position with regard to the interpretation of stellar parallax observations and that section also treats some of my criticisms that I have discussed above (Section XII). Chapter III is a revised version of a published rebuttal to an extended critique of Grünbaum by Putnam (1963).

In the course of the book Grünbaum treats aspects of the work of a large variety of historically important authors. These include Newton, Riemann, Carnap, Reichenbach, Poincaré, Einstein, Milne, Russell, Whitehead, Eddington, Clifford and of course, Duhem.

Among the interesting topics that I have not touched on there are the whole variety of issues having to do with time, the construal of continuity as an inductive framework principle of physics, a defense of the view that Reichenbach's primary use of 'universal force' is purely metaphorical, consideration of the bearing of Grünbaum's views on the general theory of relativity (and vice versa), a discussion of simultaneity in relativistic physics and some remarks on Zeno's paradoxes.

There is more. But this ought to suggest a sufficient picture of the scope of the work as a whole. Even a confirmed absolutist will appreciate that Grünbaum's work provides a solid philosophical stone against which his own axe can be ground and, no doubt, sharpened.

*University of Illinois at Chicago Circle*

## NOTES

\* I shall cite Adolf Grünbaum, *Geometry and Chronometry in Philosophical Perspective* (1968) as GCPP and my bracketed chapter and page references will be to this book. I want to take this opportunity to thank Professor Grünbaum for lively correspondence over the years and for making available to me the rough draft of a portion of some recent material of his that will contain careful and detailed reworkings of his position on space and time. Part of this material appeared in Grünbaum (1970).

[1] The phrases quoted here are from Koyré (1957, p. 252).
[2] The development by Leibniz is contained in his 'The Metaphysical Foundations of Mathematics'. This aspect of the Newton/Leibniz controversy, including the above citations, is discussed in van Fraassen (1970, Ch. 2).
[3] A similar point of view with regard to Newton is argued persuasively in Stein (1967). Stein puts more emphasis than I would on the necessity for a separable space-time structure.
[4] In making this classification I am being deliberately provocative, for Grünbaum would claim Einstein and Poincaré for his own. But then my way of marking the absolute/relative debate is different from Grünbaum's.
[5] Earman (1970) contains interesting applications of the Newtonian program with general relativity taken as the appropriate physical theory.
[6] If the reader has difficulty with 'geochronometric conventionalism' then, as I do, he may find it useful to read GC as 'Grünbaum's conventionalism'.
[7] Menger (1954) is an excellent survey of the geometrical development of Fréchet's work.
[8] For discrete space Grünbaum sometimes (e.g., pp. 154, 216) tries to restrict the admissible metrics by requiring that they give rise to physically realizable units. This demand, however, is of no avail unless one shows the theoretical necessity for recourse to just such external standards. For an intrinsic metric there is no necessity of this kind and so the question of units is merely one of convenience. Surely convenience should not be an overriding consideration in assessing conceptual constraints.
[9] These points are discussed in a preprint by John Earman, 'Are Spatial and Temporal Congruence Conventional?' Grünbaum (1970) contains a critique of this paper by Earman.
[10] Grünbaum recognizes that distance is a multiple-criteria concept. For simplicity, however, he generally focuses the discussion on criteria that involve the transport of rigid rods. I shall follow him in this.
[11] A reading, for example, of p. 149 alone might suggest that Grünbaum is making this equation, but setting this in context makes it plausible that he is following the more reasonable procedure outlined below.
[12] One should not conclude from this that I hold the reactionary view according to which science cannot discover that there are essential errors in our daily procedures and thus that these procedures should be revised. Rather my view is, roughly, that the intelligibility of science requires a continuity between scientific concepts and their 'ordinary' counterparts. Thus locally there cannot be much distinguishable difference although globally the difference may be very great indeed.
[13] Massey (1969) is a panel discussion of Grünbaum's philosophy of science which treats many of the issues considered here. I have profited by this discussion but I believe that my criticisms here and below are relevant even when Grünbaum is read in the light of his able defenders on this panel.
[14] Grünbaum (1969) is the most recent of Grünbaum's writings on the Duhemian argument. It contains references to his previous works and to the literature. See also Massey (1969), especially the paper there by Philip Quinn.
[15] I have considered here only what might be called an attempt 'directly' to check on freedom from deformations. One might, however, proceed indirectly. One might set the hypothesis that there are no deforming influences in the framework of other physical assumptions and laws, and then see how well the whole fits. But since the issue at hand concerns the separability of geometry from physics, this otherwise reasonable approach is unavailable here.

[16] Lenzen (1938, p. 12) suggests breaking out of the circle by a procedure somewhat of this sort.
[17] Wiener (1922) is an interesting and largely unnoticed set of articles.

BIBLIOGRAPHY

Alexander, H. G. (ed.), *The Leibniz-Clarke Correspondence*, Philosophical Library, New York, 1956.
Carnap, R., *Philosophical Foundations of Physics*, Basic Books, New York, 1966.
Earman, J., 'Space-Time', *Journal of Philosophy* **37** (1970), 259–78.
Fine, A., 'Physical Geometry and Physical Laws', *Philosophy of Science* **31** (1964), 156–62.
Grünbaum, A., 'Geometry, Chronometry and Empiricism', in, *Minnesota Studies in the Philosophy of Science* III, University of Minnesota Press, Minneapolis, 1962.
Grünbaum, A., *Geometry and Chronometry in Philosophical Perspective*, University of Minnesota Press, Minneapolis, 1968.
Grünbaum, A., 'Can We Ascertain the Falsity of a Scientific Hypothesis', *Studium Generale* **22** (1969), 1061–93.
Grünbaum, A., 'Space, Time and Falsifiability', *Philosophy of Science* **37** (1970), 469–588.
Koyré, A., *From the Closed World to the Infinite Universe*, Johns Hopkins Press, Baltimore, 1957.
Lenzen, V., 'Procedures of Empirical Science', in *International Encyclopedia of Unified Science* **1** (1938), 12.
Massey, G. J., 'A Panel Discussion on Grünbaum's Philosophy of Science: Toward a Clarification of Grünbaum's Conception of an Intrinsic Metric', *Philosophy of Science* **36** (1969), 331–99.
Menger, K., *Géometrie Générale*, Gauthier-Villars, Paris, 1954.
Putnam, H., 'An Examination of Grünbaum's Philosophy of Geometry', in *Philosophy of Science: The Delaware Seminar 2*, Interscience, New York, 1963.
Schlesinger, G., 'It Is False that Overnight Everything Has Doubled in Size', *Philosophical Studies* **15** (1964), 65–71.
Stein, H., 'Newtonian Space-Time', *Texas Quarterly* **10** (1967), 174–200.
van Fraassen, B., *Introduction to the Philosophy of Time and Space*, Random House, 1970.
Wiener, N., 'The Relation of Space and Geometry to Experience', *The Monist* **32** (1922), 12–60; 200–47; 364–94.

ADOLF GRÜNBAUM*

# THE ONTOLOGY OF THE CURVATURE OF EMPTY SPACE IN THE GEOMETRODYNAMICS OF CLIFFORD AND WHEELER

I. INTRODUCTION

For nearly two decades before 1972, Professor John Wheeler pursued a research program in physics that was predicated on a monistic ontology which W. K. Clifford had envisioned in 1870 and which Wheeler (1962b, p. 225) epitomized in the following words: "There is nothing in the world except empty curved space. Matter, charge, electromagnetism, and other fields are only manifestations of the bending of space. *Physics is geometry*." In an address to a 1960 Philosophy Congress (Wheeler, 1962a), he began with a qualitative synopsis of the protean role of curvature in endowing the one presumed ultimate substance, empty curved space, with a sufficient plurality of attributes to account for the observed diversity of the world. Said he:

... Is space-time only an arena within which fields and particles move about as "physical" and "foreign" entities? Or is the four-dimensional continuum all there is? Is curved empty geometry a kind of magic building material out of which everything in the physical world is made: (1) slow curvature in one region of space describes a gravitational field; (2) a rippled geometry with a different type of curvature somewhere else describes an electromagnetic field; (3) a knotted-up region of high curvature describes a concentration of charge and mass-energy that moves like a particle? Are fields and particles foreign entities immersed *in* geometry, or are they nothing *but* geometry?

It would be difficult to name any issue more central to the plan of physics than this: whether space-time is only an arena, or whether it is everything [p. 361].

For nineteen years, Wheeler and his co-workers, such as Charles Misner, developed some of the detailed physics of Clifford's ontology of curved empty space-time as an outgrowth of general relativity under the name of "geometrodynamics" (GMD). In Wheeler's parlance, "a geometrodynamical universe" is "a world whose properties are described by geometry, and a geometry whose curvature changes with time – a *dynamical geometry*" (Wheeler, 1962a, p. 361). But in a lecture at a 1972 conference,[1] Wheeler disavowed his erstwhile long quest for a reduction of all of physics to space-time geometry.[2] In a brief notice of that Conference

P. Suppes (ed.), Space, Time and Geometry, 268–295. All rights reserved
Copyright © 1973 by D. Reidel Publishing Company, Dordrecht-Holland

(*Nature* **240** [Dec. 15, 1972], 382), the pertinent part of this lecture was summarized as follows: "He [Wheeler] also developed the theme that the structure of space-time could only be understood in terms of the structure of elementary particles rather than the converse statement which he has advocated for many years."

To emphasize his new conception of the fundamental and indispensable ontological role played by entities or processes *other than* space-time, Wheeler repeatedly spoke of PREgeometry. As I understood him, he sought to emphasize in this way that he now regards space-time not as the basic stuff of a monistic ontology but rather as an *abstraction* from the events in which quantum processes are implicated. That is to say, according to Wheeler's new conception of PREgeometry, space-time is an abstraction from the constitution of physical events ONTOLOGICALLY no less than epistemologically! In Wheeler's *erstwhile* view of space-time as the only autonomous substance, space-time is absolute in the older familiar sense of being *empty*. In the perspective of Wheeler's new program of a PREgeometric ontology, his earlier all-out geometric absolutism appears as a kind of ontological chauvinism. Wheeler now wishes to supplant that absolutism by a neo-Leibnizian view which is RELATIONAL in the sense that space-time structure is only one aspect of a quantum universe whose ontological furniture cannot be constituted out of space-time. I believe that one reason for Wheeler's explicit mention of Leibniz was to emphasize the relational character of his notion of pregeometry.

In the minds of great scientists like Wheeler, there often is a subtle interplay between empirical and conceptual or philosophical promptings for abandoning no less than for espousing a major theory with its research program. It seems to me that this state of affairs need not at all betoken a gratuitous and pernicious apriorism. At least generally, there is no sharp divide between legitimate empirical and conceptual or philosophical reasons for the rejection (acceptance) of a major theory. Thus, I believe that predominantly philosophical considerations are likewise germane to the appraisal of certain facets of Wheeler's erstwhile purely geometrical ontology vis-à-vis the rival ontology adumbrated in his more recent notion of pregeometry. Hereafter, when I speak of 'geometrodynamics' or use its acronym 'GMD', I shall disregard Wheeler's basic change of view and will use this term to refer to his earlier relativistic theory of *empty* space, rather than to the ontologically more noncommittal standard version of general relativity. Even though it was *not*, of course, relativistic, Clifford's so-called 'Space-Theory of Matter' (1876) shared the essential ontological assumptions of Wheeler's GMD. Hence I shall also occasionally denote Clifford's theory-sketch by 'GMD'.

Elsewhere, in a chapter on 'Geometrodynamics and Ontology' (Grünbaum, 1973, Ch. 22), I am devoting attention to several such ontological facets of geometrodynamics, partly in response to a 1972 GMD Symposium (*Journal of Philosophy* **69**, No. 19 [Oct. 26, 1972]) which focused on Graves (1971). Here I shall confine myself to just one of these facets by concentrating on the status of curvature properties in GMD.

In my reply (Grünbaum, 1970, p. 470) to 'A Panel Discussion of Grünbaum's Philosophy of Science' (Massey *et al.*, 1969), I had questioned the Clifford-Wheeler statements of GMD in regard to "the compatibility of the theory [GMD] with the Riemannian metrical philosophy apparently espoused by its proponents." Thereafter, Clark Glymour (1972) devoted his paper 'Physics by Convention' to a critique of Grünbaum (1970). Speaking of the latter essay, Glymour (1972) writes in the opening paragraph of his paper:

> Grünbaum's replies to his critics, especially his most recent reply (1970)… involve unusually important claims which fail to be buttressed by the arguments he gives. I have in mind such claims as… that the foremost advocates of geometrodynamics, Clifford and Wheeler, were and are enmeshed in contradiction. My own view is that all of these claims are dubious or false, but I shall be less concerned with establishing their falsity than with discrediting the arguments offered for them [p. 322].

Glymour (1972) amplifies this assessment near the end of his paper as follows:

> Grünbaum charges the advocates of geometrodynamics, Clifford and Wheeler in particular, with inconsistency. Since these men have maintained that matter reduces to curved space, they must also have thought, according to Grünbaum, that curvature is an intrinsic property of space. Yet, Grünbaum argues, both Clifford and Wheeler deny that space has an intrinsic metric. But this is inconsistent, Grünbaum concludes, since "… this curvature would need to obtain with respect to a metric implicit in empty space" (Grünbaum, 1970). Now I do not think this claim especially important, partly because I am not at all convinced that Wheeler would deny that space has intrinsic metric properties, and partly because the program of geometrodynamics certainly does not require such a denial. Even so, I doubt that Grünbaum has provided, or can provide, anything like sufficient grounds for his conclusion. He gives no argument at all as to why we should think curvature properties presuppose or require or "would need to obtain with respect to" a metric. It cannot be

because the curvature tensor of space does in fact determine a *unique* Riemannian metric, for that is not true, as Grünbaum himself appears to have noted [at this point, Glymour cites Grünbaum (1963, pp. 89–105)]. Perhaps by "curvature" Grünbaum intends properties some of which are not determined by the curvature tensor alone; affine properties generally, perhaps, or sectional curvatures. But a 3-dimensional manifold fitted with a Riemannian connection does not, in general, have a *unique* compatible metric, even up to similarity (p. 338, italics added). ... Even if the properties in question include all affine properties *and* sectional curvatures it is not clear that they determine a *unique* metric (p. 339, the *second* italics are added).... So interpreted, then, Grünbaum's contention that curvature requires metric is at best moot. Of course, Grünbaum may simply have meant that curvature properties are just not the sort of thing that can exist unsupported. But he has given us no shade of reason why that might be so, let alone demonstrated that what is required for their support is a metric [p. 339].

Here Glymour raises two issues as follows:

(i) Quite *apart* from how Clifford and Wheeler conceived specifically of curved empty space as the building material of the physical world, do curvature properties of space presuppose (require) a metric, so that it would be inconsistent to claim that space is intrinsically curved but is devoid of any intrinsic metric?

(ii) What is the answer to the latter question of presupposition and inconsistency when posed in the specific context of the assumptions made by Clifford and Wheeler, and if the answer is positive, did either Clifford or Wheeler avow such an inconsistency?

## II. QUESTION (i): DO CURVATURE PROPERTIES OF EMPTY SPACE REQUIRE A METRIC?

Do curvature properties of space require or presuppose a metric, so that it would be inconsistent to claim that space is both intrinsically curved and yet devoid of an intrinsic metric? We shall see before long that the correctness of an affirmative answer to this question turns crucially on whether the given curvature properties include those specified (or determined) by the *covariant* fourth-rank Riemann-Christoffel curvature tensor or are *confined* to those furnished (determined) merely by the *mixed* Riemann-Christoffel curvature tensor (which is contravariant of rank 1 and covariant of rank 3). But before I can develop this point, I must show in some detail how Glymour misconstrued our Question (i) from the outset.

Glymour overlooked that, to begin with, the issue here is (a) whether given curvature properties require or depend on SOME Riemannian metric OR OTHER, and *not*, as Glymour would have it, (b) whether given curvature properties require or presuppose a metric which is *unique* at least to within a choice of the unit (i.e., *unique* up to a constant positive factor $k$ or 'similarity'). Glymour's conflation of (a) with (b) is important here for the following two reasons:

1. If the answer to (a) were *affirmative*, that answer to (a) would be logically *weaker* than an affirmative answer to (b), since the former merely asserts that *some* Riemannian metric *or other* is presupposed, while the latter asserts that a metric *unique* at least up to a constant positive factor $k$ is presupposed.
2. If the answer to (a) were affirmative, that *weaker* affirmation would be *sufficient* to justify the charge of *inconsistency* against the claim that space is both intrinsically curved and yet devoid of any intrinsic metric at all.

Let me refer to the claim that space is both intrinsically curved and yet devoid of *any* intrinsic metric as 'MUC' to convey its assertion of the feasibility of the Metrical Unsupportedness of intrinsic Curvature. And let me call the charge that MUC is *inconsistent* 'The Inconsistency Charge'. The strong affirmation that the answer to (b) is positive asserts that given curvature properties require a metric *unique* at least up to the factor $k$, and hence this strong assertion will hereafter be designated as 'the curvature requirement of a unique metric' (CRUM). In terms of these designations, I maintain that the two reasons I gave for rejecting Glymour's conflation of (a) with (b) show the following: It was unavailing for Glymour to have tried to undermine the inconsistency charge by wrongly assuming that this charge against MUC rests on the logically strong affirmative answer CRUM to question (b) rather than merely on the logically weaker affirmative answer to (a). As we noted above, the latter weaker affirmation asserts merely that given curvature properties require *some* Riemannian metric *or other*. On the basis of his erroneous belief that the inconsistency charge against MUC rests on the strong assertion CRUM of uniqueness, Glymour then proceeded to argue irrelevantly by means of certain technical results that this uniqueness claim is either false or gratuitous, *even if* the given curvature properties comprise some which are *not* determined by a fourth-rank curvature tensor alone.

It is instructive to note the reason which prompted Glymour to be concerned to assert the falsity or unfoundedness of CRUM with respect to a set of given curvature properties that is *wider* than one containing only those determined by the fourth-rank curvature tensor alone. In Grünbaum (1963, Ch. 3, Sec. B), I had given several examples to show that even in two-dimensional Euclidean space – where the one independent component of the curvature tensor vanishes along with the total Gaussian curvature $K$ – the vanishing curvature $K$ of that two-space *fails* to determine a Riemannian metric or metric tensor which is *unique* up to a choice of unit. Thus, I had *denied* the uniqueness claim CRUM *with respect to the Riemannian (Gaussian) curvature of two-space*.

In personal correspondence, I had called Glymour's attention to this denial of mine in response to the following statement by him in an earlier draft of his paper (1972) which he had kindly made available to me: "Even assuming he is accurate in the views he ascribes to Clifford and Wheeler, Grünbaum would *only* be correct [in his charge that MUC is inconsistent] if a curvature tensor on a differentiable manifold somehow determined or required a *unique* metric tensor; but that is not true" (italics added). As shown by my quotation above from his *published* text, Glymour modified his cited earlier statement to the extent of taking cognizance of the fact that I had denied CRUM with respect to the curvature tensor. But he persisted in his earlier misguided concern with discrediting CRUM, albeit by now widening the membership of the class of curvature properties to which this uniqueness claim is held to pertain. Instead of realizing that the inconsistency charge against MUC need not rest on any kind of uniqueness claim CRUM, and that I had never claimed it does, Glymour mistakenly remained convinced that it must. Having thus adhered to his initial misconstrual (b) of our Question (i), he therefore proceeded to inquire irrelevantly though interestingly what wider miscellaneous assortment of curvature properties might perhaps sustain the uniqueness claim CRUM. Not surprisingly, even then he found CRUM wanting or at best moot. But we saw that this finding does not at all undermine my inconsistency charge against MUC. So much for Glymour's own handling of our Question (i).

We must now deal with this latter question when properly construed as initially posing the issue (a) whether given curvature properties require (presuppose, depend on) *some* Riemannian metric *or other*. And let us propose to construe the given curvature properties as those specified or determined by the fourth-rank Riemann-Christoffel curvature tensor. We must then point out at once that in the context of our present inquiry the following fact is very important indeed: There is an ambiguity in the expression '*the* Riemann-Christoffel curvature tensor' as used nowadays, according as we mean the so-called Riemann tensor of the *first* kind – which is the *covariant* fourth-rank curvature tensor – or the so-called Riemann curvature tensor of the *second* kind, which is the *mixed* curvature tensor that is contravariant of rank 1 and covariant of rank 3.[3] Clearly, this ambiguity makes it dependent on the context whether given statements about curvature properties, construed as rendered by '*the* Riemann tensor', do refer to or involve a commitment to both kinds of Riemann tensor or to only one of them!

A few brief preliminaries will provide us with the means to define both kinds of Riemann tensor. Let us denote ordinary partial differentiation by a comma followed by the variable with respect to which we differentiate partially. Thus, the partial derivative $\partial g_{ik}/\partial x^m$ of the symmetric covariant, second-rank metric tensor $g_{ik}$ would be written as $g_{ik,m}$. The component $g$'s of this metric or 'fundamental' tensor are functions of the coordinates subject to the restriction that their determinant be nonzero. In this notation, the so-called Christoffel symbols of the first kind are defined by

$$[ij, k] = \tfrac{1}{2}(g_{ik,j} + g_{jk,i} - g_{ij,k}).$$

Furthermore, the contravariant conjugate metric tensor $g^{lk}$ is defined by the condition $g_{ml}g^{lk} = \delta_m^k$, where we sum over the repeated index $l$, and $\delta_m^k$ is the Kronecker delta: $\delta_m^k = 1$ for $k = m$, and $\delta_m^k = 0$ for $k \neq m$. The so-called Christoffel symbols of the second kind are then defined by

$$\left\{ \begin{matrix} l \\ i\ j \end{matrix} \right\} = g^{lk}[ij, k].$$

Now consider a set of functions $\Gamma_{ij}^l$ of the coordinates, such that these functions transform like the Christoffel symbols of the second kind under a change of coordinate system but do *not necessarily* share all of the other properties of these Christoffel symbols. A set of such functions

$\Gamma^l_{ij}$ is called a set of 'coefficients of affine connection'. There are affine connections $\Gamma^l_{ij}$ which have a greater number of independent components than are possessed by any of the second kind of Christoffel symbols $\begin{Bmatrix} l \\ i\ j \end{Bmatrix}$, although all affine coefficients transform just like the latter Christoffel symbols. Specifically, in a space of $n$ dimensions, there are affine connections which have as many as $n^3$ independent components. But any Christoffel symbol $\begin{Bmatrix} l \\ i\ j \end{Bmatrix}$ has *fewer* independent components, since the number of independent components possessed by any such symbol in an $n$-space is restricted by the merely $(n^2+n)/2$ independent components which the two *symmetric* fundamental conjugate tensors $g_{ij}$ and $g^{lk}$ can have both collectively and severally. Thus, whenever an affine connection $\Gamma^l_{ij}$ has a greater number of independent components than any Christoffel symbol $\begin{Bmatrix} l \\ i\ j \end{Bmatrix}$, the partial differential equations $\begin{Bmatrix} l \\ i\ j \end{Bmatrix} = \Gamma^l_{ij}$ fail to have solutions $g_{ij}$ and $g^{lk}$. Thus, *there are* affine connections (i.e., sets of affine coefficients) which are *not* obtainable (derivable), via the set of Christoffel symbols of the second kind, from the combination of a fundamental metrical tensor $g_{ij}$ with its conjugate. Hence, Christoffel's $\begin{Bmatrix} l \\ i\ j \end{Bmatrix}$ are only a particular species of affine coefficients.

Although the affine connection is not itself a tensor, the following object defined by means of it is a demonstrably mixed fourth-rank tensor, contravariant of rank 1 and covariant of rank 3:

(1) $\quad R^a_{bcd} \equiv \Gamma^a_{bd,c} - \Gamma^a_{bc,d} + \Gamma^a_{nc}\Gamma^n_{bd} - \Gamma^a_{nd}\Gamma^n_{bc}$,

where $n$ is a dummy summation index and where we have adopted one of the two sign conventions used by various authors. Furthermore, we can lower the contravariant index in Equation (1) by contracting a metric tensor $g_{an}$ with the particular mixed tensor $R^n_{bcd}$, which *corresponds* to $g_{an}$ in the following sense: $R^n_{bcd}$ is generated via Equation (1) from those functions $\Gamma^l_{an}$ that do qualify as Christoffel symbols $\begin{Bmatrix} l \\ a\ n \end{Bmatrix}$ with respect to the metric tensor $g_{an}$. Lowering the index in this way, we obtain

(2) $\quad R_{abcd} = g_{an} R^n_{bcd}$.

The *covariant* fourth-rank tensor on the left-hand side of Equation (2) is the so-called Riemann (or Riemann-Christoffel) curvature tensor of the *first* kind, while the mixed fourth-rank tensor on the left-hand side of Equation (1) is the so-called Riemann curvature tensor of the second kind.

In order to show now that the first kind of Riemann curvature tensor does presuppose a Riemannian metric while the second kind does not, let us note several prior relevant points.

(1) As we saw, there are affine connections $\Gamma$ which are not derivable from the metric tensor via the Christoffel symbols of the second kind. Furthermore, the affine connections which are so derivable do *not* depend on the metric tensor for their definition (Adler *et al.*, 1965, Ch. 2, especially p. 48). Hence we shall say that $\Gamma$ does not require (presuppose) a metric. Let us hereafter refer to any $\Gamma^l_{ij}$ which is symmetric *in its subscripts* as 'symmetric' for brevity.

(2) And consider any tensor which is *constructible* or *obtainable* from any nonsymmetric or symmetric affine connection $\Gamma^a_{bc}$ in the following sense: The components of the tensor are expressible in terms of the $\Gamma^a_{bc}$ and of their derivatives up to a certain order. Any such tensor is said to be *a differential concomitant* of the connection $\Gamma^a_{bc}$.

There is a known proof (Schouten, 1954, pp. 164–5) that the *covariant* Riemann curvature tensor $R_{abcd}$ specified by our Equation (2) is *not* a differential concomitant of any nonsymmetric or symmetric affine connection $\Gamma$ alone; instead, $R_{abcd}$ is a differential concomitant of the combination of $\Gamma$ with the metric tensor $g_{ik}$. Thus, we are entitled to say that the covariant Riemann curvature tensor is *not* a curvature tensor of the affine connection, although – as shown by Equation (1) above – the *mixed* Riemann tensor *is* such a curvature tensor. The stated nonobtainability of the covariant curvature tensor from $\Gamma$ is not vouchsafed by the mere presence of the metric tensor $g_{an}$ on the right-hand side of the equation $R_{abcd} = g_{an} R^n_{bcd}$, because that mere presence does not, of itself, prove the otherwise known fact that $R_{abcd}$ fails to be *any* kind of differential concomitant of an affine connection.

Hence we can conclude that, at least with respect to currently known resources for constructing $R_{abcd}$, the covariant curvature tensor is simply *not defined without a metric tensor*. We shall see later in this section that this conclusion will be strengthened decisively by showing that, at any given point of space (a) the components $R_{abcd}$ of the covariant curvature tensor have the physical dimension $L^2$ of the square of *length*, where length will be seen to be a *metrical property*, and (b) the components $R^a_{bcd}$ of the purely affine mixed curvature tensor have the dimensionless status of pure numbers, as we would expect from this *non*metrical object. In this clear sense, the first kind of Riemann tensor requires or presupposes a Riemannian metric.

On the other hand, since the affine connection does not presuppose a metric, the mixed Riemann tensor, which is a differential concomitant of $\Gamma$ alone in our Equation (1), does not require a Riemannian metric. Indeed, it can be shown that the class of mixed Riemann curvature tensors which are obtainable in our Equation (1) from a metric tensor via the Christoffel symbols of the second kind is only a *proper* subclass of the set of mixed curvature tensors which are obtainable from any and all nonsymmetric or symmetric connections $\Gamma$. $R^a_{bcd}$ is a rather complicated differential concomitant of $\Gamma$ in Equation (1). Hence the existence of mixed curvature tensors which are *not* obtainable from a metric tensor is not obvious from the mere fact that the Christoffel symbols $\left\{ \begin{matrix} l \\ i\ j \end{matrix} \right\}$, which are generated from the metric tensor, are only a particular *species* of affine coefficients. But the mixed curvature tensor "has certain symmetry properties only on the condition that the components of the affine connection are Christoffel symbols" of the second kind, and it then turns out that a mixed curvature tensor which is derivable from a metric tensor has fewer independent components than one which is derivable from $\Gamma$'s that do *not* qualify as Christoffel symbols $\left\{ \begin{matrix} l \\ i\ j \end{matrix} \right\}$ (Bergmann, 1946, p. 169).

Thus, the latter kind of mixed curvature tensor, which has the larger number of independent components, is not derivable from a metric tensor.

Our stated conclusions concerning the status of $R_{abcd}$ and $R^a_{bcd}$, respectively, in regard to the presupposition of a metric are borne out by the following concise statement by Adler *et al.* (1965):

Note that, although we have introduced $R^a_{bcd}$ for a Riemannian space, it is evident that our derivation holds also in a general affine space, since it involves only the coefficients of connection, and not the metric tensor itself. In the more general case, one need only replace $-\begin{Bmatrix} a \\ c\ b \end{Bmatrix}$ by $\Gamma^a_{cb}$. *However, as soon as we lower an index [by contraction with the covariant metric tensor] to form the tensor $R_{abcd}$, we commit ourselves to a metric space* [p. 136, italics added].[4]

Note that the concluding sentence of this citation says that the covariant curvature tensor requires *a* metric space, i.e., *some* metric *or other*, but *not* that the curvature properties rendered by $R_{abcd}$ presuppose a *unique* metric. It will be recalled that Glymour conflated these two different kinds of presupposition when he erroneously saddled me with needing the stronger of the two in order to establish my stated inconsistency charge.

As Kline (1972, pp. 894–5, 1125) points out in a recent monumental historical work, Riemann himself used the *covariant* tensor $R_{abcd}$ to render curvature properties in his 1861 *Pariserarbeit*, which elaborates on his foundational 1854 Inaugural Dissertation. And Kline (1972, p. 892) stresses that curvature tensor's presupposition of a metric by writing: "Strictly speaking, Riemann's curvature, like Gauss', is a property of the metric imposed on the manifold rather than of the manifold itself," a lucid declaration which is grist to my mill in more ways than one. In making this statement, Kline is well aware that the much later 1917–1918 work of Levi-Civita, Hessenberg and Weyl yielded a purely affine generalization of Riemannian metric space such that the *mixed* curvature tensor can be construed nonmetrically (Kline, 1972, pp. 1131–2, 1134). The statement that the covariant curvature tensor involves a commitment to a metric holds even for that generalization of this tensor which corresponds to the *nonsymmetric* metric tensor $g_{ik}$ employed in Einstein (1955, Appendix II).

We can now formulate the two-fold upshot of our analysis for the answer which we shall give to our Question (i). In the first place, a given *covariant* curvature tensor field $R_{abcd}$ over a particular manifold *M does require* some Riemannian metric on *M*, although one and the same such curvature properties can be conferred on *M* by metrics *differing* other than by a choice of unit. Thus, the manifold *M* on which the given $R_{abcd}$ is defined does *not* merely *happen* to be a metric space. Nor can the given $R_{abcd}$ on *M* be said to require some metric or other *just because* that curvature tensor qualifies as a differential concomitant of one or more metric tensors $g_{ik}$, respectively, corresponding to as many Riemannian metrics any one of which may have been imposed on *M*. Instead, the curvature properties on $M$ rendered by the given $R_{abcd}$ presuppose a Riemannian metric because their very definition involves some metric tensor $g_{ik}$ or other, so that these curvature properties would not be defined without having one of the metrics whose respective fundamental tensors $g_{ik}$ can each generate the given curvature properties.

In the second place, a given *mixed* curvature tensor field $R^a_{bcd}$ over a particular manifold $M$ does *not* presuppose (require) a Riemannian metric, because $R^a_{bcd}$ is *definable* on $M$ without any metric tensor $g_{ik}$ even when that curvature tensor does qualify as a differential concomitant of one or more tensors $g_{ik}$. Thus, even when a given $R^a_{bcd}$ over a manifold $M$ is a differential concomitant of the metric tensor of a metric whose imposition on $M$ has turned $M$ into a Riemann space $S$, we can say the following: The particular curvature properties $R^a_{bcd}$ no more require that $M$ be a metric space than the $n$-dimensionality of $M$ presupposes the metric whose imposition has turned $M$ into an $n$-dimensional Riemann space $S$.

These results now permit us to answer our Question (i). If the curvature properties of a manifold $M$ *include* those rendered or determined by the first (covariant) kind of Riemann curvature tensor $R_{abcd}$ and are not *confined* to those specified by the second (mixed) kind of Riemann tensor, then the curvature properties of space do require (presuppose) some Riemannian metric or other. And in that case, it would indeed be *inconsistent* to claim that space is both intrinsically curved and yet devoid of any intrinsic metric. On the other hand, if the curvature properties of a manifold $M$ are *solely* those specified or determined by a *mixed* curvature tensor, then no metric is presupposed and hence it is then obviously *not* inconsistent to assert MUC of $M$ by saying that $M$ is intrinsically curved but devoid of any intrinsic metric. In that case, the inconsistency charge against MUC would be plainly unsound. Indeed, if $M$'s curvature properties *exclude* the 'covariant' ones $R_{abcd}$ while including the 'mixed' ones $R^a_{bcd}$, then $M$ does not even qualify as a Riemann metric space but is merely an affinely connected space without a metric. The reason is that if $M$ were a Riemann space rather than just an affine space, then its metric would be *sufficient* to assure, via the metric tensor, that $M$ is endowed with 'covariant' curvature properties $R_{abcd}$ after all no less than with the 'mixed' ones $R^a_{bcd}$. Conversely, if $M$ is merely a space of affine connection without a metric rather than a Riemann space, then $M$'s curvature properties do exclude the covariant ones while including the mixed ones.

Thus, in any Riemann space, the metric of that space is both necessary and sufficient for *covariant* curvature properties $R_{abcd}$ though only sufficient for mixed ones $R^a_{bcd}$. It follows importantly that in any Riemann space, the *metric* of that space is *nontrivially required specifically by the covariant curvature properties* with which that space is *automatically* endowed, rather than being just trivially required by the *metric character* of Riemann space!

We had seen earlier that since curvature properties are being construed as rendered by 'the' Riemann curvature tensor, the ambiguity of that construal as between $R_{abcd}$ and $R^a_{bcd}$ makes it inevitably *context-dependent* whether given statements about curvature pertain to a set of properties which properly include the class $R_{abcd}$ or pertain only to the restricted class $R^a_{bcd}$. And we see now that this context-dependence of what is intended by the term 'curvature' is tantamount to whether the manifold whose curvature properties are being discussed is a Riemann space or only an affinely connected space.

Apart from his conflation of Question (a) with Question (b) above, Glymour's critique of my inconsistency charge against MUC is vitiated by his neglect of the following facts:

1. The *context* in which my inconsistency charge against MUC occurred is indispensable for determining whether the curvature properties to which I referred in that charge include the covariant ones $R_{abcd}$ or not.
2. On the very page of Grünbaum (1970, p. 523) on which I asserted – as the basis for my inconsistency charge – that curvature properties require a metric, I explicitly referred to a "Riemannian manifold" and to "the framework of Riemannian geometry" – and *not* to a merely affine space! – as the context to which my assertion of metrical presupposition pertained.
3. In the avowedly Riemannian metrical context of my inconsistency charge, the covariant curvature properties $R_{abcd}$, which do presuppose a metric, are indeed included among the curvature properties to which that charge pertains.

It follows that, contrary to Glymour, the inconsistency charge against MUC as leveled by me is true.

Furthermore, it is evident that qua intended criticism of the views I set forth in Grünbaum (1970), the following statement by Earman (1972, p. 647, n. 13) is directed against a straw man: "Clark Glymour ... has argued against the claim that curvature properties must obtain with respect to a metric." I had never denied that *there exist* a species of curvature properties which do not require a metric, namely the mixed ones $R^a_{bcd}$, since I had *not* claimed that *any and all* curvature properties must obtain with respect to a metric. Nor was it essential in the context of my argument – which avowedly pertained to "the framework of Riemannian geometry" – to point out that the particular curvature species $R^a_{bcd}$ need not have a metrical anchorage. By contrast, as shown by Glymour's discussion of the consistency of MUC, he construed curvature properties as those rendered by 'the' Riemann curvature tensor, *but* took no cognizance at all of the *relevant* fact that the covariant curvature properties $R_{abcd}$ must obtain with respect to a metric.

Finally, we can appraise the following statement of Glymour's (1972):

> Of course, Grünbaum may simply have meant that curvature properties are just not the sort of thing that can exist unsupported. But he has given us no shade of reason why that might be so, let alone demonstrated that what is required for their support is a metric [p. 339].

As is patent from our analysis, it is indeed the case, contrary to Glymour, that "curvature properties are just not the sort of thing that can exist unsupported." Any given covariant curvature properties $R_{abcd}$ over a manifold $M$ in fact cannot exist as such unsupported by some metric or other, though a multiplicity of Riemannian metrics are each capable of conferring these given properties $R_{abcd}$ on $M$. Moreover, any given mixed curvature properties $R^a_{bcd}$ over a manifold $M$ cannot exist as such unsupported by some affine connection $\Gamma$ or other, though a multiplicity of affine connections are each capable of endowing $M$ with these given properties $R^a_{bcd}$. Glymour fails to see that there is *abundant* reason – rather than "no shade of reason" – for holding that curvature properties as such cannot exist unsupported, because he is again victimized by his conflation of two different requirements as follows: On the one hand, requiring the support of a *unique* metric for $R_{abcd}$ and of a *unique* $\Gamma$ for $R^a_{bcd}$, and, on the other hand, requiring more weakly the support of respectively *some* metric *or other*, and of *some* $\Gamma$ *or other*.

Glymour (1973) has offered a rebuttal to my arguments and considers the following triplet of contentions:
(1) Space-time has, intrinsically, certain curvature properties.
(2) Space-time does not have, intrinsically, a metric.
(3) Material bodies reduce to curved space-time [Footnote 8].

Glymour (1973) then goes on to reply to me as follows:

> Suppose someone were to offer a space-time theory strictly in accord with (1)–(3), and suppose he intended to include among curvature properties some which are not obtainable from an affine connection: the chief examples of the latter are sectional curvatures. His theory, then, would perhaps contain variables for an affine connection, various scalar vector or tensor fields, and quantities, such as the [mixed] (1, 3) curvature tensor definable from these. We can also imagine that his theory contains a field quantity which takes every pair of vectors in every tangent space at every point into a real number in a smooth way, and that he calls this quantity the "sectional" curvature. We can even imagine that his theory contains a quantity which is a (0, 4) tensor [i.e., a fourth-rank *covariant* tensor], and that his axioms guarantee that this quantity has all of the symmetry properties of a (0, 4) curvature tensor, and in addition all of the properties of $R_{ijkl}$ (e.g., action on vector fields) that can be stated without use of the metric tensor. His theory does not contain any explicit quantity which is a metric tensor.
>
> Of course, no one has actually developed such a theory, but someone could I suppose, and I think it would be entirely reasonable to regard the properties he is talking about as what we ordinarily regard as curvature properties. When is it reasonable to say that someone holding such a theory is committed to the existence of a metric? My inclination is to say that he is so committed when his axioms are strong enough to permit the definition of a metric quantity from the curvature and affine quantities to which he is already committed. That was the point of my observation, against Grünbaum, that even all affine properties and sectional curvatures do not contain a unique metric, and so do not permit the definition of a metric.
>
> One can object against such a theory that it is incomplete, that there are relations among the different curvature properties that cannot be described without a metric, and that might (or, conceivably, might not) be true, but I fail to see how such a theory, coupled with claims (1)–(3), is inconsistent [Footnote 8].

This criticism gives me the opportunity to offer what I believe to be decisive further support for my claims by now showing that, at any given point, the components of the *covariant* curvature tensor (0, 4) have the physical dimension $L^2$ of the square of length – where length is a METRICAL property! – while those of the *mixed* (1, 3) curvature tensor have the dimensionless status of pure numbers. It is also going to be relevant that, as is well known, the reciprocal $L^{-2}$ of $L^2$ is the physical dimension of the Gaussian curvature $K$ at a point of a surface, and that $L^{-2}$ is also the physical dimension of the sectional curvature $K_N$ at a point with respect to a given orientation $N$. I shall now proceed to demonstrate the stated results for the two curvature tensors in order to show that they render Glymour's purported speculative counterexample to me untenable.

Let me introduce the general considerations, which are to follow, by a simple example. On the surface of a two-sphere embedded in Euclidean three-space, let the two sets of surface coordinates be respectively the colatitude $\phi$ in *radians* and the longitude $\theta$ in *radians*. By the very definition of radians, the values of $\phi$ and $\theta$ are each dimensionless pure numbers, as are the coordinate differentials $d\phi$ and $d\theta$. If $a$ is the length of the radius of the sphere, then the familiar spatial metric on the sphere is given by

$$ds^2 = a^2 \, d\phi^2 + a^2 \sin^2 \phi \, d\theta^2.$$

Since the physical dimension of $ds^2$ is $L^2$, each of the two terms in the sum on the right-hand side must have the dimension $L^2$. But since $d\phi^2$ and $d\theta^2$ are each pure numbers, it is clear that the metric coefficients $g_{11}$ and $g_{22}$ must each be of dimension $L^2$, as indeed they are because $a^2$ is the square of a length. In particular, in the case of the *unit* sphere, where $a^2$ and hence $g_{11}$ has the value *unity*, $g_{11}$ is of dimension $L^2$. Thus, at least in our example, the metric coefficients $g_{ik}$ are not only *qualitatively* of dimension $L^2$, but they contain a metric scale factor whose value is dictated by the choice of the *unit* of length! More generally, it is clear that after any particular coordinatization is introduced in an $n$-dimensional manifold, alternative metrizations which differ *only* in the choice of unit of length will issue in metric tensors $g_{ik}$ which, in the *given* coordinate system, differ only in the scale factor contained by them. Thus, the choice of metric unit of length always makes itself felt in the functions $g_{ik}$!

As shown by our simple example of the coordinate system $(\phi, \theta)$, *there are* permissible coordinate systems in which we MUST assign the dimensional status $L^2$ to the metric coefficients $g_{ik}$! Moreover, since a coordinate system is specified by a function from points into ordered $n$-tuples of pure real numbers in any space of $n$-dimensions, it is always *possible* to assign the dimensional status $L^2$ to the metric coefficients of *any* coordinate system, even in the simple case when one or more of them has the constant value unity.

Thus, to take another simple example, if $(\rho, \theta)$ are polar coordinates on the Euclidean plane, not only the values of $\theta$ in radians but also those of $\rho$ can qualify as pure (dimensionless) real numbers. In that construal, the values of $d\rho^2$, $\rho^2$ and $d\theta^2$ are each pure numbers. And we can then make perspicuous the dimensional role of unit coefficients of dimension $L^2$ in the components $g_{ik}$ by writing the metric in the form

$$ds^2 = 1 \cdot d\rho^2 + 1 \cdot \rho^2 \, d\theta^2.$$

We see that it is possible to impose the requirement that the dimensional status of the metric tensor $g_{ik}$ be *the same* in *every* coordinate system not only qualitatively but also in the quantitative sense of containing a scale factor which is dictated by the choice of unit. At least for reasons of simplicity and generality, I shall impose this requirement in the general statement which I shall give below. But we shall note at the conclusion of our analysis of the two curvature tensors that it suffices for the refutation of Glymour's purported counterexample that THERE ARE coordinate systems in which at least some of the metric coefficients not only MUST be of dimension $L^2$ but also MUST contain the metrical scale factor dictated by the choice of unit of length! Since the general formulations which are about to follow are to be applicable to four-dimensional space-time, d$s$ will need to refer to space-time 'length', and $L$ will denote the latter physical dimension. For our purposes, I shall not need to take account of the separate dimensions of spatial length and temporal duration in order to accommodate the cases in which space-time can be split up into space and time. But it is to be understood that I allow for this refinement in the general statement which I shall now proceed to give.

A coordinate system is specified by a function from points into ordered *n*-tuples of real numbers. The real number coordinates clearly have the dimensionless status of *pure* numbers. Hence, coordinate differentials $dx^i$ and $dx^k$, and products of these, are also pure numbers. But in the equation for the metric d$s$ specified by $ds^2 = g_{ik} \, dx^i \, dx^k$, $ds^2$ has the dimension $L^2$. Hence at any given point of space, the components of the metric tensor $g_{ik}$ have the dimension $L^2$ while the components of its conjugate, the contravariant metric tensor $g^{kl}$, have the dimension $L^{-2}$ (Schouten, 1951, p. 129). Partial derivatives of $g_{ik}$ with respect to the coordinates have the same dimensions as $g_{ik}$, and similarly for $g^{kl}$.

It follows from the definition of the Christoffel symbols of the second kind given earlier in this section that, at any given point of the space, the entities represented by them are each dimensionless *pure* numbers. For note that the $\begin{Bmatrix} l \\ ij \end{Bmatrix}$ are obtained by contracting the contravariant metric tensor $g^{lk}$, whose components each have dimension $L^{-2}$, with the Christoffel symbols $[ij, k]$ of the first kind, which are objects of dimension $L^2$, being one half of sums of partial derivatives of components of the tensor $g^{ik}$.

Since the dimensionless Christoffel symbols of the second kind are a particular species of affine coefficients, at any given point the coefficients $\Gamma^l_{ij}$ are pure, dimensionless numbers, regardless of whether they satisfy a set of equations $\begin{Bmatrix} l \\ ij \end{Bmatrix} = \Gamma^l_{ij}$ by being obtainable from the two conjugate metric tensors $g^{lk}$ and $g_{ik}$.

These results will now enable us to see the following via our earlier Equations (1) and (2) of this section: Whereas at any given point, the components of the *mixed* curvature tensor (1, 3) are dimensionless pure numbers, the components of the covariant curvature tensor (0, 4) each have the dimension $L^2$, where length is a *metrical property*. No wonder, therefore, that a purely affine space has the ontological resources to constitute the mixed curvature properties $R^a_{bcd}$ out of *bona fide* nonmetrical entities. But it is then likewise clear that, contrary to Glymour's allegation, the speculative (0, 4) tensor of Glymour's gleam-in-the-eye future nonmetrical theory cannot possibly render the $L^2$-dimensional properties $R_{abcd}$ which we ordinarily regard as the covariant curvature properties!

Let us now justify these claims concerning the dimensional difference between the two curvature tensors. As we saw, the coefficients $\Gamma^l_{ij}$ are dimensionless pure numbers, and therefore sums of products of these as well as partial derivatives of them are likewise pure numbers. Hence, at any given point, the components of the mixed curvature tensor $R^a_{bcd}$ defined by our Equation (1) must likewise be dimensionless *pure numbers* which are independent of the choice of any unit of length. But, since $R_{abcd} = g_{an}R^n_{bcd}$, the dimensional status of $R_{abcd}$ is exactly the same as that of the metric coefficients $g_{an}$, which not only have dimension $L^2$ but also contain the metrical scale factor dictated by the choice of unit of length. Hence $R_{abcd}$ depends on the choice of metric unit, and its $L^2$-dimensionality is metrical rather than merely qualitative. It follows that, at any given point, *the components of the* (0, 4) *curvature tensor each have dimension $L^2$, where L is a METRICAL property*!

This dimensional analysis of $R_{abcd}$ was carried out on the basis of the requirement that in every coordinate system the dimensional status of the metric tensor $g_{ik}$ not only be the same, but also that the components $g_{ik}$ contain a metric scale factor dictated by the choice of unit of length. This requirement led to conclusions about the dimensional status of $R_{abcd}$ which hold alike for all components of that tensor and in every coordinate system. But the reader will recall my simple example of the coordinatization of the two-sphere by means of colatitude $\phi$ and longitude $\theta$, both of which range over radians, which are pure numbers, so that $d\phi^2$ and $d\theta^2$ MUST be pure numbers. As I emphasized by reference to that example, my refutation of Glymour's purported counterexample does *not* depend on the imposition of the stated requirement that the dimensional status of the metric coefficients be the same in every coordinate system. For it suffices for my argument against Glymour's alleged counterexample that THERE ARE coordinate systems, such as spherical coordinates $r$, $\phi$, $\theta$ in Euclidean three-space, in which at least some of the metric coefficients MUST contain the metrical scale factor dictated by the choice of the unit of length and must be of dimension $L^2$. The existence of such coordinate systems has the consequence that there are at least some components of some covariant curvature tensors $R_{abcd}$ which depend on the choice of metric unit and have the dimension $L^2$. At least for the sake of simplicity and generality, I shall hereafter continue to employ my general formulations as based on the stated $L^2$-dimensionality of $g_{ik}$ *in every coordinate system* without thereby jeopardizing my argument against Glymour's supposed counterexample in the face of someone who does not accept the necessity for such invariance of dimensional status.

The sectional or Riemannian curvature $K_N$ at a point $P$ of a Riemann space with respect to the orientation $N$ determined by two linearly independent contravariant vectors $A^i$ and $B^i$ is given by

$$K_N = \frac{R_{abcd} A^a B^b A^c B^d}{(g_{ac}g_{bd} - g_{ad}g_{bc}) A^a B^b A^c B^d}.$$

But the (0, 4) curvature tensor in the numerator and the (0, 2) metric tensor in the denominator each has dimension $L^2$. It follows that *the scalar sectional curvature $K_N$ at the point $P$ has the dimension $L^{-2}$*. This sectional curvature $K_N$ is the Gaussian (or total) curvature at the point $P$

of the two-dimensional geodesic surface swept out by geodesics through $P$ which have directions in the 2-parameter family of directions $uA^a + vB^a$ at $P$. Thus, the Gaussian curvature of this geodesic surface at the given point $P$ is the Riemannian (or sectional) curvature of the enveloping $n$-dimensional Riemann space at $P$ with respect to the given orientation $N$. And this sectional curvature $K_N$ is of dimension $L^{-2}$!

This is as it should be, since it follows from Gauss' *theorema egregium* of *surface* theory – which gives the Gaussian curvature $K$ as a function of the metric coefficients $g_{ik}$ and their first and second partial derivatives – that $K$ has dimension $L^{-2}$. We would also expect this on elementary grounds by noting that the Gaussian curvature $K$ of a two-sphere embedded in Euclidean three-space is $r^{-2}$, where $r$ is the *length* of its radius, and thus depends numerically on the choice of the unit of length.

As will be recalled, on the basis of results *other than* the dimensional analysis of $R_{abcd}$, I had concluded earlier in this section that "at least with respect to currently known resources for constructing $R_{abcd}$, the covariant curvature tensor is simply *not defined without a metric tensor*." But now our dimensional analysis of this (0, 4) curvature tensor has shown its dimension to be $L^2$, where $L$ is a metrical property. Hence this further result obviates the proviso "at least with respect to currently known resources for constructing $R_{abcd}$" and entitles me to assert *categorically* that the (0, 4) curvature tensor properties are simply not defined without a metric!

Thus, even *if* – and it is indeed a *big* IF – the NONMETRICAL resources outlined by Glymour sufficed for the construction of a covariant fourth-rank tensor that *shares* those particular properties of the (0, 4) *Riemann curvature tensor* which he lists, I deny flatly on dimensional grounds that he is entitled to the following conclusion of his: "... it would be entirely reasonable to regard the properties he [the putative theoretician] is talking about as what we ordinarily regard as [covariant] curvature properties [$R_{abcd}$]." Glymour's putative nonmetrical (0, 4) tensor does *not* have the METRICAL dimension $L^2$, while the (0, 4) curvature tensor does have that metrically constituted dimension. Therefore, Glymour has failed to show that my inconsistency charge against claims (1)–(3) is gratuitous or unsound in the context of $R_{abcd}$-curvature and that the tenability of that charge requires the feasibility of *defining* a *unique* metric by means of the curvature properties. Indeed the demonstrated METRICAL $L^2$-dimensionality of the covariant curvature properties shows my inconsistency charge to be true.

Finally, since the sectional curvature $K_N$ has dimension $L^{-2}$, I see very little more than semantic baptism in Glymour's gambit of *calling* a quantity 'sectional CURVATURE' merely because its values are given by a smooth function from pairs of vectors in every tangent space at every point into the real numbers. Consider an analogy. For a fixed time $t$, a function from married couples (pairs) into real numbers whose values are the combined ages of the pairs in number of years (at the given time $t$) cannot be validly claimed to render the joint financial worth in dollars of the respective pairs (at time $t$) merely because the combined-age values of the former function in number of years are each given by a real number no less than the number of dollars which is the economic worth of any given married pair at time $t$. Even if, at the time $t$, the two numbers which are the values of the respective functions *happen to be the same* for any given married pair, they are each 'impure' (in Carnap's sense) by being respectively a number-of-years and a number-of-dollars.

We are now ready to deal with our Question (ii).

### III. QUESTION (ii): WERE CLIFFORD AND WHEELER CONSISTENT IN THEIR ONTOLOGY OF CURVATURE?

Our question now is whether MUC is inconsistent in the specific context of the assumptions made by Clifford and Wheeler, and if so, whether either Clifford or Wheeler did assert MUC.

Clifford (1876) gave his Cambridge lecture 'On the Space-Theory of Matter' in 1870 and died in 1879, six years after publishing his English translation of Riemann's Inaugural Dissertation of 1854.[5] Riemann elaborated technically on his foundational Inaugural Dissertation in his *Pariserarbeit* of 1861, but the latter was not published until 1876, which was ten years after Riemann's death and only three years before Clifford died (Kline, 1972, p. 889; Weber, 1953). Furthermore, Clifford's 1870 lecture 'On the Space-Theory of Matter' is both brief and purely verbal rather than mathematical. Hence it is *not* safe to assume that Clifford was familiar with Riemann's 1861 paper when he gave his lecture in 1870 or even by 1875, when he dictated the essay on 'Space' which appeared as Chapter II of his posthumously published *The Common Sense of the*

*Exact Sciences* (Clifford, 1950).[6] On the other hand, Riemann's 1854 Inaugural Dissertation was published in 1868, and in the opening paragraph of his 1870 lecture, Clifford wrote (though without giving an explicit reference) that "Riemann has shewn that... there are different kinds of space of three dimensions; and that we can only find out by experience to which of these kinds the space in which we live belongs." Hence Clifford was almost certainly familiar with Riemann's Inaugural Dissertation when he gave his 1870 lecture. Thus Clifford can be presumed to have been familiar by 1870 with the sectional two-dimensional (Gaussian) curvatures that obtain at any given point with respect to various orientations in $n$-dimensional Riemann space, since Riemann dealt with these sectional curvatures in his Inaugural Dissertation while *not* mentioning anything corresponding to a fourth-rank curvature tensor until his 1861 *Pariserarbeit*.

But even if Clifford had been aware of the contents of the 1861 *Pariserarbeit* by 1870 or by 1875, there is still ample reason to think that Clifford never had an inkling of the *mixed* Riemann curvature tensor, let alone of the feasibility of its nonmetrical construal or of the existence of any kind of nonmetrical curvature properties. In the first place, in his 1861 *Pariserarbeit*, Riemann employed only what we now call the *covariant* Riemann curvature tensor, when stating a necessary condition for the isometry of two spaces (Kline, 1972, pp. 894–5, 1125–27). Though he can thus be said to have worked with a particular *species* of tensor, Riemann did *not* have the concept of the *genus* tensor, introduced after his and Clifford's death by Ricci and Levi-Civita as an object of rank $r = l + m$, contravariant of rank $l$ and covariant of rank $m$. Nor did Riemann have knowledge of the tensor calculus as such (Kline, 1972, pp. 1122–7), and there is no indication anywhere in his work of the concept of contravariance. For these reasons, Morris Kline has expressed the view that "The question as to whether Riemann had any concept of the mixed curvature tensor must, I am quite sure, be answered in the negative.... there was no tensor analysis in 1861 or even 1870, though the subject had its beginnings in the work of Beltrami, Christoffel and Lipschitz."[7] Thus, there is good reason to think that even if Clifford did know Riemann's 1861 *Pariserarbeit* when Clifford gave his seminal 1870 GMD lecture, he had no notion then or thereafter of the *mixed* Riemann tensor as a curvature entity, even in the purely metrical construal of that tensor.

In the second place, it was not until 1917–1918, almost four decades after Clifford's death, that mathematicians first dispensed with the Riemann metric when achieving a purely affine generalization of Riemannian metric space such that the mixed curvature tensor can be construed nonmetrically (Kline, 1971, pp. 1130–4; Weatherburn, 1957). Hence there is every reason to conclude the following: The notion of curvature known to Clifford by 1870 or even by 1875 – the date of his aforementioned chapter on 'Space' – was such that Clifford, no less than Gauss and Riemann, conceived of curvature as a kind of property which is conferred on a space (at any given point) *only* by a metric (tensor). Consequently, in the context of Clifford's construal of curvature, the affirmation of the thesis MUC would be inconsistent.

Though Clifford's 1870 sketch of his 'Space-Theory of Matter' was wholly nonmathematical, it clearly enunciated the remarkably original thesis that all matter and radiation, including all physical devices which effect spatiotemporal measurements, are constituted out of *empty, curved, metric space*. Said he:

I hold in fact
(1) That small portions of space *are* in fact of a nature analogous to little hills on a surface which is on the average flat; namely, that the ordinary laws of [Euclidean] geometry are not valid in them.
(2) That this property of being curved or distorted is continually being passed on from one portion of space to another after the manner of a wave.
(3) That this variation of the curvature of space is what really happens in that phenomenon which we call the *motion of matter*, whether ponderable or etherial.
(4) That in the physical world nothing else takes place but this variation, subject (possibly) to the law of continuity.

As is evident from our prior analysis, Clifford's own construal of his program of reducing all of physics to geometry required a *metric* both nontrivially – by postulating that space is curved in the pre-1917 sense – and trivially, by postulating that the world's ultimate substance is a metric (presumably Riemannian) space. Moreover, we can now see that the *emptiness* which Clifford attributed to the curved metric space of his monistic vision required the following: Up to a constant positive factor $k$, both the curvature and the metric of space are 'implicit' in it or intrinsic to it in the sense of at least *not* being *imposed* on the continuous spatial manifold, but of being grounded solely in the very structure of that spatial manifold itself. In particular, any and all kinds of mensurational devices are themselves held to be constituted out of empty *curved* space to begin with. Hence apart from the possible exception of the scale factor $k$, the nontrivial metric and curvature of space cannot first be imposed on space or first be induced into space by the behavior of any such devices. Any and all bodies or radiation which serve as standards of metric equality qualify as such at best *only epistemically* as means of discovering the metric properties of space. And no entities other than the structure of empty space itself are *ontologically* necessary for endowing space with metric ratios or with such curvature properties as are determined by these ratios. Thus his 1870 lecture committed Clifford to the claim that space is both intrinsically curved *and* intrinsically metric (modulo a scale factor $k$).

But in his subsequent 1875 essay on 'Space', Clifford (1950) asserted that space *lacks* a (nontrivial) intrinsic metric by declaring:

> The measurement of distance is only possible when we have something, say a yard measure or a piece of tape, which we can carry about and which does not alter its length while it is carried about. The measurement is then effected by holding this thing in the place of the distance to be measured, and observing what part of it coincides with this distance... [p. 48]. The reader will probably have observed that we have defined length or distance by means of a measure which can be carried about *without changing its length*. But how then is this property of the measure to be tested? We may carry about a yard measure in the form of a stick, to test our tape with; but all we can prove in that way is that the two things are always of the same length when they are in the same place; not that this length is unaltered....
> Is it possible, however, that lengths do really change by mere moving about, without our knowing it?
> Whoever likes to meditate seriously upon this question will find that it is wholly devoid of meaning [pp. 49–50].

This 1875 statement of Clifford's constitutes a denial of the existence of an intrinsic spatial metric. To see this, note that if space were intrinsically metric to within a constant scale factor $k$ in the sense of Clifford's 1870 Lecture, then it would surely *not* be "wholly devoid of meaning," as Clifford contends in this 1875 passage, to ask whether the lengths of our familiar measuring standards "do really change by mere moving about, without our knowing it." For, as I explained in Grünbaum (1970, p. 523), "the supposition of such an unnoticed change would surely have meaning with respect to the presumed implicit metric that confers a curvature on empty space, *provided* that (1) the underlying conception of the empty space manifold has meaning, and (2) there exists a metric implicit to the latter manifold."

We can conclude that the combination of Clifford's 1870 thesis with his cited declaration of 1875 entails commitments which are inconsistent as follows: (a) there is the outright inconsistency that space is and also is not intrinsically metric (to within a scale factor $k$), and (b) there is the more subtly inconsistent commitment to MUC, whose inconsistency is due to Clifford's construal of curvature as requiring a metric. Incidentally, Clifford's 1870 assumption that continuous space is intrinsically and nontrivially metric (modulo $k$) had been in direct contradiction with a cardinal explicit tenet of Riemann's Inaugural Dissertation.

Turning to Wheeler, we find the following statement by him in the Foreword to his book *Geometrodynamics* (1962b):

> The sources of the curvature of space-time are conceived differently in geometrodynamics and in usual relativity theory. In the older analysis any warping of the Riemannian space-time manifold is due to masses and fields of non-geometric origin. In geometrodynamics – by contrast – only those masses and fields are considered which can be regarded as built out of the geometry itself.
> ... The central ideas of geometrodynamics can be easily summarized.
> The past has seen many attempts to describe electrodynamics as one or another aspect of one or another kind of non-Riemannian geometry. Every such attempt at a unified field theory has foundered. Not a change in Einstein's experimentally tested and solidly founded 1916 theory, but a closer look at it by Misner in 1956 (III), gave a way to think of electromagnetism as a property of curved empty space [p. xi].

Unlike Einstein and Schrödinger, who tried *non*metrical affine approaches to unified field theory, Wheeler cast his theoretical lot with Riemannian *metrical* theories, as he explicitly tells us here. And thus Wheeler's attempt to implement the program of GMD avowedly shared the stated assumptions of Clifford's monistic ontology of empty curved *metric* space. Hence the thesis that continuous space is *devoid* of any nontrivial metric which is intrinsic to within a scale factor $k$ is just as inconsistent with Wheeler's fundamental GMD assumptions as with those of Clifford's 1870 lecture.

Did Wheeler avow that continuous space is devoid of any intrinsic metric in the manner of Clifford's 1875 declaration? I know of no such *outright* declaration by Wheeler. But in Grünbaum (1970, p. 523), I cited from Marzke and Wheeler (1964) their conception of the *metrical ratio* of two intervals of space-time as depending ontologically – and not just epistemically – on the behavior of a metric standard along different routes of transport, rather than as depending only on the very structure of the space-time manifold itself. And I went on to point out there (p. 524) that if, in that joint paper, Wheeler had conceived of the specified metrical ratio as depending ontologically only on a metric intrinsic to empty space-time, then "the question of a possible dependence of the ratio of the measures of the two intervals on the route of intercomparison (path of transport) would not even need to arise!" If it is correct, as I think it is, to read this joint paper by Wheeler and Marzke as an implicit denial by Wheeler that empty space-time is intrinsically metric, then Wheeler was being inconsistent no less than Clifford.

Glymour gives no indication at all as to why he does not regard the textual documentation adduced by me as a cogent basis for the conclusion that Wheeler was thus being inconsistent. Instead, Glymour (1972, p. 338) contents himself with an unavailing autobiographical *obiter dictum*, saying "I am not at all convinced that Wheeler would deny that space has intrinsic metric properties."[8]

We saw that, *apart* from the stated inconsistencies, the basic GMD program of Clifford and Wheeler does require the assumption that continuous space is intrinsically (and nontrivially) metric, modulo a scale factor $k$. Yet, as I have documented in some detail (Grünbaum, 1970, pp. 515–22), this assumption was explicitly and fundamentally denied by Riemann in his Inaugural Dissertation. Thus, one of the important assumptions required by the GMD ontology of empty space is incompatible with one of the cardinal tenets of Riemann's conception of the foundations of the metric geometry of continuous space.

*University of Pittsburgh*

NOTES

* I owe warm thanks to Allen Janis, Morris Kline, Gerald Massey, John Porter, and John Stachel for the substantial benefit which this paper had from conversations or correspondence with them. Clark Glymour kindly sent me preprints cited here.

I am also indebted to the National Science Foundation for support of research.

[1] Conference on Gravitation and Quantization, held in October and November, 1972 at the Boston University Institute of Relativity Studies, directed by John Stachel. I am grateful to Professor Stachel for having given me the opportunity to attend this conference.

[2] This disavowal can now be seen to have been heralded by Wheeler's Foreword in Graves (1971, p. viii).

[3] For a statement of these designations, see, for example, Eisenhart (1949, Ch. I, Sec. 8). Speaking of the first and second kinds of Riemann curvature tensor, Kline (1972) writes: "Either form is now called the Riemann-Christoffel curvature tensor" [p. 1127].

[4] For notational convenience, I have replaced the Greek indices used by these authors by our Latin ones, and it will be noticed from the minus sign in their statement that they use rather unusual conventions such that $\left\{\begin{matrix} a \\ b\ c \end{matrix}\right\} = -\Gamma^a_{bc}$; see also their pp. 48–50, especially p. 50, Equation (2.10).

[5] See the posthumously published Clifford (1950, pp. xxx and 247).

[6] The 1875 dictation date of that chapter is given in K. Pearson's Preface to this work, p. LXIII.

[7] Private communication, quoted with the kind permission of Professor Kline.

[8] In Chapters 16–22 and in the *Appendix* of Grünbaum (1973), I have replied to other recent criticisms. Thus, §47 of this *Appendix* offers a detailed rebuttal to a key point made by Arthur Fine in his stimulating essay in the present volume.

BIBLIOGRAPHY

Adler, R., Bazin, M., and Schiffer, M., *Introduction to General Relativity*, McGraw-Hill, New York, 1965.

Bergmann, P. G., *Introduction to the Theory of Relativity*, Prentice-Hall, New York, 1946.

Clifford, W. K., 'On the Space-Theory of Matter', *Proceedings of the Cambridge Philosophical Society* **2** (1876), pp. 157–8. Reprinted in R. Tucker (ed.), *Mathematical Papers by William Kingdon Clifford*, London, 1882. Reissued, Chelsea, New York, 1968. Reprinted in J. R. Newman (ed.), *The World of Mathematics*, Vol. 1, Simon and Schuster, New York, 1956, pp. 568–9.

Clifford, W. K., *The Common Sense of the Exact Sciences*, Dover Publications, New York, 1950.

Earman, J., 'Some Aspects of General Relativity and Geometrodynamics', *Journal of Philosophy* **69** (1972), 634–47.

Einstein, A., *The Meaning of Relativity* (5th ed.), Princeton University Press, Princeton, 1955.

Eisenhart, L. P., *Riemannian Geometry*, Princeton University Press, Princeton, 1949.

Glymour, C., 'Physics by Convention', *Philosophy of Science* **39** (1972), 322–40.

Glymour, C., 'Space-Time Indeterminacies and Space-Time Structure', paper presented at the Colloquium for the Philosophy of Science, Boston, 1973.

Graves, J. C., *The Conceptual Foundations of Contemporary Relativity Theory*, The MIT Press, Cambridge, 1971.

Grünbaum, A., *Philosophical Problems of Space and Time* (1st ed.), Knopf, New York, 1963.

Grünbaum, A., 'Space, Time and Falsifiability, Part I', *Philosophy of Science* **37** (1970), 469–588. Reprinted as Chapter 16 in Grünbaum (1973).

Grünbaum, A., *Philosophical Problems of Space and Time* (2nd ed.), D. Reidel, Dordrecht and Boston, 1973. The title of Chapter 22 is 'General Relativity, Geometrodynamics and Ontology'.

Kline, M., *Mathematical Thought from Ancient to Modern Times*, Oxford University Press, New York, 1972.

Marzke, R. F. and Wheeler, J. A., 'Gravitation as Geometry, Part I: The Geometry of Space-Time and the Geometrodynamical Standard Meter', in H. Chiu and W. F. Hoffman (eds.), *Gravitation and Relativity*, Benjamin, New York, 1964.

Massey, G. J., 'A Panel Discussion of Grünbaum's Philosophy of Science'. Papers by G. J. Massey, B. C. van Fraassen, M. G. Evans, R. B. Barnett, G. Wedeking, and P. L. Quinn, *Philosophy of Science* **36** (1969), 331–99.
Schouten, J. A., *Tensor Analysis for Physicists*, Oxford University Press, London, 1951.
Schouten, J. A., *Ricci-Calculus* (2nd ed.), Springer-Verlag, Berlin, 1954.
Weatherburn, C. E., *Riemannian Geometry and the Tensor Calculus*, Cambridge University Press, England, 1957, pp. 176–8.
Weber, H. (ed.), *The Collected Works of Bernhard Riemann* (2nd ed.), Dover Publications, New York, 1953, pp. 391–404.
Wheeler, J. A., 'Curved Empty Space-Time as the Building Material of the Physical World', in E. Nagel, P. Suppes, and A. Tarski (eds.), *Logic, Methodology and Philosophy of Science*, Proceedings of the 1960 International Congress, Stanford University Press, Stanford, 1962(a).
Wheeler, J. A., *Geometrodynamics*, Academic Press, New York, 1962(b).

MICHAEL FRIEDMAN

# RELATIVITY PRINCIPLES, ABSOLUTE OBJECTS AND SYMMETRY GROUPS*

## I. THE PROBLEM AND ANDERSON'S PROGRAM

Traditionally, certain physical theories have been thought to have relativity principles associated with them. Associated with Newtonian mechanics is the principle of Galilean relativity, associated with special relativity is the special or restricted principle of relativity, associated with general relativity is the general principle of relativity, etc. Such relativity principles are often expressed in terms of groups of transformations. The principle of Galilean relativity is expressed by the 'invariance' of Newtonian mechanics under the Galilean group, that of special relativity by the 'invariance' of special relativity under the Lorentz group, and that of general relativity by the 'invariance' (or 'covariance') of general relativity under the group of all 1-1 transformations with non-vanishing Jacobian. Unfortunately, just what these various groups are groups *of*, and exactly how they are *associated* with given physical theories, has been far from clear. Similarly, the nature and role of the various relativity principles has been correspondingly unclear.

These questions typically get different answers when different theories are being considered. In the case of general relativity, the associated group is almost invariably taken to be a group of *coordinate transformations*, and the group is associated with the theory by being its *covariance group*. A covariance group is defined as follows: Let $f$ be an arbitrary coordinate transformation in the covariance group. Then, if $0_1, ..., 0_k$ are the geometrical objects (i.e., tensor fields or affine connexions – cf. Anderson, 1967, pp. 14–6; Trautman, 1965, pp. 84–7) of the theory as expressed in one coordinate system and $f0_1, ..., f0_k$ are the transforms of these objects in the transformed coordinate system, $0_1, ..., 0_k$ satisfy the equations of the theory in the original coordinate system iff $f0_1, ..., f0_k$ satisfy the equations of the theory in the transformed coordinate system. General relativity is then thought to be at least partially defined by the 'principle of general covariance': the requirement that its covariance

*P. Suppes (ed.), Space, Time and Geometry, 296–320. All rights reserved*
*Copyright © 1973 by D. Reidel Publishing Company, Dordrecht-Holland*

group be the group of *all* admissible coordinate transformations. However, since this is just the requirement that the equations of the theory are tensor equations, it will be satisfied by *any* theory in tensor form. And it is well known by now that both Newtonian mechanics and special relativity are also expressible in four-dimensional tensor form.[1] The 'principle of general covariance' cannot, therefore, be used to classify or single out general relativity.

When we come to special relativity, on the other hand, we find that the Lorentz group is sometimes considered as a group of coordinate transformations and sometimes as a group of automorphisms – i.e., mappings of the space-time manifold onto itself. Sometimes it is treated as a covariance group – e.g., in discussions of relativistic electrodynamics, but sometimes it is treated as an 'invariance' group – e.g., it is sometimes claimed that the laws or equations of special relativity are 'invariant' under Lorentz transformations (Newtonian mechanics and Galilean transformations are treated similarly). This notion of 'invariance', however, is seldom precisely defined. It is still more difficult to see how Lorentz 'invariance' is supposed to relate to the general *covariance* claimed for general relativity – how the latter is supposed to be a generalization of the former, for example.

Anderson (1967) tries to clarify these questions. He proposes that the groups under consideration are associated with their corresponding physical theories in the following way. First, he divides the geometrical objects postulated by the theory into two classes, the *absolute* objects and the *dynamical* objects. The absolute objects are thought to be those objects that are not affected by the interactions described by the theory; they are 'independent' of the dynamical objects – part of the fixed 'background framework' within which interaction takes place. Examples of absolute objects are the metric of special relativity and the absolute time of Newtonian mechanics. Examples of dynamical objects are the metric of general relativity (which is affected by the mass-energy distribution) and the electromagnetic field (which is affected by the charge-current distribution). Thus, absolute objects are the kinds of things that Einstein criticized as "contrary to the mode of thinking in science," since they affect other things but are not affected in turn (cf. Einstein, 1956, pp. 55–6). Indeed, the desire to eliminate such absolute objects from physical theory was one of Einstein's main motivations in developing general relativity. Anderson then proposes that the various groups under consideration are associated with their respective theories by being the *symmetry groups* of the absolute objects of the theories. He proceeds to argue that the symmetry group of Newtonian mechanics is the Galilean group, that of special relativity the Lorentz group, and that of general relativity the group of all admissible transformations.

However, the central notions Anderson uses in his program are not very clearly defined. First, it is often unclear whether he is talking about coordinate transformations or point transformations; second, it is consequently unclear what a *symmetry* of an object amounts to; and finally and most importantly, the definition of 'absolute object' given (Anderson, 1967, p. 83) is exceedingly obscure – it is especially difficult to see how to use Anderson's definition in *proving* that any particular object is absolute or dynamical. In this paper I try to carry out Anderson's program in a precise way. I provide a precise definition of 'absolute object' which is (I hope) in the spirit of Anderson's discussion. I then use this notion to derive the symmetry groups of Newtonian mechanics, special relativity, and general relativity – showing in each case that the derived group is the intuitively correct one. Finally, I try to relate this notion of symmetry to traditional claims about the relativity of motion.

## II. SYMMETRY GROUPS

Let $M$ be a $C^\infty$ manifold.[2] A *differentiable transformation* of $M$ is a 1-1 $C^\infty$ mapping $h$ from $M$ onto itself such that $h^{-1}$ is also $C^\infty$ (i.e., a diffeomorphism from $M$ onto itself). Such a differentiable transformation $h$ induces a mapping $h*$ which associates the geometrical objects defined on $M$ with other such geometrical objects as follows:

(1) If $f \in T^{(0,0)}(M)$, $h*f(hp) = f(p)$
(2) If $X \in T^{(1,0)}(M)$, $h*X_{hp} h*f = X_p f$
(3) If $w \in T^{(0,1)}(M)$, $h*w_{hp} h*X = w_p X_p$
(4) If $\theta \in T^{(r,s)}(M)$, $h*\theta_{hp}(w_1, ..., w_r, X_1, ..., X_s) =$
$= \theta_p(h*^{-1} w_1, ..., h*^{-1} X_s)$
(5) If $D$ is an affine connexion on $M$, $(h*D_{h*X} h*Y)_{hp} = (D_X Y)_p$

where $p \in M$, $T^{(0,0)}(M)$ is the set of real-valued $C^\infty$ functions on $M$, $T^{(1,0)}(M)$ is the set of contra vector fields on $M$, $T^{(0,1)}(M)$ is the set of covector fields on $M$, and $T^{(r,s)}(M)$ is the set of tensor fields of type $(r, s)$ on $M$. Since $h*$ is 1-1, all these objects are well defined. It follows from the above definition that if $\{x_i\}$ is a local coordinate system and $g$ is a geometrical object, then the components of $g$ with respect to $\{x_i\}$ at $p$ are equal to the components of $h*g$ with respect to $\{h*x_i\}$ at $hp$. A differentiable transformation $h$ is a *symmetry* of a geometrical object $g$ (or $h$ leaves $g$ *invariant*) iff $h*g = g$. So $h$ is a symmetry of $g$ iff the components of $g$ with respect to $\{x_i\}$ at $p$ are equal to the components of $g$ with respect to $\{h*x_i\}$ at $hp$.

A *Lie group* is a set $L$ such that

(1)      $L$ is a group
(2)      $L$ is a $C^\infty$ manifold
(3)      The mapping $\circ : L \times L \to L$ determined by the group operation $\circ$ is $C^\infty$ on the product manifold $L \times L$.

A Lie group $L$ is said to be a Lie group of differentiable transformations on a $C^\infty$ manifold $M$ if to each pair of elements $p \in M$, $l \in L$ there corresponds an element $lp$ of $M$ such that

(1)      The mapping $(l, p) \to lp$ of $M \times L$ into $M$ is $C^\infty$
(2)      For fixed $l \in L$, $lp$ is a differentiable transformation of $M$
(3)      $lp = p$ for all $p \in M$ iff $l = e$, the identity element of $L$
(4)      $k(lp) = (k \circ l) p$ for all $p \in M$, $k, l \in L$.

If the dimension of $L$ is $d$, $L$ is called a *d-parameter* group of differentiable transformations. Any $d$-parameter group of differentiable transformations $L$ is a subgroup of the group $\mathcal{M}$ of all differentiable transformations on $M$, with composition as the group operation. For if $l \in L$ there is an $h \in \mathcal{M}$ such that $lp = h(p)$ for all $p \in M$; and by (4) above the group operation $\circ$ on $L$ can be viewed as composition of functions on $M$.

Let $L$ be a $d$-parameter group of differentiable transformations on a $C^\infty$ $n$-manifold $M$. Let $\{y_i\}$ be a local coordinate system around $e \in L$ and $\{x_j\}$ be a local coordinate system around $p \in M$. We can represent the action of $L$ on $M$ by the relations

$$x_i(lq) = f_i(y_1(l), \ldots, y_d(l), x_1(q), \ldots, x_n(q))$$

where $f_i$ is $C^\infty$ for $l$ in a neighborhood of $e$ and $q$ in a neighborhood of $p$. Form the $d + n$ functions

$$g^i{}_j = \frac{\partial f_i}{\partial y_j}\Big|_{l=e}$$

and the $d$ vector fields defined on a neighborhood of $p$

$$X_j = g^i_j \frac{\partial}{\partial x_i}.$$

The fundamental theorem of the theory of Lie groups of differentiable transformations is that the $d$ vector fields $X_j$ span a $d$-dimensional subspace of $T^{(1,0)}(A)$ (where $A$ is a neighborhood of $p$); conversely, every such set of $d$ independent vector fields $X_j$ determines a (local) $d$-parameter Lie group (cf. Cohn, 1968, Ch. 5, pp. 66–8). Hence, the vector fields $X_j$ are called the *generators* of the Lie group $L$.

The *Lie derivative* of a tensor field $\theta$ of type $(r, s)$ on $M$ with respect to the vector field $X$ on $M$ is a tensor field $L_X\theta$ of type $(r, s)$ on $M$ such that

(1)    If $f \in T^{(0,0)}(M)$, $L_X f = X f$
(2)    If $Y \in T^{(1,0)}(M)$, $L_X Y = [X, Y] = XY - YX$
(3)    If $w \in T^{(0,1)}(M)$, $L_X w(Y) = Xw(Y) - w(L_X Y)$
(4)    If $\theta \in T^{(r,s)}(M)$, $(L_X\theta)(w_1, \ldots, w_r, X_1, \ldots, X_s) =$
$= L_X(\theta(w_1, \ldots, w_r, X_1, \ldots, X_s)) - \theta(L_X w_1, \ldots, w_r, X_1, \ldots, X_s)$
$- \ldots - \theta(w_1, \ldots, w_r, X_1, \ldots, L_X X_s).$

It can be shown that $L_X\theta = 0$ is a necessary and sufficient condition for the (local) 1-parameter group generated by $X$ to be a *symmetry group* of the object $\theta$ (cf. Bishop and Goldberg, 1968, pp. 128–31). Here, a symmetry group of differentiable transformations is such that every transformation in the group is a symmetry of $\theta$. So if $h_t$ is a 1-parameter group generated by $X$, then $h_t * \theta = \theta$ for every $t$ iff $L_X \theta = 0$.

I will say that *the* symmetry group $S$ of an object $\theta$ is the largest subgroup of $\mathcal{M}$ which is a symmetry group of $\theta$. If the symmetry group $S$ of an object $\theta$ is a $d$-parameter group we can use the Lie derivative to find an explicit form for the $d$ 1-parameter subgroups of $S$ in a neighborhood of $p \in M$ with respect to a local coordinate system $\{x_i\}$ around $p$. First we solve $L_{X_j}\theta = 0$ with respect to $\{x_i\}$ to find an explicit form for the components of $d$ independent generators $X_j$. Suppose the components of

$X_j$ are $a^i_j$. We then solve the equations

$$\frac{\partial (x_i \circ h_t)}{\partial t} = a^i_j$$

subject to the initial conditions $x_i \circ h_t = x_i$ for $t=0$, to find the functions $x_i \circ h_t$. This gives us an explicit representation in a neighborhood of $p$ of the 1-parameter subgroup of $S$ generated by $X_j$.

If we can supply a precise definition of 'absolute object' we are now in a position to carry out Anderson's program. We can define *the symmetry group of a theory* $T$ as the largest group of differentiable transformations which is a symmetry group of all the absolute objects of $T$. We can then use the above process to find an explicit form for the symmetry groups of various theories – showing in each case that the symmetry group as so defined is the intuitively correct one. In the next section I provide a definition of 'absolute object'. In Section IV I show that this definition gives the correct results.

### III. SPACE-TIME THEORIES AND ABSOLUTE OBJECTS

I view the three theories here under consideration as *space-time* theories. Each theory describes a four-dimensional differentiable manifold $M$, space-time. It postulates various geometrical objects defined on the space-time manifold, and requires that these objects satisfy certain relations called the *field equations* of the theory in question. With the help of the objects appearing in the field equations the theory singles out a privileged class of *curves* on the space-time manifold – a class of curves which is intended to represent the trajectories of some privileged class of physical particles; e.g., free particles, or particles affected only by gravitational forces, or charged particles subject to an external electromagnetic field. Such a class of curves is picked out by the *equations of motion* of the theory. The simplest example of this kind of theory is one that postulates one geometrical object, a flat affine connexion, on the space-time manifold, and requires that free particles follow the geodesics of the postulated connexion. In this case the field equations are simply $R^i_{jkl}=0$, where $R^i_{jkl}$ is the Riemann-Christoffel curvature tensor. The equation of motion is just

$$\frac{d^2x_i}{du^2}+\Gamma^i_{jk}\frac{dx_j}{du}\frac{dx_k}{du}=0.$$

Often, one of the purposes of a space-time theory is to describe a particular form of *interaction* – gravitational, electromagnetic, etc. Such a theory postulates two types of geometrical objects in its field equations: *source variables*, which represent the sources of the interaction in question (mass density, charge density, etc.); and *field variables*, which appear in the equations of motion and represent the forces arising from the interaction.

I will illustrate the above ideas by formulating Newtonian mechanics, special relativity, and general relativity in this way. (More detailed formulations within this framework can be found in Anderson, 1967; Friedman, 1972; Havas, 1964; Trautman, 1965, 1966.)

## A. *Newtonian Mechanics*

Newtonian kinematics postulates three objects: a flat affine connexion $\Gamma^i_{jk}$, a covector field $t_i$ representing absolute time, and a symmetric tensor field $g^{ij}$ of type (2, 0) and signature (0, 1, 1, 1) which induces a Euclidean metric on the instantaneous three-dimensional 'spaces' of $M$. The field equations of Newtonian kinematics are

$$R^i_{jkl}=0$$
$$t_{i;j}=0$$
$$g^{ij}_{;k}=0$$
$$g^{ij}t_it_j=0,$$

while the equation of motion is the geodesic law

$$\frac{d^2x_i}{du^2}+\Gamma^i_{jk}\frac{dx_j}{du}\frac{dx_k}{du}=0.$$

It can be shown that there exists a $C^\infty$ function $t$ such that $t_i=\partial t/\partial x_i$, and that $g^{ij}$ induces a Euclidean metric on the hypersurfaces $t=$constant. Also, it can be shown that $t$ is an affine parameter, so the equation of motion can be written

$$\frac{d^2x_i}{dt^2}+\Gamma^i_{jk}\frac{dx_j}{dt}\frac{dx_k}{dt}=0.$$

In *inertial* coordinate systems – where $\Gamma^i_{jk}=0$, $(t_i)=(1, 0, 0, 0)$, and

$(g^{ij}) = \text{diag}(0, 1, 1, 1)$ – the law of motion reduces to Newton's first law

$$\frac{d^2 x_i}{dt^2} = 0.$$

Newtonian gravitation theory can be incorporated into this framework by introducing a scalar field $\varphi$ representing the gravitational potential and a scalar field $\rho$ representing the mass density. We add the field equation

$$g^{ij}\varphi_{;i;j} = 4\pi k \rho$$

and change the equation of motion to

$$\frac{d^2 x_i}{dt^2} + \Gamma^i_{jk} \frac{dx_j}{dt} \frac{dx_k}{dt} = -g^{ir}\varphi_{;r}.$$

In inertial coordinate systems these reduce to the Poisson equation

$$\nabla^2 \varphi = 4\pi k \rho$$

and Newton's second law

$$\frac{d^2 x_i}{dt^2} = -\frac{\partial \varphi}{\partial x_i}.$$

B. *Special Relativity*

The kinematics of special relativity is simple almost to the point of triviality. We have two geometrical objects: a flat connexion $\Gamma^i_{jk}$ and a symmetric tensor field $g_{ij}$ of type $(0, 2)$ and signature $(1, -1, -1, -1)$; and two field equations

$$R^i_{jkl} = 0$$
$$g_{ij;k} = 0.$$

Our equation of motion is again the geodesic law

$$\frac{d^2 x_i}{du^2} + \Gamma^i_{jk} \frac{dx_j}{du} \frac{dx_k}{du} = 0.$$

A contra vector field $a^i$ is called *timelike* iff $g_{ij}a^i a^j > 0$. A curve is said to be timelike iff its tangent vector field is timelike. We can define the *length* $\tau$ of any timelike curve $\sigma(u)$ by the formula

$$\tau(u) = \int_a^u \sqrt{g_{ij} \frac{dx_i}{du} \frac{dx_j}{du}}\, du$$

where $x_i = x_i \circ \sigma$. $\tau$ (proper time) is an affine parameter, so we can rewrite our equation of motion as

$$\frac{d^2 x_i}{d\tau^2} + \Gamma^i_{jk} \frac{dx_j}{d\tau} \frac{dx_k}{d\tau} = 0.$$

In *inertial* coordinate systems $x_0, x_1, x_2, x_3$ – where $\Gamma^i_{jk} = 0$ and $(g_{ij}) = \text{diag}(1, -1, -1, -1)$ – the law of motion is just

$$\frac{d^2 x_i}{d\tau^2} = 0.$$

So $x_0 = t$ (coordinate time) is an affine parameter and our law of motion finally becomes

$$\frac{d^2 x_i}{dt^2} = 0.$$

Electrodynamics can be formulated in this context by adding an antisymmetric tensor field $F^{ij}$ of type (2, 0) representing the electromagnetic field and a contra vector field $J^i$ representing the charge-current density. Our additional field equations are

$$F^{ij}_{;j} = -4\pi J^i$$
$$F_{[ij;k]} = 0$$

where $F_{ij} = g_{ia} g_{jb} F^{ab}$ and [ ] indicates antisymmetrization. Our new law of motion is

$$m \left[ \frac{d^2 x_i}{d\tau^2} + \Gamma^i_{jk} \frac{dx_j}{d\tau} \frac{dx_k}{d\tau} \right] = q F^i_j \frac{dx_j}{d\tau}$$

where $F^i_j = g_{ja} F^{ia}$, for a particle of (rest) mass $m$ and charge $q$. In inertial coordinate systems we obtain the Maxwell-Lorentz equations

$$\nabla \cdot \mathbf{E} = 4\pi\sigma$$
$$\nabla \times \mathbf{B} - \frac{\partial \mathbf{E}}{\partial t} = 4\pi\mathbf{j}$$

and

$$\nabla \cdot \mathbf{B} = 0$$
$$\nabla \times \mathbf{E} + \frac{\partial \mathbf{B}}{\partial t} = 0$$

where $\mathbf{E}$ is the electric field intensity, $\mathbf{B}$ the magnetic induction, $\sigma$ the charge density, and $\mathbf{j}$ the current density, from our two field equations. Similarly, the equation of motion yields the Lorentz equation

$$\frac{d\mathbf{p}}{dt} = q(\mathbf{E} = \mathbf{v} \times \mathbf{B})$$

with $\mathbf{p}$ the relativistic three-momentum and $\mathbf{v}$ the (coordinate) velocity.

C. *General Relativity*

General relativity has three geometrical objects: a (not necessarily flat) connexion $\Gamma^i_{jk}$, a symmetric tensor field $g_{ij}$ of type (0, 2) and signature $(1, -1, -1, -1)$, and tensor field $T^{ij}$ of type (2, 0) representing the mass-energy density. The field equations are well known:

$$g_{ij;k} = 0$$
$$R^{ij} - \tfrac{1}{2} g^{ij} R = -8\pi k T^{ij},$$

and we again have the geodesic law of motion

$$\frac{d^2 x_i}{d\tau^2} + \Gamma^i_{jk} \frac{dx_j}{d\tau} \frac{dx_k}{d\tau} - 0$$

with $\tau$ the proper time as above. General relativity, unlike our other space-time theories, is standardly presented in this four-dimensional tensor form.

If $T$ is a space-time theory a *dynamically possible model* (dpm) for $T$ is an $n$-tuple $\langle M, g_1, \ldots, g_{n-1} \rangle$, where $M$ is a four-dimensional differentiable manifold and $g_1, \ldots, g_{n-1}$ are the geometrical objects postulated by $T$, such that the objects $g_1, \ldots, g_{n-1}$ satisfy the field equations of $T$.

Thus, a dpm of Newtonian kinematics takes the form $\langle M, \Gamma^i_{jk}, t_i, g^{ij}\rangle$, where $\Gamma^i_{jk}$ is a flat connexion, $t_i$ is a covector field, and $g^{ij}$ is a symmetric contra tensor of signature (0, 1, 1, 1). Such a dpm is called a *Galilean manifold* or *Galilean space-time*. A dpm of special relativity kinematics takes the form $\langle M, \Gamma^i_{jk}, g_{ij}\rangle$, where $\Gamma^i_{jk}$ is a flat connexion and $g_{ij}$ is a semi-Riemannian metric tensor of signature $(1, -1, -1, -1)$. Such a dpm is called a *Minkowski* manifold or *Minkowski* space-time. Similarly, a dpm of Newtonian gravitation theory takes the form $\langle M, \Gamma^i_{jk}, t_i, g^{ij}, \varphi, \rho\rangle$, those of (relativistic) electrodynamics the form $\langle M, \Gamma^i_{jk}, g_{ij}, F^{ij}, J^i\rangle$, and those of general relativity the form $\langle M, \Gamma^i_{jk}, g_{ij}, T^{ij}\rangle$.

I would now like to discuss the problem of dividing up the geometrical objects postulated by a theory $T$ into absolute objects and dynamical objects, the problem of giving a precise definition of 'absolute object'. As we saw above, the absolute objects are intended to be those objects – e.g., the metric of special relativity – which are intuitively independent or not affected by interaction with the other objects of the theory. The dynamical objects are those objects – e.g., the metric of general relativity – which are affected by interaction with other objects – in this case, the mass-energy distribution $T^{ij}$. Since the absolute objects are thus independent of the other objects, the field equations of our theory $T$ determine them in a way in which they do not determine the other objects. For example, in the case of electrodynamics as formulated on pp. 304–5, $g_{ij}$ is an absolute object because it is determined by the field equations $R^i_{jkl}=0$ and $g_{ij;k}=0$ independently of $F^{ij}$ and $J^i$. On the other hand, $F^{ij}$ is a dynamical object because the field equations do not determine it – it can only be fixed when $J^i$ is determined. Our problem is to give a precise sense to an object's being *determined* by the field equations of a theory $T$.

An obvious strategy in finding a definition of 'absolute object' is therefore to look for a natural equivalence relation between geometrical objects, and to demand that the field equations of $T$ determine the absolute objects up to equivalence under this relation. I think there is such an equivalence relation. Consider the Minkowski metric tensor, for example. Let $g$ be a Minkowski tensor on a 4-manifold $M$ and let $h$ be a second Minkowski tensor on a 4-manifold $N$. Let $p$ be a point of $M$ and $q$ a point of $N$. There is a neighborhood $A$ of $p$, a neighborhood $B$ of $q$, and a diffeomorphism $f$ from $A$ onto $B$ such that $h=f*g$ on $B$. This can be shown by the following argument.

Let $\varphi$ be a chart around $p \in M$ such that $(g_{ij}) = \text{diag}(1, -1, -1, -1)$ and let $\psi$ be a chart around $q \in N$ such that $(h_{ij}) = \text{diag}(1, -1, -1, -1)$ – i.e., $\varphi$ and $\psi$ are inertial coordinate systems. Let $x_i = u_i \circ \varphi$, $y_i = u_i \circ \psi$ – where $u_i$ is the 'slot' function on $R^4$: $u_i(a_1, a_2, a_3, a_4) = a_i$. Suppose $(a_1, a_2, a_3, a_4)$, $(b_1, b_2, b_3, b_4)$ are the coordinates of $p$ and $q$, respectively; i.e., $x_i(p) = a_i$, $y_i(q) = b_i$. Change to new coordinates $x_i^* = x_i - a_i$, $y_i^* = y_i - b_i$. Clearly, these transformations leave the components of $g$ and $h$ unchanged, so $(g_{ij})$ and $(h_{ij})$ take the form $\text{diag}(1, -1, -1, -1)$ in $\varphi^*$ and $\psi^*$ also. Let $R_{\varphi^*}^4$ be the range of $\varphi^*$, $R_{\psi^*}^4$ the range of $\psi^*$. Consider the set $C = R_{\varphi^*}^4 \cap R_{\psi^*}^4$. $C$ is nonempty, and $\varphi^{*-1}(C)$ is a neighborhood of $p$, $\psi^{*-1}(C)$ a neighborhood of $q$. Now consider the function $f: \varphi^{*-1}(C) \to \psi^{*-1}(C)$ defined by $f(r) = \psi^{*-1} \circ \varphi^*(r)$ for $r \in \varphi^{*-1}(C)$.

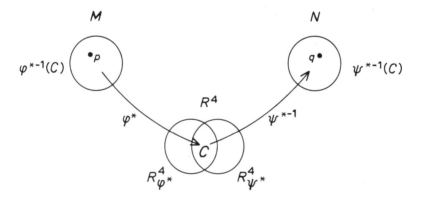

The function $f$ is a diffeomorphism since it is the composition of two diffeomorphisms $\varphi^*$ and $\psi^{*-1}$. Also, $y_i^* = f * x_i^*$ since $y_i^*(fr) = x_i^*(r)$. By hypothesis, the components of $h$ with respect to $y_1^*, y_2^*, y_3^*, y_4^*$ are the same as the components of $g$ with respect to $x_1^*, x_2^*, x_3^*, x_4^*$. It follows that $h = f * g$ on $\psi^{*-1}(C)$.

More generally, given two arbitrary geometrical objects, $g$ and $h$, on two arbitrary manifolds, $M$ and $N$, I will say that $g$ and $h$ are *locally d-equivalent* around $p \in M$, $q \in N$ iff there is a neighborhood $A$ of $p$, a neighborhood $B$ of $q$, and a diffeomorphism $f$ from $A$ onto $B$ such that $h = f * g$ on $B$. Thus, two Minkowski tensors are locally d-equivalent around every pair of points $p \in M$, $q \in N$. I will say that $g$ and $h$ are *d-equivalent simpliciter* if the neighborhoods on which they are locally

308                    MICHAEL FRIEDMAN

$d$-equivalent can be pieced together to cover the whole manifold in the following way:

DEFINITION 1: Two geometrical objects, $g$ and $h$, on $M$ and $N$, respectively are *d-equivalent* iff there is an open covering $\{A_i\}$ of $M$ and a family of diffeomorphisms $\{f_i\}$ such that $\{f_iA_i\}$ covers $N$, and for every $i$, $h=f_i*g$ on $f_iA_i$.

Since any two Minkowski tensors are locally $d$-equivalent around *every* pair of points $p\in M$, $q\in N$, they are obviously $d$-equivalent. $D$-equivalence is trivially reflexive and symmetric. Transitivity can be shown by the following argument.[3]

Let $g$, $h$, and $m$ be three geometrical objects on $M$, $N$, and $S$, respectively. Suppose that $g$ and $h$ are $d$-equivalent and that $h$ and $m$ are $d$-equivalent; i.e., there is a covering $\{A_i\}$ of $M$ and a family of diffeomorphisms $\{f_i\}$ such that $\{f_iA_i\}$ covers $N$ and $h=f_i*g$ on $f_iA_i$. Similarly, there is a covering $\{B_j\}$ of $N$ and a family of diffeomorphisms $\{g_j\}$ such that $\{g_jB_j\}$ covers $S$ and $m=g_j*h$ on $g_jB_j$.

$$\begin{array}{ccc} M & N & S \\ \{A_i\} \to f_i \to \{f_iA_i\} & & \\ & \{B_j\} \to g_j \to \{g_jB_j\} & \end{array}$$

Consider the family of sets $\{f_iA_i\cap B_j\}$ on $N$. For fixed $j$ the family $\{f_iA_i\cap B_j\}$ covers $B_j$, so the family $\{g_j(f_iA_i\cap B_j)\}=g_jB_j$. With varying $j$ the family $\{g_jB_j\}$ covers $S$. Therefore $\{g_j(f_iA_i\cap B_j)\}$ covers $S$. Similarly, the family $\{f_i^{-1}(f_iA_i\cap B_j)\}$ covers $M$. Therefore, there is a covering of $M$, $\{f_i^{-1}(f_iA_i\cap B_j)\}$, and a family of diffeomorphisms, $\{g_j\circ f_i\}$, such that $\{(g_j\circ f_i)(f_i^{-1}(f_iA_i\cap B_j))\}$ covers $S$, and for every $\langle i,j\rangle$, $m=(g_j\circ f_i)*g$ on $(g_j\circ f_i)(f_i^{-1}(f_iA_i\cap B_j))$.

An analogous condition is not satisfied by the electromagnetic field tensor; i.e., it is not the case that any two antisymmetric tensor fields of type (2, 0) satisfying the electrodynamic field equations are $d$-equivalent. For there can be two such tensor fields, $F_1$ and $F_2$, such that $F_1$ is everywhere zero and $F_2$ is not everywhere zero. Let $p$ be an arbitrary point in the domain of $F_1$, and let $q$ be a point at which $F_2$ is not zero; i.e., the components of $F_2$ with respect to an arbitrary chart at $q$ do not all vanish. Suppose there were a diffeomorphism $f$ from a neighborhood of $p$ onto a neighborhood of $q$ such that $F_2=f*F_1$ on that neighborhood. Then if

$\{x_i\}$ is an arbitrary coordinate system around $p$, the components of $F_2$ with respect to $\{f * x_i\}$ at $q (= fp)$ would be the same as the components of $F_1$ with respect to $\{x_i\}$ at $p$. But this is impossible since these latter components are all zero. So there is no point $p$ in the domain of $F_1$ such that $F_1$ and $F_2$ are locally $d$-equivalent around $p, q$. It follows that $F_1$ and $F_2$ are not $d$-equivalent. Similarly, the metric tensor of general relativity does not satisfy such a requirement either. For there can be both everywhere flat and not everywhere flat solutions to Einstein's field equations; and if $q$ is a point at which the metric is not flat, there is no point $p$ in the domain of a flat metric tensor such that the two objects are locally $d$-equivalent around $p, q$.

Thus, I require that absolute objects be determined up to $d$-equivalence:

DEFINITION 2. A geometrical object $g_i$ is an *absolute object* of a theory $T$ iff for any two dpm's of $T$, $\langle M, g_1, ..., g_{n-1} \rangle$ and $\langle N, h_1, ..., h_{n-1} \rangle$, $g_i$ and $h_i$ are $d$-equivalent.

To show that there are enough absolute objects note that the argument on pp. 307-8 can be easily generalized to yield the following theorem:

THEOREM. Let $g$ and $h$ be two geometrical objects of the same type defined on the $n$-manifolds $M$ and $N$, respectively. Suppose that for every point $p \in M$ there is a chart $\varphi$ around $p$ such that the components of $g$ with respect to $\varphi$ are some fixed set of constants $A_1, ..., A_r$ (the *same* set for every point); and suppose that the same condition holds for $h$, for the same set of constants. Then $g$ and $h$ are $d$-equivalent.

Thus, for example, two flat connexions are always $d$-equivalent since there are coordinate systems around every point in which the components of a flat connexion are all zero. Similarly, the metric tensor of a flat space is an absolute object since around every point there is a coordinate system in which the matrix of components is diagonal with diagonal elements 1 or $-1$. The Minkowski metric is just the special case where $(g_{ij}) = \text{diag}(1, -1, -1, -1)$. We see, therefore, that the existence of absolute objects is closely tied to the existence of 'privileged' or 'standard' coordinate systems; coordinate systems in which the components of some geometrical object take on a preassigned set of values.

I think that the above definition of 'absolute object' divides the absolute and dynamical objects up in the intuitively correct way and captures the sense in which the absolute objects are determined by a theory's field equations. In the next section I will show that my definition leads to the intuitively correct symmetry groups for our three classical space-time theories.

## IV. SOME EXAMPLES

We are now in a position to use the method outlined on pp. 300–1 to find an explicit form for the symmetry group of a given space-time theory.

### A. *Special Relativity*

As we saw above, a dpm of special relativity takes the form $\langle M, \Gamma^i_{jk}, g_{ij}\rangle$ where $\Gamma^i_{jk}$ and $g_{ij}$ satisfy the field equations

$$R^i_{jkl} = 0$$
$$g_{ij;k} = 0.$$

It follows from these equations that there is a coordinate system around every point $p \in M$ in which $\Gamma^i_{jk} = 0$ and $(g_{ij}) = \text{diag}(1, -1, -1, -1)$. Thus, both $\Gamma^i_{jk}$ and $g_{ij}$ are absolute objects of special relativity. Therefore, the symmetry group $L$ of special relativity must leave $\Gamma^i_{jk}$ and $g_{ij}$ invariant. I will derive an expression for the 1-parameter subgroups of $L$ in an inertial coordinate system. First of all, we know that a generator $X$ of a 1-parameter subgroup of $L$ must satisfy $L_X g = 0$. If the components of $X$ are $(a^0, a^1, a^2, a^3)$ this yields Killing's equation

$$a_{i;j} + a_{j;i} = 0$$

where $a_i = g_{ik}a^k$. Furthermore, in an inertial coordinate system covariant differentiation reduces to ordinary differentiation, and we have

$$a_{i;j;k} - a_{i;k;j} = 0.$$

Permuting indices in the above two equations we obtain

$$a_{i;j;k} = 0.$$

So in an inertial coordinate system we have the conditions

$$\frac{\partial a^k}{\partial x_i} g_{kj} + \frac{\partial a^k}{\partial x_j} g_{ki} = 0$$

and

$$\frac{\partial^2 a^p}{\partial x_k \partial x_j} g_{pi} = 0$$

on the components $a^i$ of our generators. The $a^i$ must therefore satisfy

$$a^i = b_{ij} x_j + c_i$$

where the $b_{ij}$ and $c_i$ are constants, $b_{ij} = -b_{ji}$ $(i, j = 1, 2, 3)$, $b_{i0} = b_{0i}$, and $b_{00} = 0$. This equation has ten undetermined constants. We can therefore obtain ten independent generators by successively setting one constant equal to 1 and the rest equal to 0. In this way we obtain the following ten generators.

| Translations | Rotations | Lorentz Transformations |
|---|---|---|
| $(1, 0, 0, 0)$ | $(0, x_2, -x_1, 0)$ | $(-x_1, -x_0, 0, 0)$ |
| $(0, 1, 0, 0)$ | $(0, 0, x_3, -x_2)$ | $(-x_2, 0, -x_0, 0)$ |
| $(0, 0, 1, 0)$ | $(0, x_3, 0, -x_1)$ | $(-x_3, 0, 0, -x_0)$ |
| $(0, 0, 0, 1)$ | | |

To find a representation of the 1-parameter subgroups of $L$ generated by these generators we solve the equations

$$\frac{\partial x_i^*}{\partial s} = a^i$$

where $x_i^* = x_i \circ h_s$.

For example, the translation $(1, 0, 0, 0)$ yields the equations

$$\frac{\partial x_0^*}{\partial s} = 1 \quad \frac{\partial x_1^*}{\partial s} = 0 \quad \frac{\partial x_2^*}{\partial s} = 0 \quad \frac{\partial x_3^*}{\partial s} = 0,$$

which have the solution

$$x_0^* = x_0 + s$$
$$x_1^* = x_1$$
$$x_2^* = x_2$$
$$x_3^* = x_3;$$

i.e., it generates a translation along the $x_0$ (coordinate time) axis. The rotation $(0, x_2, -x_1, 0)$ yields the equations

$$\frac{\partial x_0^*}{\partial s} = 0 \qquad \frac{\partial x_1^*}{\partial s} = x_2^* \qquad \frac{\partial x_2^*}{\partial s} = -x_1^* \qquad \frac{\partial x_3^*}{\partial s} = 0$$

which have the solution

$$x_0^* = x_0$$
$$x_1^* = x_1 \cos(s) + x_2 \sin(s)$$
$$x_2^* = x_2 \cos(s) - x_1 \sin(s)$$
$$x_3^* = x_3 \,;$$

i.e., it generates a rotation around the $x_3$-axis. Finally, the Lorentz transformation $(-x_1, -x_0, 0, 0)$ yields the equations

$$\frac{\partial x_0^*}{\partial s} = -x_1^* \qquad \frac{\partial x_1^*}{\partial s} = -x_0^* \qquad \frac{\partial x_2^*}{\partial s} = 0 \qquad \frac{\partial x_3^*}{\partial s} = 0$$

which have the solution

$$x_0^* = x_0 \cosh(s) - x_1 \sinh(s)$$
$$x_1^* = x_1 \cosh(s) - x_0 \sinh(s)$$
$$x_2^* = x_2$$
$$x_3^* = x_3 \,.$$

Changing to a new parameter $v = \tanh(s)$ we obtain the more familiar form

$$x_0^* = \frac{x_0 - v x_1}{\sqrt{1-v^2}}$$

$$x_1^* = \frac{x_1 - v x_0}{\sqrt{1-v^2}}$$

$$x_2^* = x_2$$
$$x_3^* = x_3$$

of a Lorentz transformation along the $x_1$-axis.[4]

Thus, the symmetry group of the Minkowski metric $g$ is a 10-parameter group consisting of spatial and temporal translations, spatial rotations, and Lorentz transformations. Since in this coordinate system invariance of the connexion $\Gamma^i_{jk}$ adds no new constraints on our symmetry group – it merely means that the representation of our full symmetry group $L$ must be linear – this 10-parameter group is the symmetry group of our theory. Finally, note that the symmetry group of (relativistic) electrodynamics is also the Lorentz group $L$. This is because both the electromagnetic field $F^{ij}$ and the charge-current density $J^i$ are dynamical objects (by the argument on pp. 308–9).

B. *Newtonian Mechanics*

A dpm of Newtonian kinematics takes the form $\langle M, \Gamma^i_{jk}, t_i, g^{ij} \rangle$ where $\Gamma^i_{jk}, t_i,$ and $g^{ij}$ satisfy the field equations

$$R^i_{jkl} = 0$$
$$t_{i;j} = 0$$
$$g^{ij}_{;k} = 0$$
$$g^{ij} t_i t_j = 0.$$

It follows from these equations that there is a coordinate system around every point $p \in M$ in which $\Gamma^i_{jk} = 0$, $(t_i) = (1, 0, 0, 0)$, and $(g^{ij}) = \text{diag}(0, 1, 1, 1)$. Thus, $\Gamma^i_{jk}, t_i,$ and $g^{ij}$ are absolute objects of Newtonian kinematics. Therefore, the symmetry group $G$ of Newtonian mechanics must leave $\Gamma^i_{jk}, t_i,$ and $g^{ij}$ invariant. I will derive an expression for the 1-parameter subgroups of $G$ in an inertial coordinate system. Expanding the Lie derivatives of $g^{ij}$ and $t_i$ in an inertial coordinate system we find that

$$\frac{\partial a^i}{\partial x_k} g^{kj} + \frac{\partial a^j}{\partial x_k} g^{ki} = 0$$

and

$$\frac{\partial a^k}{\partial x_i} t_k = 0$$

where the $a^i$ are the components of a generator of a 1-parameter subgroup of $G$. Using reasoning just like the above treatment of special relativity we find that the $a^i$ take the form

$$a^i = b_{ij} x_j + c_i$$

where the $b_{ij}$ and $c_i$ are constants, $b_{ij} = -b_{ji} (i, j = 1, 2, 3)$, and $b_{0j} = 0$.

Again, this equation has ten undetermined constants. Substituting these into our formula one at a time we again obtain ten independent generators.

| Translations | Rotations | Galilean Transformations |
|---|---|---|
| (1, 0, 0, 0) | (0, $x_2$, $-x_1$, 0) | (0, $-x_0$, 0, 0) |
| (0, 1, 0, 0) | (0, 0, $x_3$, $-x_2$) | (0, 0, $-x_0$, 0) |
| (0, 0, 1, 0) | (0, $x_3$, 0, $-x_1$) | (0, 0, 0, $-x_0$) |
| (0, 0, 0, 1) | | |

(Compare the generators of the Galilean transformations with those of the Lorentz transformations on p. 311). This group differs from the Lorentz group $L$ only by containing the Galilean transformations in place of the Lorentz transformations. To see that the above generators really do generate the Galilean transformations, consider the generator (0, $-x_0$, 0, 0). It yields the equations

$$\frac{\partial x_0^*}{\partial v}=0 \quad \frac{\partial x_1^*}{\partial v}=-x_0^* \quad \frac{\partial x_2^*}{\partial v}=0 \quad \frac{\partial x_3^*}{\partial v}=0$$

which have the solution

$$x_0^*=x_0$$
$$x_1^*=x_1-vx_0$$
$$x_2^*=x_2$$
$$x_3^*=x_3;$$

i.e., it generates a Galilean Transformation along the $x_1$-axis.

Thus, the symmetry group of $t_i$ and $g^{ij}$ is a 10-parameter group consisting of spatial and temporal translations, spatial rotations, and Galilean transformations. Since in this coordinate system invariance of $\Gamma^i_{jk}$ again adds no new constraints on our symmetry group, this 10-parameter group is the full symmetry group of our theory. Finally, note that the symmetry group of Newtonian gravitation theory is also the Galilean group $G$ – because the gravitational potential $\varphi$ and the mass density $\rho$ are dynamical objects.

C. *General Relativity*

A dpm of general relativity takes the form $\langle M, \Gamma^i_{jk}, T^{ij}\rangle$ where $\Gamma^i_{jk}$, $g_{ij}$, and $T^{ij}$ satisfy the field equations

$$g_{ij;k}=0$$
$$R^{ij}-\tfrac{1}{2}g^{ij}R = -8\pi k T^{ij}.$$

From these equations it is clear (by the argument on pp. 308–9) that $\Gamma^i_{jk}$, $g_{ij}$, and $T^{ij}$ are all dynamical objects. General relativity has no absolute objects. Thus, the symmetry group of general relativity is a degenerate case – it is just the group $M$ of all differentiable transformations.

## V. RELATIVITY PRINCIPLES

Anderson's program involves construing the traditional relativity principles associated with our three space-time theories as requirements on the symmetry groups of these theories. In this interpretation the principle of Galilean relativity says that the symmetry group of Newtonian mechanics is the Galilean group, the special principle of relativity says that the symmetry group of special relativity is the Lorentz group, and the general principle of relativity says that the symmetry group of general relativity is the group $\mathcal{M}$ of all differentiable transformations. Such an interpretation of traditional relativity principles has several virtues. First, it gives a clear and precise sense to the traditional principles – a sense that enables us to classify and distinguish the three classical space-time theories. This contrasts with the construal of relativity principles as covariance requirements, since all three theories satisfy the 'principle of general covariance'. Second, it emphasizes the fact that our three theories deal with space-time structure. The relativity principles, construed as symmetry requirements, put constraints on the space-time structures described by their respective theories. Thus, the Galilean group $G$ is essentially the symmetry group of Galilean space-time $\langle M, \Gamma^i_{jk}, t_i, g^{ij}\rangle$, and the Lorentz group $L$ is essentially the symmetry group of Minkowski space-time $\langle M, \Gamma^i_{jk}, g_{ij}\rangle$. The general principle of relativity, construed as the requirement that our symmetry group include *all* differentiable transformations, has as a consequence the dynamical nature of general relativistic space-time. It means in effect that our theory can have no absolute objects; for the only geometrical objects which are left invariant under all differentiable transformations are constant-valued scalars. Thus, it implements Einstein's anti-absolutist motivations alluded to on

p. 297. Finally, this interpretation of relativity principles allows us to view the principle of general relativity as a generalization of the principle of special relativity (again, unlike the 'principle of general covariance').

In most discussions of relativity principles the notion of a reference frame plays a central role. In these discussions a relativity principle is thought to express the 'equivalence' of various classes of reference frames. Thus, the special principle of relativity expresses the 'equivalence' of all inertial frames, and the general principle of relativity expresses the alleged 'equivalence' of all (inertial *and* accelerated) frames. This sense of 'equivalence' is sometimes explained by the claim that two frames are equivalent for a given theory if the laws of the theory take the same 'form' in the two frames. Finally, relativity principles are sometimes thought to be connected with a relational or relativistic view of motion: a view that all motion should be interpreted as the motion of some bodies in relation to other bodies. For example, the special principle of relativity is thought to express the relative, nonabsolute character of all uniform inertial motion, and the general principle of relativity is thought to express the relative character of all motion. Indeed, one of Einstein's principal motivations for the general theory of relativity was his desire to implement such a relational view of motion. He wanted to extend such a view of motion, which is consistent with special relativity only as applied to uniform inertial motion, to nonuniform accelerated motion as well. Thus, Einstein was dissatisfied with special relativity (and classical physics) because of effects like the following: of two bodies rotating in relation to each other, only the one that is absolutely rotating (i.e., moving noninertially with respect to the affine structure of Minkowski (or Galilean) space-time) will experience distorting centrifugal and Coriolis forces. This is unsatisfactory, according to Einstein, because such force effects have an unobservable cause – i.e., the affine structure of space-time – rather than an observable cause – i.e., rotation relative to a second physical body (cf. Einstein, 1952, pp. 112–3). Consequently, he wanted to modify special relativity so that only relative acceleration produced such effects. How many of these traditional claims about relativity principles can we make sense of in the present framework?

First of all, what is a reference frame? I think of reference frames as coordinate systems associated with a given body in a particular state of motion. I will say that a coordinate system $x_0, x_1, x_2, x_3$ is *associated* with a particle with trajectory $\sigma$ iff $\sigma$ has the equations $x_i = 0$ ($i = 1, 2, 3$) in $x_0, x_1, x_2, x_3$. Thus, a particle is at rest at the origin of its associated coordinate system. A *reference frame* will be a pair consisting of a given particle and an associated coordinate system. A reference frame has trajectory (state of motion) $\sigma$ iff the particle it is associated with has trajectory $\sigma$. An *inertial* frame will be a reference frame whose coordinate system is an inertial coordinate system. This class of frames will vary from theory to theory. For example, in Newtonian mechanics an inertial frame is characterized by the conditions $\Gamma^i_{jk} = 0$, $(t_i) = (1, 0, 0, 0)$, and $(g^{ij}) = \mathrm{diag}(0, 1, 1, 1)$, while in special relativity an inertial frame is characterized by $\Gamma^i_{jk} = 0$ and $(g_{ij}) = \mathrm{diag}(1, -1, -1, -1)$. In both theories inertial frames execute inertial motions; i.e., an inertial coordinate system can be associated with a particle with trajectory $\sigma$ iff $\sigma$ is an affine geodesic. In general relativity, of course, there are no privileged inertial frames.

In the cases of Newtonian mechanics and special relativity we can connect our symmetry groups, which are groups of *point* transformations, with groups of *coordinate* transformations which relate different inertial frames. Both the Galilean and Lorentz groups have the general linear form

$$x_i^* = A_{ij} x_j + B_i$$

in inertial coordinates. We can associate this linear point transformation with a linear coordinate transformation given by

$$y_i = A_{ij} x_j + B_i.$$

Such coordinate transformations are easily seen to take inertial coordinate systems onto inertial coordinate systems; i.e., the Galilean transformations preserve the conditions $\Gamma^i_{jk} = 0$, $(t_i) = (1, 0, 0, 0)$, and $(g^{ij}) = \mathrm{diag}(0, 1, 1, 1)$, the Lorentz transformations preserve the conditions $\Gamma^i_{jk} = 0$ and $(g_{ij}) = \mathrm{diag}(1, -1, -1, -1)$. Furthermore, in the cases of the Galilean and Lorentz groups, considered as groups of coordinate transformations, it can be shown that the group parameter '$v$' denotes the relative velocity of the two related inertial frames; i.e., if frame I is related by a Galilean (Lorentz) transformation to frame II, the particle associated with frame II moves with constant velocity $v$ with respect to frame I (cf. Friedman, 1972, pp. 78–9; 116–8).

Now what is the connection between the relativity of motion and principles which talk about the 'forms' of laws in different reference frames? Presumably, it is something like the following: in order for a certain class of motions (e.g., inertial motions) to be interpretable as relative motions it must be the case that any two reference frames performing motions from that class (e.g., inertial frames) are physically equivalent (or physically indistinguishable). For otherwise some states of motion would possess features not explainable in terms of relative motion alone. Thus, it is a necessary condition for the relativity of motion within a certain class of motions that all reference frames performing motions from this class be physically equivalent (indistinguishable). It is then suggested that two reference frames are physically equivalent just in case the laws of nature have the same 'form' in both of them.

Something close to the above line of thought underlies Einstein's treatment of the general principle of relativity, as well as numerous later expositions of the theory. I think the problem with it is that the notion of *the* 'form' of the laws of nature is not precise. In general, the same laws of a given theory can be written in several different forms. Thus, the laws of (relativistic) electrodynamics can be written in tensor form with respect to an arbitrary coordinate system or in the more familiar three-dimensional notation with respect to an inertial frame. The fact that these laws have the same three-dimensional form in every inertial system does not show that all inertial systems are physically equivalent any more than the fact that they take the same four-dimensional tensor form in every coordinate system shows that *all* reference systems are physically equivalent. Similarly, the fact that Newtonian mechanics can be written in four-dimensional tensor form does not show that all reference frames are equivalent in Newtonian mechanics. Therefore, the fact that the laws of general relativity take the same form in all reference systems should not be taken as showing that all reference systems are physically equivalent according to general relativity. A more sensible criterion for the physical equivalence of two reference systems for a given theory is that the geometrical objects of the theory take the same form in the two systems. Thus, all inertial frames are equivalent in Newtonian kinematics; for in any such frame the objects of our theory have the form $\Gamma^i_{jk} = 0$, $(t_i) = (1, 0, 0, 0)$, and $(g^{ij}) = \text{diag}(0, 1, 1, 1)$. And all inertial frames are equivalent in the kinematics of special relativity; for in any such frame the objects of our theory have the form $\Gamma^i_{jk}=0$ and $(g_{ij})=\text{diag}(1, -1, -1, -1)$. Therefore, the form of the *objects* of a given theory provides a better test for the equivalence of reference systems than the form of its *laws*.

How does general relativity square with the relativity of motion in our sense, on which the members of a class of motions are interpretable as relative motions only if any two reference systems performing motions from that class are physically equivalent? In order to satisfy the Machian relationalist motiviations mentioned above, the class of motions in question must include all motions. So the general principle of relativity would have to imply that all reference frames are physically equivalent. But this is absurd. First, it is obvious that the geometrical objects of general relativity take different forms in different frames. Second, in general, accelerating frames (e.g., rotating frames) are physically distinguishable (e.g., by centrifugal and Coriolis type force effects) from non-accelerating frames (cf. Earman, 1970, pp. 300–2). Thus, contrary to Einstein's motivations, general relativity does not successfully implement a Machian or relational view of motion. It does not generalize the relativity' is not an appropriate name for our theory, as Fock (1964) has physics) to noninertial motions. In one sense, therefore, 'general relativity', is not an appropriate name for our theory, as Fock (1964) has forcefully argued. What happens in general relativity is this: we do not widen the class of physically equivalent reference frames to include *all* frames; rather, there is *no* privileged class of frames within which such equivalence holds. We no longer have a privileged state of inertial motion, nor, consequently, a privileged class of inertial frames. But this elimination of a privileged state of inertial motion does not imply a generalized relativity of motion, since it does not imply a generalized equivalence of refrence frames.

*Harvard University*

NOTES

\* The basic ideas of this paper grew out of conversations with Clark Glymour. I am also indebted to John Earman for valuable discussions and advice. Most of the material in the paper is drawn from Friedman (1972).
[1] Compare, e.g., Anderson (1967), Friedman (1972), Havas (1964), and Trautman (1965, 1966). Such a treatment will be sketchily outlined in Section III.

[2] Most of the notions from differential geometry used here are explained in Bishop and Goldberg (1968) or Hicks (1965).

[3] Transitivity follows more simply if we note that Definition 1 is equivalent to the following:

DEFINITION 1'. Two geometrical objects, $g$ and $h$, on $M$ and $N$, respectively, are $d$-equivalent iff for every $p \in M$ there is a point $q \in N$ such that $g$ and $h$ are locally $d$-equivalent around $p$, $q$; and for every $q \in N$ there is a point $p \in M$ such that $g$ and $h$ are locally $d$-equivalent around $p$, $q$.

This was pointed out to me by Clifton McIntosh and Dan Friedan.

[4] This technique for finding the symmetry group is taken largely from Robertson and Noonan (1968, pp. 321-3).

## BIBLIOGRAPHY

Anderson, J. L., *Principles of Relativity Physics*, Academic Press, New York, 1967.
Bishop, R. and Goldberg, S., *Tensor Analysis on Manifolds*, Macmillan, New York, 1968.
Cohn, P., *Lie Groups*, Cambridge University Press, 1968.
Earman, J., 'Who's Afraid of Absolute Space?', *Australasian Journal of Philosophy* **48** (1970), 287–319.
Einstein, A., 'The Foundations of the General Theory of Relativity', in *The Principle of Relativity*, Dover, New York, 1952, pp. 109–64.
Einstein, A., *The Meaning of Relativity*, Princeton University Press, Princeton, 1956.
Fock, V., *The Theory of Space, Time, and Gravitation*, Pergamon Press, New York, 1964.
Friedman, M., *Foundations of Space-Time Theories*, Unpublished Ph.D. dissertation, Princeton University, Princeton, 1972.
Havas, P., 'Four-Dimensional Formulations of Newtonian Mechanics and Their Relation to the Special and the General Theory of Relativity', *Reviews of Modern Physics* **36** (1964), 938–65.
Hicks, N. J., *Notes on Differential Geometry*, Van Nostrand, Princeton, 1965.
Robertson, H. P. and Noonan, T. W., *Relativity and Cosmology*, W. B. Saunders, Philadelphia, 1968.
Trautman, A., 'Foundations and Current Problems of General Relativity', in S. Deser and K. Ford (eds.), *Lectures on General Relativity*, Prentice-Hall, Englewood Cliffs, 1965.
Trautman, A., 'Comparison of Newtonian and Relativistic Theories of Space-Time', in B. Hoffman (ed.), *Perspectives on Geometry and Relativity*, Indiana University Press, Bloomington, 1966.

ROBERT W. LATZER

# NONDIRECTED LIGHT SIGNALS AND THE STRUCTURE OF TIME

**Abstract.** 'Temporal betweenness' in space-time is defined solely in terms of light signals, using a signalling relation that does not distinguish between the sender and the receiver of a light signal. Special relativity and general relativity are considered separately, because the latter can be treated only locally. We conclude that the (local) coherence of time can be described if we know only which pairs of space-time points are light-connected. Other consequences in the case of special relativity: (1) a categorical axiom system exists in terms of nondirected light connection alone, with neither 'particle' nor 'time order' as a primitive concept, though we do not actually present the axioms; (2) any concept definable by coordinates is also definable in terms of nondirected light signals if and only if it is invariant under Lorentz transformations, translations, dilations, space reflections, and time reflections; and (3) any transformation of space-time (not necessarily continuous) which preserves nondirected light connection is a product of transformations just listed above. The bulk of the paper is devoted to proving that the definitions we give correspond to their intended interpretations in the usual space-time continua.

## I. INTRODUCTION

Temporal order, in some form, has been taken as fundamental to virtually every attempt to describe the structure of space-time in terms of simple primitive relations. In this paper, however, I will show that the structure of the Minkowski space of special relativity can be completely described in terms of a single primitive relation that does not involve order in any way: the *symmetric* light relation $\lambda$, where $a \lambda b$ means that the space-time points $a$ and $b$ can be connected by a light signal, without specifying anything as to which point is the origin and which is the receiver of the light signal. This conclusion will follow from the work of Robb (1914) and others, once I have shown that a notion of temporal betweenness can be defined in Minkowski space $M$ solely on the basis of knowing which pairs of points in $M$ can be connected by light signals.

For general relativity, I will prove that the *local* temporal structure can be completely described in terms of *local* nondirected light connection. Some kind of localization is to be expected here, in view of the complicated nature of the global geometry that can be encountered in general relativity. Furthermore, the need to have a suitable notion of locality before

*P. Suppes (ed.), Space, Time and Geometry,* 321–365. *All rights reserved*
Copyright © 1972 *by Robert W. Latzer, Stanford*

proceeding will lead to introducing a topology on $M$, whereas the work for special relativity is completely free of topological notions, such as continuity, limit point, etc. Still, the fact that a local notion of temporal betweenness can be obtained on the basis of this rather meager information – a topological structure on $M$ and a knowledge of the point pairs that are light-connected in each locality – certainly suggests that it might be possible to completely pin down the metric structure of $M$, and hence the entire geometry, as can be done for special relativity. However, this is not done here, since the available proofs that temporal order determines the geometry of space-time (Robb, 1914; Zeeman, 1964) depend strongly on the behavior of lines near infinity, and thus do not go over to general relativity in any obvious way.

This work bears directly upon the question of what would be a suitable set of *primitive concepts* for studying relativistic kinematics. Let us restrict our attention for the time being to special relativity. Certainly many concepts underlie what one wants to say in this subject, such as event, particle, observer, light signal, distance, past, future. However, if it can be shown that some of these can be completely described in terms of others, then a firmer base for our theory has been laid. For example, a *particle* can be described as a subset of the events of space-time (namely, its worldline) which satisfies certain properties. If I can state these properties in terms of other concepts which are available to me, then I no longer need to take 'particle' as primitive. If my concept of particle should change (for example, by ideas that particles can or cannot move faster than light, that they may occasionally be reflected in time, or that they need not have continuous existence), then the logical framework on which my theory is based is not attacked, and only the particular description given to the word 'particle' within this framework is different. Or, to give a more extreme example, suppose that someone objects to the use of real numbers to measure distance. Within the usual framework of relativistic kinematics, this sort of question cannot even be discussed, because to do so would involve first throwing out the idea of coordinate systems, on which the whole mathematical framework is based. If only 'event' and 'light connection' are taken as primitive, however, then at least we can approach the problem. After the concepts 'line', 'parallel', and 'between on a line' have been defined in terms of these primitives, one can make statements such as, "Distance (or length) is a property of event pairs one can make. It is an equivalence relation. If $c$ is between $a$ and $b$ on a line, then the length of $ab$ depends only on that of $ac$ and $cb$. Opposite sides of a parallelogram have the same length. All sides of a parallelogram have the same length if both pairs of opposite corners can be light connected." If these statements are agreed to, then a theory of distance can be begun, and if not, then the reasons for disagreement can be discussed.

The most important step toward discussing relativity theory in terms of simple primitives was made by Robb (1914), who presented an axiom system based on the two primitives *element* ( = space-time event) and *after*, where the physical interpretation of '*a after b*' is that a signal can go from $a$ to $b$ with velocity less than or equal to that of light. He carried out the development of this system up to the introduction of coordinates, thereby showing that any system with these two primitives which satisfies his twenty-one postulates must be isomorphic to $R^4$ with the usual geometry of Minkowski space. Later axiom systems have generally started out with much more structure. Carathéodory (1924) developed relativity using only light signals, but he introduced *space points* (his "materielle Punkte") as a new primitive, and assumed that these space points could be imbedded in $R^3$, and that the events *observed at* (another primitive) a given point were totally ordered. A number of other authors (Reichenbach, 1924; Schnell, 1938; Walker, 1959) have also had a primitive notion of space point or of particle. In each case, the events occurring at a point or particle are taken to be totally ordered, and various signaling relations are generally assumed as well. (Walker gives a special role to a certain distinguished particle, the observer, and takes the order of events as primitive only for this one particle.) Basri (1966) used particle chains as one way of defining temporal order, but in the absence of at least a local notion of 'before' and 'after', it is hard to see how the distinction between the appearance and disappearance of a particle, upon which this definition depends, can be maintained. Carnap (1958, Chapter G) described several axiom systems, all of them depending on some form of time order. Part of Robb's basic result was rediscovered from a more modern point of view by Zeeman (1964), who studied the relations $a < b$ (a signal can go from $a$ to $b$ with velocity less than that of light) and $a \lessdot b$ (a signal can go from $a$ to $b$ with velocity equal to that of light). He assumed that space-time had the standard Minkowski space structure, and then showed that any one-to-one map which preserved either of these time-order relations had to be given by an element of the Lorentz group, plus perhaps a translation, a dilation, and a spatial reflection.

Let $M$ stand for the set of all events in space-time. Since we are interested in the geometry of $M$, let us refer to its elements as *points*. This is one of our primitives. The other is the *symmetric light relation* $\lambda$, where $a\lambda b$ means that a light signal can go from $a$ to $b$ (written $a < b$) *or* that a light signal can go from $b$ to $a$ (written $b < a$). There are several reasons for preferring this as a starting point for space-time geometry. One is that the same results are derived on the basis of less initial structure. For example, any of the three binary relations of time order considered above (*after*, $<$, and $\ll$) can easily be defined in terms of any of the others (cf. Zeeman, 1964, p. 491), and the relation $\lambda$ can be defined in terms of them as well, but it is not possible to define any of them in terms of $\lambda$, so that $\lambda$ appears to involve less information. Another reason is that $\lambda$ is clearly imbedded in a very fundamental way in the manner in which space-time is put together. Thus, particles whose velocity is precisely equal to that of light have invariance properties which give them a special role, and lead to that velocity appearing so prominently in relativistic physics. Slower-than-light particles, on the other hand, have no such special role, nor do faster-than-light particles (which are not *logically* excluded by the theory.) Also, $\lambda$ avoids using the time asymmetry, which all of the relations above make use of, but which is completely absent from the local geometry of space-time. And $\lambda$, being a binary relation defined on *pairs* of events, reflects the essentially discrete nature of light transmission. Whereas any particle with a nonzero rest mass can be transformed to zero velocity by a change of coordinates, and thus appears to have a continuous existence, a photon is never at rest, can never be accelerated or decelerated, but passes without change and without other evidence of its existence directly from the point of emission to the point of absorption (see Zeeman, 1967, for further comments on the peculiar nature of light transmission). But perhaps the most compelling reason for wanting to build up space-time geometry from $\lambda$ lies in what it says about the *coherence* of time. Suppose, for the moment, that we look on $M$ as being a completely unstructured set of events. The only information available about it is a list of unordered pairs of events, such that a photon can go from one member of each pair to the other member. Then, on the basis of this information alone, it becomes possible to deduce all of the geometric interrelationships of the elements of $M$, and even to assign a temporal order throughout $M$, given only one arbitrary choice as to the direction of time for the whole space $M$. Thus, a global concept of time can be pieced together from this limited information that is completely independent of time order.

In order to accomplish the objectives stated above for special relativity, it is sufficient to do the following two things: (a) give definitions in terms of $\lambda$ for each of the primitive concepts (except *point*) in some collection of primitives that has been shown to be adequate for the description of Minkowski space, and (b) prove that when our set of points $M$ is Minkowski space and $\lambda$ is the symmetric light-connection relation on $M$, then each of these definitions that we have given in terms of $\lambda$ in fact coincides with the intended interpretation of the corresponding primitive concept in $M$. (We assume that *point*, i.e. element of $M$, will be a primitive in all systems, so we exclude it from specific redefinition.) If these two things are done, then by (a) it becomes possible to translate any definition that may have been given in terms of the old set of primitives into one in terms of $\lambda$. By (b), the new definition will be satisfied by precisely the same things as the old, whenever $M$ is Minkowski space and $\lambda$ is the symmetric light relation on $M$. It also becomes possible in the same way to translate an axiom system in terms of the old primitives into one involving $\lambda$ instead.

In point of fact, we cannot carry out this program in exactly the form outlined above. The reason is obvious: all of the previous sets of primitives have involved the direction of time, and $\lambda$ does not. Indeed, since $\lambda$ is invariant under time reversal, any property defined in terms of $\lambda$ must also be invariant under time reversal. In effect, an arbitrary choice must be made as to the direction of time. Here is one way of proceeding:

(1) We define in terms of $\lambda$, in Section II of this paper, the relation *temporally between*, and show that when $M$ is Minkowski space and $\lambda$ is the symmetric light relation on $M$, then $c$ is temporally between $a$ and $b$ if and only if either $a < c < b$ or $b < c < a$, where $<$, as previously, is the *before* relation that uses slower-than-light signals.

(2) This 'temporally between' relation contains all information concerning the time order of events, except for the arbitrary choice between past and future. We use it to define *temporally parallel*: an ordered pair

$(a, b)$ is temporally parallel to $(c, d)$ if there exists an $x$ such that $a$ is temporally between $x$ and $b$, and $c$ is temporally between $x$ and $d$. Clearly, this will be true if and only if either (i) $a < b$ and $c < d$, or (ii) $a > b$ and $c > d$. We may also use it to define *timelike interval* (which can also be defined directly from $\lambda$): $(a, b)$ is a timelike interval if there exists an $x$ such that $x$ is temporally between $a$ and $b$. For the present discussion, the statement that $(a, b)$ is a timelike interval is abbreviated by $T(a, b)$.

(3) We introduce the notation $a <_{cd} b$ as another way of saying that $(a, b)$ is temporally parallel to $(c, d)$. By fixing $c$ and $d$, this defines a binary relation $<_{cd}$ on $M$. If $c < d$, then $<_{cd}$ is the same as $<$. If $c > d$, then $<_{cd}$ is the same as $>$. Finally, if $T(c, d)$ is false, then $<_{cd}$ is the empty relation (i.e., false for all pairs of points in $M$).

(4) In terms of $<_{cd}$, we define the corresponding binary relations $\leqslant_{cd}$ and $before_{cd}$: $a\ before_{cd}\ b$ means that for all $x$, $x <_{cd} a$ implies $x <_{cd} b$, and $a \leqslant_{cd} b$ means that $a\ before_{cd}\ b$ is true but $a <_{cd} b$ is false. This extends the time order to velocity-of-light intervals. $a\ after_{cd}\ b$ is defined to mean $b\ before_{cd}\ a$.

(5) Let $P(a_1, ..., a_n, <)$ be a propositional function having $<$ as its only primitive relation and $a_1, ..., a_n$ as its free variables. From the remarks in (3), it is clear that

(∗)   $(\exists c)(\exists d)[T(c, d) \wedge P(a_1, ..., a_n, <_{cd})]$

will be true if and only if either $P(a_1, ..., a_n, <)$ or $P(a_1, ..., a_n, >)$ is true. Thus any propositional function $P(a_1, ..., a_n, <)$ whose value is invariant under time reversal can be replaced by the equivalent expression (∗), which has the same free variables and has $\lambda$ as its only primitive relation. The same is of course true if we use $\leqslant$ or *after* instead of $<$. The variables $a_1, ..., a_n$ need not be individual points, but could range over sets of points, or higher types, if desired.

(6) As a consequence of (5), we see that any property which is definable in Robb's system and is invariant under time reversal can be defined in terms of $\lambda$. In particular, elementary geometric concepts such as line, plane, hyperplane, orthogonal, parallel, equal length, between on a line, etc., can be defined in terms of $\lambda$. They can then be used to define *reference frame* (as a kind of function from $M$ to $R^4$), provided the definition is made broad enough to admit an arbitrary choice of origin, unit of length, and set of four mutually orthogonal directions, all

of which choices may differ from reference frame to reference frame. Finally, the resulting definition of reference frame can be used to define any concept which is definable in terms of coordinates (for example, differentiability), provided that this concept is invariant under translations, dilations, and Lorentz transformations (including reflections in space and time), and hence does not depend on the arbitrary choices made in defining the reference frame.

(7) Robb's axioms can be written in the form of a (second-order) sentence **Robb**(*after*), which has no free variables and has *after* as its only primitive relation. Let $(a \lambda b) \equiv F(a, b, after)$ be a definition for the relation $\lambda$ in terms of *after*. Then

(∗) $\quad (\exists c)(\exists d)[T(c, d) \wedge \textbf{Robb}(after_{cd})$
$\quad \quad \wedge (\forall a)(\forall b)((a \lambda b) \equiv F(a, b, after_{cd}))]$

is clearly satisfied by Minkowski space. On the other hand, if $M$ is any set and $\lambda$ is any binary relation on $M$ such that $\{M, \lambda\}$ satisfies this sentence, then for some $c$ and $d$ the relation $after_{cd}$ satisfies Robb's axioms. Hence $M$ is isomorphic to Minkowski space with $after_{cd}$ as its *after* relation. The clause

$$(\forall a)(\forall b)((a \lambda b) \equiv F(a, b, after_{cd}))$$

guarantees that $\lambda$ corresponds to the symmetric light relation of Minkowski space under this isomorphism. Therefore the sentence (∗) constitutes a *categorical* axiom system for Minkowski space.

(8) Let $f: M \to M$ be any permutation of the points of Minkowski space which preserves the relation $\lambda$ (that is, $f(a) \lambda f(b)$ if and only if $a \lambda b$). Then $f$ preserves temporal parallelism, so that it either preserves the order of every timelike interval, or reverses the order of every timelike interval. By the work of Zeeman (1964), it follows that $f$ consists of a Lorentz transformation, plus perhaps a translation, a dilation, and a reflection of space and/or time. Alternatively, we can reach the same conclusion by observing that $f$ must transform reference frames to reference frames (provided that we allow the same arbitrary choices in our definition of 'reference frame' that were used above in (6)). This is because 'reference frame' can be defined in terms of $\lambda$ if these arbitrary choices are allowed, and $f$ preserves $\lambda$. Note that no continuity hypotheses of any kind are required on the transformation $f$.

When we turn to general relativity, our task becomes somewhat more subtle. The fact that we need to work within local regions should not be surprising, because the whole structure of general relativity is defined through local (when not actually infinitesimal) properties. Indeed, it is clear that we cannot avoid this at some stage in the development, because we are trying to determine the temporal structure of space-time, and there exist general relativistic manifolds which do not admit a globally defined time coordinate.

Our approach will be to take as primitive the following things: a topological space $M$, a family $\mathscr{F} = \{U_\alpha\}$ of open sets $U_\alpha$ which covers $M$, and a symmetric binary relation $\lambda_\alpha$ defined on each $U_\alpha$. The intention is that the sets $U_\alpha$ are to be 'local regions' within $M$, and that the relations $\lambda_\alpha$ are to tell when it is possible to connect two points of $U_\alpha$ by a light signal whose path lies entirely within $U_\alpha$. The main difficulty here is in describing how small a 'local region' must be.

Consider a four-dimensional relativistic torus obtained in the following way: take $R^4$ with the usual geometry of special relativity, take a rectangular lattice within $R^4$ such that the four spacing distances between adjacent lattice points for each of the four coordinate directions are independent irrational numbers, and then identify points whose relative location with respect to the surrounding lattice points is the same. This is a compact four-dimensional manifold which is locally isometric to Minkowski space, yet there are light rays whose paths within it are actually dense subsets of the manifold. Light connection on such a manifold would certainly be strange. For example, given any point $a$, one could find a sequence of points converging to $a$ from any desired direction, all of them light connected to $a$ and to each other. What is wrong here? Simply the fact that a light signal can go 'around the universe' many times and still return arbitrarily close to the place and the time of its origin. When we ask whether two nearby points are light connected, this is not the sort of connection we have in mind.

We will give a definition in Section III for 'small open set' within any general relativistic manifold, and the only fact we will need to know concerning the open covering $\mathscr{F}$ will be that it contains a subcovering consisting of small open sets. This would be satisfied, for example, by simply requiring $\mathscr{F}$ to be a basis for the topology of $M$. However, it is easy to see that the amount of information which is given by knowing the $\lambda_\alpha$'s does not really depend on how fine the open covering $\mathscr{F}$ is, as long as the 'local regions' $U_\alpha$ are small enough to avoid the situation described above, where light paths can cross themselves or return arbitrarily close to previous points on the path. The reason is that one could in principle define a 'local light ray' within an open set $U$ to be a connected subset of $U$ such that any two members of this subset are $\lambda_\alpha$-connected for each $U_\alpha$ that contains $U$ (note the essential use here of the topological structure of $M$ in introducing concepts like connectedness). This would make it possible to define a light-connection relation within $U$, provided that $U$ is contained within some $U_\alpha$ in which light paths are well behaved. One would then be able to define a light relation on any open set, by appropriately piecing together the relations just defined for the smaller open sets such as $U$.

Our objective is to describe the local temporal structure of $M$ in terms of the $\lambda_\alpha$'s by somehow giving a local definition of temporal betweenness. The most basic requirement, before we can even give an adequate interpretation of what this should mean, is that we must have some satisfactory notion of 'time' on the underlying manifold, which is valid within each of the local regions we wish to consider. Essentially, this means that it must be possible, given any one of these local regions, to put the metric in it into a form that separates one coordinate (to be called the time coordinate, appearing quadratically with one sign) from the other three coordinates (to be called the space coordinates, appearing quadratically with the opposite sign). This is precisely what is done by a *Gaussian coordinate system*, in which

$$ds^2 = (dx^0)^2 + g_{ij}(x^0, x^1, x^2, x^3)\, dx^i\, dx^j \quad (i,j = 1, 2, 3),$$

where $g_{ij}dx^i dx^j$ is a *negative* definite quadratic form at every point of the coordinate region. Such a coordinate system exists throughout some region about any given point (see Adler *et al.*, 1965, pp. 59–62). On the other hand, the cosmological solution of Gödel (1949) provides an example of a metric in which this type of separation of 'time' from 'space' is not possible over coordinate regions which are too large.

In the case of special relativity, the definition of '$c$ temporally between $a$ and $b$' amounts to the following: each pair of the three elements $a$, $b$, $c$ can be connected by a slower-than-light signal, i.e., each pair lies on a timelike line, and the time coordinate of $c$ is between that of $a$ and $b$ in some (and therefore every) reference frame. Given an open set $U$ in a general relativistic manifold, we use the concept of Gaussian coordinate system (so that a time coordinate is available) to say what we mean by 'locally temporally between' in $U$. Specifically, we require that the following conditions be satisfied if a definition of 'locally temporally between' is to adequately reflect what we mean by this concept: (a) given three points $a$, $b$, and $c$ connected by timelike geodesics in $U$, exactly one of them must be locally temporally between the other two, and (b) whenever a point is locally temporally between two others, it must be the case that its time coordinate is between that of the other two in every Gaussian coordinate system for $U$. (In situations where the time coordinate is between that of the other two in every Gaussian coordinate system, but there are not connecting timelike geodesics that lie inside of $U$, we make no stipulation as to whether or not temporal betweenness is to be defined.) In general, we cannot expect there to be such a definition for an arbitrary open set $U$, but in Section III we will be able to define in terms of the given primitives (a) a class of *admissible* open sets, which will include all open sets that are small enough to admit a Gaussian coordinate system, and (b) a definition of temporal betweenness which satisfies the requirements above on each of the admissible open sets. These definitions will be made in terms of the collection of possible trajectories that a slower-than-light particle can follow (such trajectories can be defined in terms of the given primitives). Therefore they will have a fairly clear meaning even if the admissible set should be too large to admit a Gaussian coordinate system, though it is not clear whether that can ever happen for an admissible set.

## II. SPECIAL RELATIVITY

This part describes the method for defining 'temporal betweenness' within the set $M$ of points of the space-time continuum of special relativity, using only the primitive relation $\lambda$. Certain other useful relations in $M$ will also be defined in terms of $\lambda$. The starting point is the predicate $Q(abc)$:

DEFINITION 2.1. *If $a, b, c \in M$, we write* **Q(abc)** *to mean that*
 (i) *$a, b$, and $c$ are distinct;*
 (ii) *$a\lambda b$, $a\lambda c$, and $b\lambda c$ are all false; and*

(iii) *there is no point d which satisfies all of a$\lambda$d, b$\lambda$d, and c$\lambda$d.*

The main idea here is condition (iii): the light cones of $a$, $b$, and $c$ do not intersect simultaneously. We mean to use this as a test for collinearity. In fact, we shall prove that (iii) holds if and only if $a$, $b$, and $c$ are collinear, *provided that the mutual separations of a, b, and c are all spacelike*. This leads us to a first approximation for the concept of line:

DEFINITION 2.2. *A* **path** *is a set $P$ such that* **Q(abc)** *holds for all distinct triples $a$, $b$, $c \in P$.*

However, when pairs of points whose separation is timelike appear in the path $P$, then configurations other than a collinear set are possible. In order to exclude this possibility, we are led to our second approximation to the concept of line:

DEFINITION 2.3. *A* **spacepath** *is a path $S$ satisfying*

(i) *$S$ has at least four elements; and*

(ii) *if $T$ is a path which consists of four points, three of them also belonging to $S$, then $S \cup T$ is a path.*

By adding the requirement that any three points are enough to constrain a fourth also to lie on the path, we have by the above definition excluded the extraneous configurations. Indeed, a spacepath turns out to be precisely a collinear set which has at least four elements and is oriented in a spacelike direction. An alternate method for building up spacetime geometry would then be the following: from the definition above one gets lines which are oriented in a spacelike direction, and from them it is possible to generate arbitrary planes and hyperplanes. By intersecting the planes, one can define arbitrary lines. Using figures which involve parallel lines, one can next define betweenness on a line and equality of intervals on parallel lines. Finally, since the collection of lightlike lines is known within the collection of all lines, figures can be drawn which lead to a definition of equality of the Lorentz length for line segments, even in nonparallel directions. This introduces the metric structure and makes it possible to describe the geometry completely.

DEFINITION 2.4. *A pair of distinct points $a$, $b$ is called*

(i) **spacelike** *if $a$, $b \in S$ for some spacepath $S$ that consists of four elements;*

(*ii*) **lightlike** *if $a\lambda b$; and*
(*iii*) **timelike** *if it is neither spacelike nor lightlike.*

(The restriction in (i) that $S$ have only four elements does not affect the meaning of the definition, and it is only included so that the definition can be translated in a straightforward way into an expression involving $\lambda$ in the *first-order* predicate calculus, if that is desired.)

DEFINITION 2.5. *A point $c$ is **temporally between** $a$ and $b$ if*
(*i*) *the three pairs $ab$, $ac$, and $bc$ are all timelike; and*
(*ii*) *given any point $d$ such that $cd$ is timelike, then at least one of the pairs $ad$ and $bd$ is timelike.*

Our task is to show that Definition 2.5 has the intended meaning when $M$ is Minkowski space, the four-dimensional space-time continuum of special relativity, and $\lambda$ is the symmetric light-connection relation on $M$.

We distinguish $M$ from $R^4$, the real four-dimensional Cartesian space. There is a family of functions from $M$ to $R^4$ known as 'reference frames', such that if any one of them is given, then the complete family is obtained by following that given function by the various members of the Lorentz group.

Let $K((x_0, x_1, x_2, x_3), (y_0, y_1, y_2, y_3)) = (x_0 - y_0)^2 - \sum_{i=1}^{3}(x_i - y_i)^2$ be the characteristic 'skew metric' defined on $R^4 \times R^4$. Since it is not changed by elements of the Lorentz group, $K$ is defined on $M \times M$.

In terms of $K$, the given binary relation $\lambda$ is such that $a\lambda b$ holds if and only if $a \neq b$ and $K(a, b) = 0$. We define additional relations $\sigma, \tau, <$, and $\lessdot$ on $M$: if $a, b \in M$, and if $a_0$ and $b_0$ are the respective time coordinates of $a$ and $b$ in some reference frame, then

$a\sigma b$ means that $K(a, b) < 0$;
$a\tau b$ means that $K(a, b) > 0$;
$a < b$ (or $b > a$) means that $a_0 < b_0$ and $a\tau b$;
$a \lessdot b$ (or $b \gtrdot a$) means that $a_0 < b_0$ and $a\lambda b$.

When $a\sigma b$, $a\lambda b$, or $a\tau b$, then we refer to the line $ab$ as being *spacelike*, *lightlike*, or *timelike*, respectively. These words will not be used in the sense of Definition 2.4, until we are ready to show that Definition 2.4 in fact agrees with this terminology.

We note that the meaning of $<$ and $\lessdot$ is independent of the particular coordinate system used in the definition (except that they are reversed when the sign of the time coordinate is reversed). The following elementary lemmas and corollaries summarize some of the properties of these relations that we shall find useful. The first of them expresses the transitivity of time order, but the strict conclusion ($a < c$, never $a \ll c$, when the points are noncollinear) has other uses as well:

LEMMA 2.6. *Let $a, b, c$ be noncollinear points, and suppose that $a < b$ or $a \ll b$, and that $b < c$ or $b \ll c$. Then $a < c$.*

*Proof*: In some reference frame, let $a_0, b_0, c_0$ be the respective time coordinates of $a, b, c$. Introducing the *Euclidean* metric $d(x, y)$ in that reference frame, we see that the hypotheses are that $a, b, c$ are noncollinear and satisfy $\sqrt{2}(b_0 - a_0) \geq d(a, b)$ and $\sqrt{2}(c_0 - b_0) \geq d(b, c)$, where the $\sqrt{2}$ is due to the fact that time appears with *reversed* sign in the Euclidean metric (rather than just being absent). Adding these inequalities and using the strict triangle inequality for noncollinear points, we have $\sqrt{2}(c_0 - a_0) > d(a, c)$, the required conclusion. ∎

COROLLARY 2.7. *If $a \lambda b$, $a \lambda c$, and $b \lambda c$, then $a, b$, and $c$ are collinear.*

*Proof*: If they were noncollinear, then one of the pairs would have to be related by the $<$ relation (rather than $\ll$), by Lemma 2.6. ∎

COROLLARY 2.8. *Let $a, b, c$ be noncollinear points. If $a \ll b$ and $c \ll b$, or if $b \ll a$ and $b \ll c$, then $a \sigma c$.*

*Proof*: Any of the other possibilities ($a < c$, $a \ll c$, $c < a$, or $c \ll a$), if combined with the hypotheses $a \ll b$ and $c \ll b$, would contradict Lemma 2.6. The same is true if they are combined with the hypotheses $b \ll a$ and $b \ll c$, which differ from the preceding ones only in the direction of the relations $<$ and $\ll$. ∎

LEMMA 2.9. *If $a < b$, $a < c$, and $b \sigma c$, then there is a point $d$ satisfying $a < d$, $d \ll b$, and $d \ll c$. Similarly, if $a > b$, $a > c$, and $b \sigma c$, then there is a point $d$ satisfying $a > d$, $d \gg b$, and $d \gg c$.*

*Proof*: Suppose that $a < b$, $a < c$, and $b \sigma c$. Then there is a point $e$ such that $be$ is lightlike and $a < e < c$, because if we take $e$ on the line segment $ac$ and move it from $a$ to $c$, $be$ must change from timelike to spacelike at some point, and it will be lightlike there. Note that $e \ll b$, because $b \ll e$ would imply $b < c$. Similarly, by moving a point $d$ from $e$ to $b$, we get a position where $cd$ is lightlike and $e \ll d \ll b$. Here $a < d$

by transitivity, and $d \ll c$ because $c \ll d$ would imply $c < b$, so this is the required point.

The case in which $a > b$, $a > c$, and $b\sigma c$ is obtained from the preceding case by reversing the direction of the relations $<$ and $\ll$. ∎

Given any point $a \in M$, we shall need the following cones:

$$S_a = \{x: x\sigma a\},$$
$$L_a = \{x: x\lambda a\} \cup \{a\},$$
$$T_a = \{x: x\tau a\},$$
$$LA_a = \{x: x \gtrdot a\} \cup \{a\},$$
$$LB_a = \{x: x \lessdot a\} \cup \{a\},$$
$$TA_a = \{x: x > a\},$$
$$TB_a = \{x: x < a\}.$$

These are, respectively, the space, light, time, light-after, light-before, time-after, and time-before cones. If we give $M$ the usual topology from $R^4$, then $S_a$, $T_a$, $TA_a$, $TB_a$ are open sets, and $L_a$, $L_a$, $LA_a$, $LB_a$ are their respective boundaries. If $a < b$, then $TA_a \cap TB_b$ is a nonempty *bounded* open set.

Take three points $a$, $b$, $x$, and let their representations in some reference frame be $(a_0, a_1, a_2, a_3)$, $(b_0, b_1, b_2, b_3)$, and $(x_0, x_1, x_2, x_3)$. If $x \in L_a \cap L_b$, then

$$(x_0 - a_0)^2 - \sum_{i=1}^{3} (x_i - a_i)^2 = 0$$

and

$$(x_0 - b_0)^2 - \sum_{i=1}^{3} (x_i - b_i)^2 = 0,$$

which implies

$$(a_0 - b_0)\left(x_0 - \frac{a_0 + b_0}{2}\right) - \sum_{i=1}^{3} (a_i - b_i)\left(x_i - \frac{a_i + b_i}{2}\right) = 0.$$

Thus $L_a \cap L_b$ lies in a hyperplane which is orthogonal to the line $ab$ and intersects this line at the midpoint between $a$ and $b$, provided that orthogonality is interpreted with respect to the nondegenerate bilinear form $x_0 y_0 - x_1 y_1 - x_2 y_2 - x_3 y_3$. Within that hyperplane, the intersection is a quadric surface, since its equation is of the second degree.

The following lemma may seem geometrically obvious, but we prove it anyway, partly because it will be useful in a somewhat different context in Section III:

LEMMA 2.10. *If $a < b$, then $L_a \cap L_b = LA_a \cap LB_b$. On the other hand, if $a\sigma b$, then $L_a \cap L_b = (LB_a \cap LB_b) \cup (LA_a \cap LA_b)$, where the union is disjoint and the two parts are topologically separated from one another.*

*Proof*: Since $L_a = LB_a \cup LA_a$ and $L_b = LB_b \cup LA_b$, we have $L_a \cap L_b = (LB_a \cap LB_b) \cup (LB_a \cap LA_b) \cup (LA_a \cap LB_b) \cup (LA_a \cap LA_b)$. Furthermore, the union is disjoint, because neither $a$ nor $b$ is in $L_a \cap L_b$, and any other point belongs to at most one of the sets $LB_a$ and $LA_a$, and to at most one of the sets $LB_b$ and $LA_b$. If $c$ is an arbitrary point in $L_a \cap L_b$, observe that $c \in LA_a \cap LB_b$ implies $a < b$ or $a \lessdot b$ by transitivity, that $c \in LB_a \cap LA_b$ implies $a > b$ or $b \lessdot a$ for the same reason, and that $c \in LB_a \cap LB_b$ or $c \in LA_a \cap LA_b$ implies that $a\sigma b$ by Corollary 8. Hence the only parts of the union that can be nonempty are $LA_a \cap LB_b$ in the case $a < b$, and $LB_a \cap LB_b$ and $LA_a \cup LA_b$ in the case $a\sigma b$. Finally, the fact that $LB_a \cap LB_b$ and $LA_a \cap LA_b$ are topologically separated in the case $a\sigma b$ follows from the fact that they are both closed and are disjoint, because it is then trivial that neither one can intersect the closure of the other. ∎

It is straightforward to actually calculate what the quadric surface $L_a \cap L_b$ is, but we note that this is already implied by information at hand. For, when $a\sigma b$, then $LB_a \cap LB_b$ and $LA_a \cap LA_b$ are both non-empty (proved by taking a point very far in the past and a point very far in the future, and applying Lemma 2.9). But a disconnected quadric surface is a hyperboloid of two sheets. Furthermore, because the disconnection has been explicitly exhibited in Lemma 2.10, we know that $LB_a \cap LB_b$ and $LA_a \cap LA_b$ are the two sheets. On the other hand, when $a < b$, then $L_a \cap L_b = LA_a \cap LB_b$ lies on the boundary of the bounded open set $TA_a \cap TB_b$, and can be seen to be nonempty in a variety of ways. But a bounded quadric surface is an ellipsoid.

We are now ready to examine the predicate $Q(abc)$. We take first the case in which $ab$, $ac$, and $bc$ are all spacelike:

PROPOSITION 2.11. *Let $a$, $b$, $c$ be three distinct points such that $ab$, $ac$, and $bc$ are all spacelike. Then $Q(abc)$ is true if and only if $a$, $b$, and $c$ are collinear.*

*Proof*: Suppose first that $a$, $b$, and $c$ are collinear. Then $L_a \cap L_b$ and $L_a \cap L_c$ are contained in hyperplanes which are orthogonal to the line of $a$, $b$, and $c$ and which cut this line at different points. Consequently, these hyperplanes are parallel, and $L_a \cap L_b$ and $L_a \cap L_c$ cannot meet. Therefore $L_a \cap L_b \cap L_c$ is empty, and $Q(abc)$ holds.

Conversely, suppose that $a$, $b$, and $c$ are not collinear. We must show that $Q(abc)$ is false, by finding a point $d$ such that $a\lambda d$, $b\lambda d$, and $c\lambda d$ hold. We choose a reference frame in which $a$ and $b$ are simultaneous, which is possible because $ab$ is spacelike, and we let $A$, $B$, $C$ be the respective space positions and $t_A$, $t_B$, $t_C$ the respective times of $a$, $b$, $c$ in this reference frame. We shall restrict ourselves to the plane in which $A$, $B$, and $C$ lie. Let $h$ be the perpendicular bisector of $AB$ in this plane. By relabeling $A$ and $B$ if necessary, we may assume that $A$ and $C$ are on the same side of $h$ (unless $C$ lies on $h$). The plan is to take a point $D$ on $h$, and send light signals toward $D$ from $A$ at time $t_A$, from $B$ at time $t_B = t_A$, and from $C$ at time $t_C$. The signals from $A$ and $B$ will arrive simultaneously at some time $t_D$. If there are some points $D$ on $h$ for which the signal from $C$ is earlier than this, and other points for which it is later, then by the intermediate value theorem there will be a point at which it arrives at exactly the same time, and the resulting pair $(t_D, D)$ is the required space-time point $d$.

*Case I.* $C$ is closer to $h$ than $A$ is. By reversing the direction of time if necessary, we assume that $t_C$ is later than $t_A$. If $D$ is a distant point of $h$ on the side that is closer to $A$ than to $C$ (or if $D$ is a distant point of $h$ in either direction when $AC \perp h$), then the signal from $C$ will be late. On the other hand, if $D$ is at the intersection of $AC$ with $h$, then the signal from $C$ will be early. (Note the use here of the fact that $ac$ is spacelike.)

*Case II.* $C$ is farther from $h$ than $A$ is. This time, we assume that $t_C$ is earlier than $t_A$. If $D$ is a distant point on $h$ on the side that is closer to $C$ than to $A$ (or either side if $AC \perp h$), then the signal from $C$ will be early, but if $D$ is at the intersection of $AC$ with $h$, then the signal from $C$ will be late.

*Case III.* $C$ and $A$ are the same distance from $h$. If $D$ is a distant point of $h$ on the side that is closer to $A$ than to $C$, then the signal from $C$ will be late, but if $D$ is a distant point of $h$ in the opposite direction, then the signal from $C$ will be early. The time difference between $t_A$ and $t_C$ is not enough to make any difference, because $ac$ is spacelike. ∎

Next, we examine the cases in which not all of $ab$, $ac$, and $bc$ are spacelike. Since $Q(abc)$ is trivially false if any of them are lightlike, there are only three cases to be examined:

PROPOSITION 2.12. *Let $a$, $b$, $c$ be three distinct points. Suppose that each of $ab$, $ac$, and $bc$ is either spacelike or timelike. Then*:
 (i) *if exactly one of them is timelike, $Q(abc)$ is true;*
 (ii) *if exactly two of them are timelike, $Q(abc)$ is false;*
 (iii) *if all three of them are timelike, $Q(abc)$ is true.*

*Proof*: (i) Without loss of generality, we may assume $a < b$, $a\sigma c$, $b\sigma c$. If $d$ satisfied $a\lambda d$, $b\lambda d$, and $c\lambda d$, then we would have $a \lessdot d$ and $d \lessdot b$, because all other cases would either imply $a > b$ or $a\sigma b$. We must have either $c \lessdot d$ or $d \lessdot c$. But $c \lessdot d \lessdot b$ contradicts $b\sigma c$, and $a \lessdot d \lessdot c$ contradicts $a\sigma c$, so no such point $d$ can exist, and $Q(abc)$ is true.

(ii) We may assume $a < b$, $a < c$, $b\sigma c$, because $a < b$, $a > c$, $b\sigma c$ would contradict the transitivity of $<$. By Lemma 2.9, there is a point $h$ such that $a < h$, $h \lessdot b$, and $h \lessdot c$. Take any point $g$ such that $b < g$. Then $h \in TA_a \cap TB_g$, which is *bounded*, and $h \in LB_b \cap LB_c$, which is a sheet of a hyperboloid, and therefore *connected* and *unbounded*. It follows that there is some point $d \in LB_b \cap LB_c$ which lies on the boundary of $TA_a \cap TB_g$. Since $d$ is on the boundary, we must have $d \in LA_a$ or $d \in LB_g$. But $d \lessdot g$ and $d \lessdot b$ would contradict $b < g$, so it must be the case that $d \in LA_a$. Hence $d \in L_a \cap L_b \cap L_c$, and $Q(abc)$ is false.

(iii) In this case, $<$ totally orders the points $a$, $b$, $c$, and we may assume $a < b < c$. If $d$ satisfied $a\lambda d$, $b\lambda d$, and $c\lambda d$, then we would have $a \lessdot d \lessdot b$ and $b \lessdot d \lessdot c$. But $d \lessdot b$ and $b \lessdot d$ is a contradiction, so $d$ cannot exist and $Q(abc)$ is true. ∎

The next two propositions concern paths:

PROPOSITION 2.13. *On a path, ($\tau$ or $=$) is an equivalence relation.*

*Proof*: Let $P$ be a path. Reflexivity and symmetry are trivial, and transitivity is trivial unless the three points involved are distinct. By Definition 2.1, no two points of $P$ can be related by $\lambda$, so the only way that transitivity could fail is to have $a\tau b$, $b\tau c$, and $a\sigma c$. But this possibility is excluded by Proposition 2.12. ∎

PROPOSITION 2.14. *Let $P$ be a path. If there are more than two equivalence* classes in $P$ with respect to ($\tau$ or $=$), then all the equivalence classes are singletons and $P$ is a subset of a spacelike line. Conversely, any subset of a spacelike line is a path.

*Proof*: Suppose, on the contrary, that there are more than two equivalence classes, but that not all of them are singletons. We can then find $a$ and $b$ in the same equivalence class, and $c$ and $d$ each belonging to different equivalence classes. Since elements of the same class are related by $\tau$ and elements of different classes are related by $\sigma$, we have $a\tau b$, $a\sigma c$, $a\sigma d$, $b\sigma c$, $b\sigma d$, and $c\sigma d$. Proposition 2.11 then implies that $a$, $c$, and $d$ are collinear and that $b$, $c$, and $d$ are collinear. Thus $a$ and $b$ both lie on the spacelike line $cd$, which contradicts $a\tau b$. Thus all the equivalence classes of $P$ with respect to ($\tau$ or $=$) are singletons. By Proposition 2.11, it then follows that any three distinct points of $P$ are collinear (on a spacelike line), which establishes that $P$ is a subset of a spacelike line.

The converse, that any subset of a spacelike line is a path, follows directly from Proposition 2.11. ∎

We conclude by showing that Definitions 2.3, 2.4, and 2.5 yield exactly what was claimed above:

PROPOSITION 2.15. *A set $S$ is a spacepath if and only if it has at least four points and is a subset of a spacelike line.*

*Proof*: Let $S$ be a spacepath. Then $S$ has at least four points. By Proposition 2.14, it suffices to show that $S$ has at least three equivalence classes with respect to ($\tau$ or $=$). Thus suppose, on the contrary, that it had only one or two equivalence classes. We choose four elements $a$, $b$, $c$, $d$ from $S$, taking as many as possible (in the order just written) from one equivalence class, and taking the remaining elements (if any) from the other equivalence class. Then either $a\tau b$, $a\tau c$, $a\tau d$, $b\tau c$, $b\tau d$, and $c\tau d$ (all four in the same class), or $a\tau b$, $a\tau c$, $a\sigma d$, $b\tau c$, $b\sigma d$, $c\sigma d$ (three and one decomposition), or $a\tau b$, $a\sigma c$, $a\sigma d$, $b\sigma c$, $b\sigma d$, $c\tau d$ (two and two decomposition). Now $a$ belongs to $T_b$, and to either $T_c$ or $S_c$, and to either $T_d$ or $S_d$. Since each of these sets is open, there will be a neighborhood $U$ of $a$ that lies entirely within each of the sets above that $a$ belongs to. We choose $a' \in U$ in a spacelike direction from $a$ (thus $a\sigma a'$). Then $a'$ satisfies the same $\sigma$ and $\tau$ relations with $b$, $c$, and $d$ as $a$ does. Because no three of these four elements have mutually spacelike separations, it is easy to verify from Proposition 2.12 that $\{a', b, c, d\}$ is a path. By Definition 2.3, it follows that $S \cup \{a', b, c, d\} = S \cup \{a'\}$ is a path. But this is impossible by Proposition 2.12, since $a\sigma a'$, $a\tau b$, and $a'\tau b$.

Conversely, suppose that $S$ is a subset of a spacelike line and has at least four points. Let $T$ be any path which consists of four points, three of them (call them $a$, $b$, and $c$) in common with $S$. Since $a, b, c \in S$, we have $a\sigma b$, $a\sigma c$, $b\sigma c$. Since they are also in $T$, we see that $T$ has at least three equivalence classes with respect to ($\tau$ or $=$), and is therefore a subset of a spacelike line. Because of the three points in common, the two spacelike lines are the same, which shows that $S \cup T$ is a path. It follows that $S$ is a spacepath. ■

COROLLARY 2.16. *In Definition 2.4, spacelike means spacelike, timelike means timelike, and lightlike means lightlike.*

PROPOSITION 2.17. *A point $c$ is temporally between $a$ and $b$ (in the sense of Definition 2.5) if and only if either $a < c < b$ or $b < c < a$.*

*Proof*: Suppose first that $c$ is temporally between $a$ and $b$. Then $ab$, $ac$, and $bc$ are all timelike, so $a$, $b$, and $c$ are totally ordered by $<$. We must show that $a < b < c$, $b < a < c$, $c < a < b$, and $c < b < a$ cannot occur. By switching $a$ with $b$ and/or reversing the direction of time, it suffices to consider $a < b < c$. As remarked previously, $L_a \cap L_b$ is nonempty when $a < b$ (in fact, it is an ellipsoid). Take $d \in L_a \cap L_b$. Then $a < b < c$ implies $a \lessdot d \lessdot b < c$. Hence $a\lambda d$, $b\lambda d$, and $c\tau d$, which contradicts the second part of Definition 2.5.

Conversely, suppose that $a < c < b$ (the same reasoning with order reversed will apply to $b < c < a$). Then all three pairs are timelike. Take any $d$ such that $cd$ is timelike. If $c < d$, then $a < c < d$ shows that $ad$ is timelike. On the other hand, if $d < c$, then $d < c < b$ shows that $bd$ is timelike. Therefore the definition is satisfied. ■

### III. EXTENSION TO GENERAL RELATIVITY

Our primitives now consist of a topological space $M$, a family $\mathscr{F} = \{U_\alpha\}$ of open sets $U_\alpha$ which covers $M$, and a symmetric binary relation $\lambda_\alpha$ defined on each open set $U_\alpha$ in $\mathscr{F}$. We shall mimic what was done in the case of special relativity insofar as possible, but many significant changes are necessary. At any rate, our starting point is the same:

DEFINITION 3.1. *If $a$, $b$, $c \in U_\alpha$, we write $\mathbf{Q}_\alpha(\mathbf{abc})$ to mean that*
(i) *$a$, $b$, and $c$ are distinct;*
(ii) *$a\lambda_\alpha b$, $a\lambda_\alpha c$, and $b\lambda_\alpha c$ are all false; and*
(iii) *there is no point $d$ which satisfies all of $a\lambda_\alpha d$, $b\lambda_\alpha d$, and $c\lambda_\alpha d$.*

However, $Q_\alpha(abc)$ will not give us any effective information about collinearity. This is because the previous test for collinearity depended upon the existence or nonexistence of certain distant points of intersection, and we can no longer precisely describe behavior at a distance. Nevertheless, we proceed with Definition 3.2, to which a topological condition has been added:

DEFINITION 3.2. *An **α-path** is a set $P \subset U_\alpha$ such that*
(i) *$P$ is connected and contains at least two points; and*
(ii) *$Q_\alpha(abc)$ holds for all distinct triples $a$, $b$, $c \in P$.*

Because of the changes in the behavior of $Q_\alpha$, it is no longer possible to get a concept of 'spacepath' by the method used before. Instead, we take advantage of the topology to pick out paths which are oriented in a timelike direction:

DEFINITION 3.3a. *An **α-timepath** is an α-path $P$ such that if we are given a point $a \in P$, an open set $U_\beta \in \mathscr{F}$, and a β-path $R$ satisfying $P \cap R = \{a\}$, then there is a point $b \in R \cap U_\alpha$ and a neighborhood $V \subset U_\alpha$ of $a$ for which $Q_\alpha(abc)$ holds for all $c \in P \cap V - \{a\}$.*

DEFINITION 3.3b. *A **local timepath** is a set that is an α-timepath for some $U_\alpha \in \mathscr{F}$.*

Local timepaths may have to be very small, because each one has to lie within some $U_\alpha$. The following definition eliminates the dependence on a particular $U_\alpha$:

DEFINITION 3.4. *A **timepath** is a nonempty connected set which has an open covering (in its relative topology) by local timepaths.*

Unlike the spacepaths of Section II, these timepaths do not turn out to be straight lines, though they do, at least, turn out to be continuous paths. In fact, they are essentially the paths in space-time that a slower-than-light particle can follow (subject to certain restrictions, however. For example, the path may not cross itself, although in some manifolds it can be circular. Thus a person whose world line is a timepath would never return to a place and time of his own past, unless he were to repeat his past forever afterward.)

The definitions that we seek are the following:

DEFINITION 3.5. *Given an open set U, we say that a point c is* **U-locally temporally between** *a and b if there exist P and R such that*
(i)  *P, R, and P ∪ R are timepaths contained in U;*
(ii) *P contains c and a but not b; and*
(iii) *R contains c and b but not a.*

DEFINITION 3.6. *Given an open set U, we say that two points of U are* **U-locally time connected** *if they both belong to a timepath that is contained within U.*

DEFINITION 3.7. *An open set U is called* **admissible** *if whenever we are given three distinct points, any two of which are U-locally time connected, then exactly one of them is U-locally temporally between the other two.*

No claim is made that the definitions above are the most efficient possible from a mathematical point of view. A number of pages of proof could be saved, for example, if we required in the definitions that 'paths' and 'timepaths' be paths in the usual sense (that is, continuous images of the real line), rather than simply saying that they are topologically connected and then having to prove that a parametric representation is possible. In a way, though, this would be begging the question, because we are trying to give definitions that are as simple as possible from a logical point of view, and the inclusion of the real line within the set of primitive concepts would certainly defeat this purpose. Even our requirement that they be connected seems somewhat unsatisfactory. However, although connectedness is used in our proofs in several other ways, we make no really essential use of the connectedness of paths and timepaths until we get to Proposition 3.29, where the parameterization of timepaths is finally established. Thus a significant portion of what we do could be modified to avoid making this requirement in the definitions.

We now specialize to the situation that we have in mind. First we need the following pair of definitions, which are suggested in part by the discussion at the end of Section I:

DEFINITION 3.8a. *In a four-dimensional manifold M provided with a Riemannian metric of signature* $(1, -1, -1, -1)$, *an open set U will be called* **nice** *if*

(i) *U is geodetically convex*: *given two points in U, there is at least one geodesic that connects them within U;*

(ii) *U has unique geodesics*: *given two points in U, there is at most one geodesic that connects them within U; and*

(iii) *there is a Gaussian coordinate system, with metric* $ds^2 = (dx^0)^2 + g_{ij}dx^i dx^j$ $(i, j = 1, 2, 3)$, *which is valid at every point of U* (see Adler et al., 1965, pp. 59–62).

DEFINITION 3.8b. *An open set will be called* **small** *if it is a subset of a nice open set.*

Such a manifold $M$ can always be covered by nice open sets (see Adler *et al.*, 1965, Chapter 2 and Helgason, 1962, Chapter 1). Actually, we use the slightly stronger fact that the nice open sets form a basis for $M$, which follows immediately from this (because any open set in $M$ can be covered by nice open sets by regarding it as a manifold in its own right).

The only assumptions that we need to make regarding the given primitives are:

(i)   $M$ is a four-dimensional real manifold provided with a Riemannian metric of signature $(1, -1, -1, -1)$.

(ii)  The open covering $\mathscr{F} = \{U_\alpha\}$ is such that each point of $M$ belongs to some $U_\alpha$ which is 'small' in the sense of the definition above.

(iii) The relation $\lambda_\alpha$ on each $U_\alpha$ holds for a pair of points $a, b \in U_\alpha$ if and only if there is a segment of a lightlike geodesic (= null-geodesic) which passes through $a$ and $b$ and lies entirely within $U_\alpha$.

Our terminology regarding $M$, such as *past* or *future* direction with respect to a coordinate system, or *lightlike* geodesic (rather than null-geodesic), or *time-order relation*, is in accordance with the idea that $M$ is a space-time continuum. Formally, however, $M$ is just a manifold, and we do not make any physical assumptions about it, nor do we make use of the Einstein field equations nor any of the simplifying assumptions often associated with their solution. The fact that our definitions do what has been claimed will be demonstrated by the following theorems. First, regarding timepaths, we will prove:

THEOREM 3.9. *Let P be a timepath and U an open set with a Gaussian coordinate system. Then any component of $P \cap U$, expressed in this coordinate system, is a set of the form* $\{(x, f(x)) : x \in I\}$, *where I is an interval of real numbers (open, half-open, or closed; finite or infinite) and f is a continuous function from I to $R^3$*.

THEOREM 3.10. *Let P be a set which is connected and has at least two points. Then P is a timepath if and only if, given a point $a \in P$ and a neighborhood V of a, there is an open set W with $a \in W \subset V$ such that any two points of $P \cap W$ can be connected by a segment of a timelike geodesic which is contained within W*.

These two theorems establish the continuous and locally timelike nature of timepaths. From them we will obtain:

THEOREM 3.11. *An open set U which admits a Gaussian coordinate system is admissible.*

THEOREM 3.12. (a) *Suppose that U is an admissible open set. Then a and b are U-locally time connected if they are connected by a segment of a timelike geodesic within U.*

(b) *Suppose that U is a nice open set. Then a and b are U-locally time connected if and only if they are connected by a segment of a timelike geodesic within U.*

It should be noted here that being U-locally time connected is in general not the same as being connected by a timelike geodesic within U. Basically, the reason is that if the open set U is small but has holes in it, then a timelike geodesic from a to b might pass over one of these holes and therefore go outside of U, whereas a timepath can curve around the hole and therefore remain in U. These differences do not occur for nice open sets, because such sets are geodetically convex.

THEOREM 3.13. *In any open set U with a Gaussian coordinate system, a point c is U-locally temporally between a and b if and only if each two of*

$a, b, c$ are *U-locally time connected and the time coordinate of $c$ is between that of $a$ and $b$.*

This final theorem, combined with the preceding results, demonstrates that Definition 3.7 correctly defines local temporal betweenness, as discussed in Section I. The remainder of Section III will be devoted to proving Theorems 3.9 through 3.13.

Almost everything we need to do can be done by topological methods, and our use of differential geometry will be minimal. We make the convention that a neighborhood of a point is always open (rather than just containing the point in question as an interior point), and that a basis always consists of open sets. Also, we speak of a *segment* of a geodesic (and of *closed* geodesic segments, *open* geodesic segments, etc.) in analogy with *interval* on a line. For basic topological terminology (closed, connected, compact, boundary, component, open map, etc.), refer to Hocking and Young (1961) or Kelley (1955).

Three special types of neighborhood will be useful: *nice open sets* (as defined above), *diamonds*, and *wedges*. The main advantage of the nice open sets is that they allow us to reintroduce the notation used in Section II. Given a fixed nice open set $U$, and some particular Gaussian coordinate system defined on it, we now describe how this is done, and summarize the needed topological information concerning $U$. All definitions and properties (closed set, geodesic, etc.) are to be relativized to $U$.

Given two distinct points $a, b \in U$, there is exactly one geodesic which passes through them, which we denote by $ab$. If $a, b, c$ are three points that lie on the same geodesic, we say that $a, b, c$ are *collinear*. The geodesics of $U$ can be divided into three disjoint classes – those which are respectively timelike, lightlike, or spacelike at each of their points – and we write $a\tau b$, $a\lambda b$, or $a\sigma b$ according to the respective nature of the geodesic $ab$. If $a = (a^0, a^1, a^2, a^3)$ and $b = (b^0, b^1, b^2, b^3)$ in the specified Gaussian coordinate system, we write $a < b$ (or $b > a$) to mean that $a\tau b$ and $a_0 < b_0$, and we write $a \ll b$ (or $b \gg a$) to mean that $a\lambda b$ and $a_0 < b_0$. If $a\sigma b$, we do not define the order. From the special properties of the Gaussian coordinate system, it is easy to see that $a_0 = b_0$ cannot happen when $a\tau b$ or $a\lambda b$, so that $<$ and $\ll$ are total order relations when restricted respectively to a single timelike or single lightlike geodesic. In addition, these total order relations give rise to the same topology on a geodesic that the geodesic inherits as a subset of $U$. We use the various relations on $U$ to define the cones $S_a$, $L_a$, $T_a$, $LA_a$, $LB_a$, $TA_a$, and $TB_a$, as in Section II.

Given any point $a \in U$, let $M_a$ be the space of tangent vectors at $a$, and let $M'_a$ be the collection of all geodesics through $a$. Give $M'_a$ the quotient topology, and let $\theta: U - \{a\} \to M'_a$ be the projection map which takes any point $b \in U - \{a\}$ to the unique geodesic $ab$ containing it. Then the map $\theta$ is continuous and open. It is clear that $M'_a$ may also be obtained by identifying the tangent vectors in $M_a$ which differ only by a nonzero factor. Thus $M'_a$ is homeomorphic to projective 3-space $P^3$. Since each geodesic through $a$ lies entirely within one of the cones $S_a$, $L_a$, or $T_a$, the respective images $S'_a$, $L'_a$, or $T'_a$ of these cones in $M'_a$ form a decomposition of $M'_a$ into three disjoint parts. What these parts are can easily be calculated in the tangent space $M_a$ using the metric at the point $a$, but it really depends only on the signature $(1, -1, -1, -1)$ of the metric. Thus $S'_a$ and $T'_a$ are open subsets of $M'_a$, and $L'_a$ is a closed subset which forms the boundary of each of them. Further, we shall find use for the fact that $L'_a$ is homeomorphic to the 2-sphere $S^2$ (taking a coordinate system such that $g_{ij}$ has the form diag$(1, -1, -1, -1)$ when evaluated at the point $a$, then each tangent vector in a lightlike direction has the form $(dx^0, pdx^0, qdx^0, rdx^0)$ with $p^2 + q^2 + r^2 = 1$, and this sets up a one-to-one correspondence between the lightlike geodesics and the points of $S^2$). Taking inverse images under the open continuous map $\theta$, we see that $S_a$ and $T_a$ are open sets in $U$, and that $L_a$ is a closed set which is the boundary for $S_a$ and $T_a$. Using the coordinate surface $x^0 = $ constant which passes through $a$, it is clear that we can disconnect $T_a$ into the two open sets $TA_a$ and $TB_a$, and that $LA_a$ and $LB_a$ are their respective boundaries.

The next kind of special neighborhood is the *diamond*, so called because it can be pictured as the intersection of two cones pointing in opposite directions. We remarked in Section II that intersections of the form $TA_a \cap TB_b$, where $a < b$, are nonempty bounded open sets. If we are working within a nice open set $U$, it no longer need be the case that such an intersection be 'bounded', that is, have compact closure within $U$ (the intersection might hit the edge of $U$). However, when there *is* compact closure within $U$, we call the set a diamond. Given any $a \in U$, and any neighborhood $V$ of $a$, it is clear that we can always find a diamond which contains $a$ and has compact closure within $V$ – this is certainly true by linear arguments if we map some neighborhood of $a$ into the tangent space $M_a$ by means of the inverse of the exponential map, and then make the approximation that geodesics near $a$ correspond to straight lines in $M_a$, with parallel lines being of the same type (timelike, lightlike, etc.) – and that approximation can be made as accurate as we please by making the neighborhood small enough. The formal definition is somewhat complicated by the need to make the nature of the localization precise:

DEFINITION 3.14a. *An* **α-diamond** *consists of*
 (i) *a set* $U_\alpha \in \mathscr{F}$;
 (ii) *a nice open set* $U \subset U_\alpha$;
 (iii) *a Gaussian coordinate system for* $U$;
 (iv) *a pair of points* $a, b \in U$ *with* $a < b$, *where* $<$ *is defined relative to the coordinate system in* (iii); *and*
 (v) *the open set* $D = \{c : a < c < b\}$;
satisfying the additional condition that the closure $D^-$ of $D$ is compact and contained in $U$.

More loosely, we speak of the set $D$ as being the α-diamond, and of $U$ as being the nice open set *associated with* (or *containing*) $D$. Alternatively, we say that $D$ is an α-diamond *within U* (with respect to the specified coordinate system for $U$). $a$ and $b$ are called the *endpoints* of $D$. $D$ is necessarily a small set, but we sometimes need still more:

DEFINITION 3.14b. *A* **local** *α-diamond is an α-diamond for which* $U_\alpha$ *is small.*

When we do not wish to emphasize the particular set $U_\alpha \in \mathscr{F}$ which appears in these two definitions, we simply say that $D$ is a *diamond* or a *local diamond*. Given any $U_\alpha \in \mathscr{F}$, the comments above imply that the family of all α-diamonds is a basis for $U_\alpha$, and that the family of all local diamonds is a basis for $M$.

Finally, we have *wedges*, which are cones truncated by a surface of constant time:

DEFINITION 3.15a. *An* **A-wedge** *consists of*
 (i) *a diamond $D$ having endpoints $a$ and $b$, $a < b$;*
 (ii) *a real number $u$ satisfying $x^0(a) < u$, where $x^0(a)$ denotes the time coordinate of $a$ in the coordinate system of $D$; and*

(iii) *the open set* $K = \{c : a < c \text{ and } x^0(c) < u\}$,
*where K and D satisfy the additional condition* $K \subset D$.

DEFINITION 3.15b. *A* **B-wedge** *consists of*
  (i) *a diamond D having endpoints a and b,* $a < b$;
  (ii) *a real number u satisfying* $u < x^0(b)$, *where* $x^0(b)$ *denotes the time coordinate of b in the coordinate system of D; and*
  (iii) *the open set* $K = \{c : u < x^0(c) \text{ and } c < b\}$,
*where K and D satisfy the additional condition* $K \subset D$.

More loosely, we regard $K$ as being the wedge, $D$ as being the diamond associated with it, $u$ as being the cutoff value of $K$, etc. Or, we say that $K$ is an $A$-wedge or $B$-wedge in $D$ with cutoff $u$. The critical part of this definition is the requirement that $K \subset D$. What this says is that the $A$-wedge $\{c : a < c \text{ and } x^0(c) < u\}$ must be small enough that it lies entirely within the cone $TB_b = \{c : c < b\}$ (and analogously for $B$-wedges). It is easy to see that the set $\{c : a < c \text{ and } x^0(c) < u\}$ can be made arbitrarily small if we take $u$ sufficiently close to $x^0(a)$, by making the same approximation to the linear situation that was used above for diamonds. Since $TB_b$ ($TA_a$ for $B$-wedges) is open, this establishes that every diamond does have both $A$-wedges and $B$-wedges.

Let $U$ be a fixed nice open set. By fixing a particular Gaussian coordinate system $(x^0, x^1, x^2, x^3)$ for $U$, we can introduce the notation of Section II in the manner described above, and recover many of our earlier results. For the time being, all points considered will belong to $U$. Only the generalization of Lemma 2.6 requires any computation:

LEMMA 3.16. *Let a, b, c be noncollinear points, and suppose that* $a < b$ *or* $a \lessdot b$, *and that* $b < c$ *or* $b \lessdot c$. *Then* $a < c$.

*Proof*: Since we know from the definitions that the $x^0$-coordinate of $c$ is greater than that of $a$, we only need show that $a \tau c$. Let $d$ be a variable point on the geodesic $bc$, and parameterize the geodesic segment $ad$ by $q$, normalizing $q$ (for each position of $d$) so that $q = 0$ at $a$ and $q = 1$ at $d$. Let

$$T = \tfrac{1}{2}\left(\frac{dx^0}{dq}\right)^2 + \tfrac{1}{2}g_{ij}(x^0, x^1, x^2, x^3)\left(\frac{dx^i}{dq}\right)\left(\frac{dx^j}{dq}\right)$$
$$= \tfrac{1}{2}(\dot{x}^0)^2 + \tfrac{1}{2}g_{ij}(x^0, x^1, x^2, x^3)\dot{x}^i\dot{x}^j,$$

where $i$ and $j$ are summed over the space-coordinates 1, 2, and 3, and $\dot{x}^i$ means $dx^i/dq$. Then $I(d) = \int_0^1 T dq$, integrated along the segment $ad$, will be positive if $ad$ is timelike, zero if $ad$ is lightlike, and negative if $ad$ is spacelike. We shall show that if $ae$ is lightlike for a point $e$ on the closed geodesic segment $bc$, then in some interval about $e$, $ad$ is spacelike for $d$ on the past side of $e$, and timelike for $d$ on the future side of $e$. Thus, as $d$ moves from $b$ to $c$, $ad$ is timelike somewhere in the interval (either at $b$, or immediately thereafter if it is lightlike at $b$). Further, once it has become timelike, it remains so thereafter, because otherwise, by continuity of $I(d)$, there would have to be a first point at which $I(d)$ stopped being positive and was zero, but by the result just mentioned, $I(d)$ would have to have been negative ($ad$ spacelike) immediately prior to that point. This shows that $ac$ must be timelike.

To get the result mentioned above, we let $d$ move away from $e$, and compute the *sign* of the variation

$$\delta\left(\int T dq\right) = \int \left(\frac{\partial T}{\partial x_i} \delta x^i + \frac{\partial T}{\partial \dot{x}^i} \delta \dot{x}^i\right) dq.$$

$$= \left[\frac{\partial T}{\partial \dot{x}^i} \delta x^i\right]_{q=0}^{q=1} + \int \left[\frac{\partial T}{\partial x^i} - \frac{d}{dq}\left(\frac{\partial T}{\partial \dot{x}^i}\right)\right] \delta x^i \, dq.$$

Since $\partial T/\partial x^i - d/dq(\partial T/\partial \dot{x}^i) = 0$ on a geodesic, and since $\delta x^i$ (the displacement from the geodesic $ae$ to the geodesic $ad$) vanishes at $q = 0$, we have

$$\delta\left(\int T dq\right) = \frac{\partial T}{\partial \dot{x}^i} \delta x^i \Big|_{q=1}$$

$$= \dot{x}^0 \delta x^0 \Big|_{q=1} + g_{ij} \dot{x}^i \delta x^j \Big|_{q=1}.$$

Here $(\dot{x}^0, \dot{x}^1, \dot{x}^2, \dot{x}^3)$ and $(\delta x^0, \delta x^1, \delta x^2, \delta x^3)$, both of them evaluated at $q = 1$, are proportional to the tangent vectors at $e$ for the geodesics $ae$ and $ed$, the sense of the first one being from $a$ to $e$, and the sense of the second being from $e$ to $d$. Using the fact that $ae$ is oriented in a lightlike direction, the fact that $ed$ (being the same geodesic as $bc$) is oriented in a time- or lightlike direction, and Schwarz' inequality as applied to the negative definite quadratic form $g_{ij}$, one can show that $g_{ij} \dot{x}^i \delta x^j$ (at $q = 1$) is less than or equal to $\dot{x}^0 \delta x^0$ (at $q = 1$) in absolute value, with equality occurring only when the signs are the same (equality with opposite signs implies that $ae$ and $ed$ are on the same geodesic, which contradicts the noncollinearity of $a$, $b$, and $c$). Hence the *sign* of $\delta(\int Tdq)$ is the same as the sign of $\dot{x}^0 \delta x^0$ (at $q = 1$). Furthermore, this is the sign of $I(d) = \int Tdq$, because the integral is zero at $d = e$. Since $e$ is on the closed geodesic segment $bc$, our hypotheses imply that $x^0(a) < x^0(b) \leqslant x^0(e)$, which shows that $\dot{x}^0$ is positive. Thus $I(d)$ becomes negative if $\delta x^0$ is negative (when $d$ moves toward the past), and positive if $\delta x^0$ is positive (when $d$ moves toward the future). ■

Lemma 3.16 establishes the transitivity of time in a strict sense: even if $a \lessdot b$ and $b \lessdot c$, we can have $a \lessdot c$ only when the three points are collinear. In all other cases, $c$ must be in the interior of the after-cone of $a$, not on the surface.

The next four lemmas and corollaries generalize from Section II with no change in statement or proof:

COROLLARY 3.17. *If $a\lambda b$, $a\lambda c$, and $b\lambda c$, then $a$, $b$, and $c$ are collinear.*

COROLLARY 3.18. *Let $a$, $b$, $c$ be noncollinear points. If $a \lessdot b$ and $c \lessdot b$, or if $b \lessdot a$ and $b \lessdot c$, then $a\sigma c$.*

LEMMA 3.19. *If $a < b$, $a < c$, and $b\sigma c$, then there is a point $d$ satisfying $a < d$, $d \lessdot b$, and $d \lessdot c$. Similarly, if $a > b$, $a > c$, and $b\sigma c$, then there is a point $d$ satisfying $a > d$, $d \gtrdot b$, and $d \gtrdot c$.*

LEMMA 3.20. *If $a < b$, then $L_a \cap L_b = LA_a \cap LB_b$. On the other hand, if $a\sigma b$, then $L_a \cap L_b = (LB_a \cap LB_b) \cup (LA_a \cap LA_b)$, where the union is disjoint and the two parts are topologically separated from one another.*

In Section II, we began our study of the predicate $Q(abc)$ by showing that it gave a test of collinearity in the case in which all three of $ab$, $ac$, and $bc$ are spacelike (Proposition 2.11). In the present context, we will obtain no information about that case, but when at least one of $ab$, $ac$, or $bc$ is timelike, then the previous results (Proposition 2.12) still hold up, provided that we use diamonds to localize them correctly. However, the proof of part (ii) of Proposition 2.12 depended in an essential way on the fact that $L_a \cap L_b$ was a quadric surface within a certain hyperplane (we reasoned that $L_a \cap L_b$, being topologically disconnected, was a hyperboloid of two sheets and therefore unbounded). The next two lemmas show that we can recover this seemingly special property of second-degree surfaces in the much more general setting in which we now find ourselves. We are still working within the fixed nice open set $U$, and all topological concepts (point, closed set, etc.) are therefore relativized to $U$.

LEMMA 3.21. *If $a\sigma b$, then $L_a \cap L_b$ is homeomorphic to an open subset of the 2-sphere $S^2$.*

*Proof*: Recall that $L'_a$ is homeomorphic to $S^2$, and that the projection map $\theta: U - \{a\} \to M'_a$ is continuous and open. Further, since $\theta(L_a \cap L_b) \subset \theta(L_a) = L'_a$, we can define a restriction $\varphi: L_a \cap L_b \to L'_a$. If we can show that $\varphi$ is continuous, open, and one-to-one, then we will have established that it is a homeomorphism onto an open subset of $L'_a$, and the lemma will be proved. Of course, the relative topology must be used for both $L_a \cap L_b$ and $L'_a$.

Continuity is trivial, since $\varphi$ is a restriction of $\theta$. One-to-oneness is easy (if $\varphi(x) = \varphi(y)$, then $x$ and $y$ are on the same lightlike geodesic through $a$. But then we would have $x\lambda y$, $x\lambda b$, and $y\lambda b$, which would show that $x$, $y$, and $b$ are collinear. Then $a$ and $b$ are both on the lightlike geodesic $xy$, which contradicts $a\sigma b$.) To show that $\varphi$ is open, take $x \in L_a \cap L_b$, and let $L_a \cap L_b \cap V$ be an arbitrary neighborhood of $x$ in the the relative topology of $L_a \cap L_b$, where $V$ is open in $U$. We must find an open set $W$ in $M'_a$ such that $\varphi(x) \in L'_a \cap W \subset \varphi(L_a \cap L_b \cap V)$.

Let $X$ be a nice open set containing $x$, with $X \subset V$. Let $y, z \in X$ be two points on the lightlike geodesic $ax$, such that $a \ll y \ll x \ll z$ if $a \ll x$, and such that $a \gg y \gg x \gg z$ if $a \gg x$. In either case, it follows from $b\sigma a$ and $b\lambda x$ that we have $b\sigma y$ and $b\tau z$ (by Lemma 3.16, Corollary 3.18, and the fact that $b \gg x$ if $a \gg x$ and $b \ll x$ if $a \ll x$), so that $y \in S_b \cap X$ and $z \in T_b \cap X$. I claim that the open set $W = \theta(S_b \cap X) \cap \theta(T_b \cap X)$ is the required open set in $M'_a$. We have $\varphi(x) = ax = \theta(y) = \theta(z) \in \varphi(L_a \cap L_b) \cap \theta(S_b \cap X) \cap \theta(T_b \cap X) = \varphi(L_a \cap L_b) \cap W \subset L'_a \cap W$, and we must show that $L'_a \cap W \subset \varphi(L_a \cap L_b \cap V)$. Take $w \in L'_a \cap W = L'_a \cap \theta(S_b \cap X) \cap \theta(T_b \cap X)$. Thus $w$ has inverse images $u \in S_b \cap X$ and $v \in T_b \cap X$. By definition of $\theta$, these lie on the same geodesic through $a$, which is lightlike since $w \in L'_a$. Since $u \in S_b$ and $v \in T_b$, there must be some $r \in L_b$ which lies between $u$ and $v$ on this geodesic. Then $r \in L_a$ because the geodesic is lightlike, and $r \in X$ because $u, v \in X$ and the nice open set $X$ is geodetically convex. This shows that $w = \varphi(r)$ is in $\varphi(L_a \cap L_b \cap X) \subset \varphi(L_a \cap L_b \cap V)$, as required. ∎

LEMMA 3.22. *Suppose that $e < a < f$ and $e < b < f$, and that $a\sigma b$. Then no component of $LB_a \cap LB_b$ nor of $LA_a \cap LA_b$ is compact.*

*Proof*: By Lemma 3.20, we have $L_a \cap L_b = (LB_a \cap LB_b) \cup (LA_a \cap LA_b)$, where the union is disjoint and the two parts are topologically separated. By Lemma 3.19, we can find a point $c$ such that $e < c$, $c \lessdot a$, and $c \lessdot b$, and also a point $d$ such that $f > d$, $d \gtrdot a$, and $d \gtrdot b$. Hence $LB_a \cap LB_b$ and $LA_a \cap LA_b$ are both nonempty. Unfortunately, we do not know whether they are connected, so we cannot assert that they are the components of $L_a \cap L_b$. All we can say is that there are at least two components, and that the decomposition of $L_a \cap L_b$ into components is subordinate to the decomposition $L_a \cap L_b = (LB_a \cap LB_b) \cup (LA_a \cap LA_b)$ (that is, the components of $L_a \cap L_b$ consist of the components of $LB_a \cap LB_b$ plus the components of $LA_a \cap LA_b$). We must now show that no component of the larger set, $L_a \cap L_b$, is compact.

Embed $L_a \cap L_b$ in $S^2$, using the homeomorphism whose existence is asserted by Lemma 3.21. Since $L_a \cap L_b$ is open in $S^2$ and $S^2$ is locally connected, the components of $L_a \cap L_b$ are also open in $S^2$. If any of them were compact, it would also be closed in $S^2$, and hence either empty or all of $S^2$. But no component of $L_a \cap L_b$ can be empty, because $L_a \cap L_b$ is not empty, and no component can be all of $S^2$, because then $L_a \cap L_b$ would be connected, and it is not. ∎

We are now ready to begin our analysis of Definitions 3.1–3.7. Part (ii) of Proposition 2.12 can be recovered for α-diamonds, and the rest of it will be recovered for local α-diamonds:

PROPOSITION 3.23. *Suppose that $D$ is an α-diamond within the nice open set $U$, and that $a$, $b$, $c$ are distinct points of $D$, each pair of which can be connected by a spacelike or timelike geodesic within $U$. If exactly two of these three connecting geodesics are timelike, then $Q_\alpha(abc)$ is false.*

*Proof*: Working within the nice open set $U$ and using the notation associated with it, we shall find a point $d$ that is light connected to all three of $a$, $b$, and $c$ within $U$. Since $U \subset U_\alpha$, it follows that $d$ is light connected to each of them within $U_\alpha$, so that $Q_\alpha(abc)$ is false.

Our proof is based on that of Proposition 2.12(ii). Write out an explicit expression $D = TA_f \cap TB_g$ for the α-diamond $D$ within $U$. Then we have $b < g$, so this $g$ can play the role of the $g$ used in the earlier proof. We then find that the point $h$ belongs to the (possibly smaller) α-diamond

$TA_a \cap TB_g$, and that $h$ is in some component of $LB_b \cap LB_c$. But this component is *closed* in the relative topology of $U$ (because it is a component of a closed set), is *connected* (because it is a component), and is *not compact* (by Lemma 3.22). Since $TA_a \cap TB_g$ has *compact closure* contained in $U$ (because $D$ does by Definition 3.14a, and $TA_a \cap TB_g \subset D$), it follows that there is some point $d$ in this component, and therefore in $LB_b \cap LB_c$, which lies on the boundary of $TA_a \cap TB_g$. This point $d$ has all the properties stated in the earlier proof, and hence $d$ is in $L_a \cap L_b \cap L_c$. ∎

PROPOSITION 3.24. *Suppose that $D$ is a local α-diamond in the nice open set $U$, and that $a$, $b$, $c$ are distinct points of $D$, each pair of which can be connected by a spacelike or timelike geodesic within $U$. Then*

(i) *if exactly one of these three connecting geodesics is timelike, $Q_\alpha(abc)$ is true;*

(ii) *if exactly two of these three connecting geodesics are timelike, $Q_\alpha(abc)$ is false;*

(iii) *if all three of these connecting geodesics are timelike, $Q_\alpha(abc)$ is true.*

*Proof*: $D$ is a *local* α-diamond, so $U_\alpha$ is small. Thus, by Definition 3.8b and Definition 3.14a, we have $D \subset U \subset U_\alpha \subset V$, where both $U$ and $V$ are nice. We pick a coordinate system for $V$, and introduce the notation above relative to $V$, not $U$. (ii) has already been proved, and it remains to be shown that $Q_\alpha(abc)$ is true for cases (i) and (iii). Since $U_\alpha \subset V$, it suffices to show that no point $d \in V$ can be light connected within $V$ to all three of $a$, $b$, and $c$. This is done by the proof of Proposition 2.12, parts (i) and (iii). ∎

As a corollary, we get an existence theorem for α-paths:

COROLLARY 3.25. *Let $D$ be a local α-diamond, and let $P \subset D$ be a connected set of at least two points, such that any two points of $P$ can be connected by a segment of a timelike geodesic within $D$. Then $P$ is an α-path.*

*Proof*: Since any two points of $P$ can be connected by a timelike geodesic within $D$, they can surely be connected within the nice open set $U$ containing $D$. The result then follows from Definition 3.2 and Proposition 3.24. ∎

COROLLARY 3.26. *Any segment of a timelike geodesic that is contained in a local α-diamond is an α-path.*

The main result concerning α-paths is that, locally, they must go in either a spacelike or a timelike direction:

PROPOSITION 3.27. *Let $P$ be an α-path, and take $a \in P$. Then there is an α-diamond $D$, with $a \in D$, such that either*

*(i) for all $b$, $c \in P \cap D$, there is a segment of a timelike geodesic in $D$ that connects $b$ and $c$; or*

*(ii) for all $b \in P \cap D$, there is a segment of a spacelike geodesic in $U$ that connects $a$ and $b$, where $U$ is the nice open set associated with $D$.*

*Proof*: Since $P$ is an α-path and $a \in P$, we have $a \in U_\alpha$. Let $E$ be any α-diamond containing $a$, let $U$ be the nice open set associated with $E$, and introduce the notation relative to $U$. We then consider two cases: (i) there exists a point $d \in P \cap E$ such that $a\tau d$, or (ii) there does not exist such a point.

(i) if such a $d$ exists, suppose without loss of generality that $a < d$, and let $D = E \cap TB_d$. Then $a \in D$, and it is easy to see that $D$ is an α-diamond, with the same nice open set $U$ associated with it. If $b, c \in P \cap D = P \cap E \cap TB_d \subset TB_d$, then we have $b\tau d$ and $c\tau d$. Also, we have $Q_\alpha(bcd)$ since $P$ is a path. Applying Proposition 3.23 to the α-diamond $D$, we conclude that $b\tau c$ must hold. We now need to show that the timelike geodesic $bc$ connects $b$ and $c$ within $D$, not just within $U$. In other words, given a point $e$ on $bc$ satisfying $b < e < c$ or $c < e < b$, we must show that $e \in D$. This follows from the fact that $D$ is a diamond and the transitivity of $<$.

(ii) If no such point $d$ exists, then $a\tau b$ is false for every $b \in P \cap E$. Since $a\lambda b$ is excluded by the definition of path, we have $a\sigma b$, so that $a$ and $b$ are connected by a segment of a spacelike geodesic in $U$. Letting $D = E$, we have the required conclusion. ∎

We are now ready to consider timepaths. The preceding proposition helps us to extend Corollary 3.25 to an existence theorem for local timepaths:

PROPOSITION 3.28. *Let $D$ be a local diamond, and let $P \subset D$ be a connected set of at least two points, such that any two points of $P$ can be connected by a segment of a timelike geodesic within $D$. Then $P$ is a local timepath.*

*Proof*: $D$ is an α-diamond for some $U_\alpha$ which is small. By Corollary 3.25, $P$ is an α-path for that $U_\alpha$. Take any point $a \in P$, any open set

$U_\beta \in \mathscr{F}$, and any $\beta$-path $R$ such that $P \cap R = \{a\}$. We must find a point $b \in R \cap U_\alpha$ and a neighborhood $V \subset U_\alpha$ such that $Q_\alpha(abc)$ holds for all $c$ distinct from $a$ in $P \cap V$.

Let $E$ be the $\beta$-diamond whose existence is asserted by Proposition 3.27, as applied to $R$ at the point $a$. Then $D \cap E$ is an open set containing $a$ (note that $D \cap E$ might not be a diamond, because $D$ and $E$ are diamond-shaped with respect to different coordinate systems, and we have not proved coordinate independence). Hence we can find a nice open set $U$ and a diamond $F$ within $U$ such that $a \in F \subset U \subset D \cap E$. Let $X$ and $Y$ be the respective nice open sets associated with $D$ and $E$. Then we have $a \in F \subset U \subset D \cap E \subset D \subset X \subset U_\alpha$ and $a \in F \subset U \subset D \cap E \subset E \subset Y \subset U_\beta$. Introduce notation with respect to $U$, and note that $F$ is both an $\alpha$-diamond and a $\beta$-diamond. In fact, $F$ is a local $\alpha$-diamond, because $U_\alpha$ is small. If two points $e, f \in U$ are connected by a timelike (respectively spacelike) geodesic in $D$ or $E$ or $X$ or $Y$, then we can write $e\tau f$ (respectively $e\sigma f$) in $U$, because they must be connected by some kind of geodesic in $U$, and the two geodesics are the same by uniqueness of geodesics in $X$ or in $Y$. This makes it possible to express statements regarding timelike or spacelike connection (such as those in Proposition 3.27) in terms of the notation of $U$, so long as the points involved belong to $U$.

Let $b$ be any point in $R \cap F$ which is distinct from $a$ (if no such point existed, then $a$ would be an isolated point of $R$, contradicting the connectedness of $R$). If 3.27(i) holds, let $V = F \cap T_b$, and if 3.27(ii) holds, let $V = F \cap S_b$. Then $V$ is a neighborhood of $a$. Take any $c$ in $P \cap V$ which is distinct from $a$. If 3.27(i) holds, then $a\tau b$ (by 3.27(i) and the remarks above), $a\tau c$ (by the hypothesis on $P$), and $b\tau c$ (because $c \in V \subset T_b$). Similarly, if 3.27(ii) holds, then $a\sigma b$ (by 3.27(ii)), $a\tau c$ (by the hypothesis on $P$), and $b\sigma c$ (because $c \in V \subset S_b$). In either case, $Q_\alpha(abc)$ is true, by Proposition 3.24 applied to the local diamond $F$. ∎

The next proposition gives a proof of Theorem 3.10, plus a localized version of Theorem 3.9. Among other things, by giving a necessary and sufficient condition for a set $P$ to be a timepath, it shows that our definition of 'timepath' (as well as 'U-locally temporally between', 'U-locally time connected', and 'admissible', which are all defined in terms of 'timepath') is independent of the particular open covering $\mathscr{F} = \{U_\alpha\}$ for which the local light connections $\lambda_\alpha$ were given.

PROPOSITION 3.29. *Let P be a set which is connected and has at least two points. Then* (i) *and* (ii) *are equivalent, and imply* (iii):

  (i)  *P is a timepath.*

  (ii)  *Given a point $a \in P$ and a neighborhood V of a, there is an open set W with $a \in W \subset V$ such that any two points of $P \cap W$ can be connected by a segment of a timelike geodesic within W.*

  (iii)  *Given a point $a \in P$, a neighborhood V of a, and a Gaussian coordinate system $(x^0, x^1, x^2, x^3)$ defined on V, there is an open set W with $a \in W \subset V$, a finite interval I of real numbers (open or half-open), and a continuous function $f: I \to R^3$ such that $P \cap W = \{(x, f(x)) : x \in I\}$ in the given coordinate system for V.*

*Proof*: We will prove that (i) implies (ii), that (ii) implies (iii), and that (ii) and (iii) jointly imply (i). The proof that (ii) implies (iii) is by far the most difficult.

*Proof that* (i) *implies* (ii): Take $a \in P$ and a neighborhood $V$ of $a$. By Definition 3.4, $P$ has an open covering by local timepaths. Hence there is a set $S$ such that (1) $a \in S \subset P$, (2) $S$ is a local timepath, i.e., an $\alpha$-timepath for some $U_\alpha \in \mathscr{F}$, and (3) $S$ is open in $P$, i.e., $S = P \cap X$ for some open set $X$ of $M$. Since $S$ is an $\alpha$-path, there is an $\alpha$-diamond $D$ which contains $a$ and satisfies either 3.27(i) or 3.27(ii) (with $S$ replacing $P$ in their statements). Let $U$ be the nice open set associated with $D$, and introduce notation relative to $U$. Let $W$ be a diamond (with respect to the nice open set $U$) such that $a \in W \subset V \cap X \cap D \subset U$.

Suppose that 3.27(i) holds, and take $b, c \in P \cap W \subset P \cap V \cap X \cap D \subset P \cap X \cap D = S \cap D$. Then $b\tau c$. We must show that the segment of the timelike geodesic $bc$ that lies between $b$ and $c$ is contained in $W$. As in the proof of 3.27(i), this follows from the fact that $W$ is a diamond and the transitivity of $<$.

If we can show that 3.27(ii) cannot occur, the proof will be complete. To do this, let $E$ be any local diamond containing $a$, and let $R$ be a segment of a timelike geodesic that lies within $D \cap E$ and passes through $a$. By Corollary 3.26, $R$ is a $\beta$-path for some $U_\beta \in \mathscr{F}$. If 3.27(ii) were true, then we would have $S \cap R = \{a\}$ ($a$ is in $S \cap R$, and any other element $d$ in $S \cap R$ would have to satisfy $a\tau d$ by the choice of $R$ and $a\sigma d$ by 3.27(ii), which is impossible). Hence there is an element $b \in R$ and a neighborhood $Y$ of $a$ such that $Q_\alpha(abc)$ holds for all $c \in (S \cap Y) - \{a\}$. Note that $a\tau b$ by the choice of $R$. Hence $a$ is in the open set $Y \cap D \cap T_b$.

Now take $c$ is $S \cap Y \cap D \cap T_b$ and distinct from $a$, which is possible because the connected set $S$ can have no isolated points. Then $c\tau b$. Finally, $a\sigma c$ by 3.27(ii). But then $Q_\alpha(abc)$ is false by Proposition 3.23 and true by the choice of $b$ and $Y$, which shows that 3.27(ii) cannot occur.

*Proof that (ii) implies (iii)*: Let $V$ be the given neighborhood of $a$, and let $U$ be any nice open set satisfying $a \in U \subset V$. Set up the notation relative to $U$, using the restriction to $U$ of the Gaussian coordinate system for $V$. Applying (ii) to $U$, we get an open set $X$ with $a \in X \subset U \subset V$, such that $b, c \in P \cap X$ implies that $b\tau c$. Let $D$ be a diamond with respect to $U$ and the given coordinate system, with $a \in D \subset X \subset U \subset V$, and choose $D$ to be small enough that some point of $P$ lies outside of $D$. Let $J = \{x^0(c) : c \in P \cap D\}$, where $x^0$ is the function that assigns to each point of $U$ its time coordinate in the given coordinate system. Note that $x^0(a) \in J$. Three possibilities exist regarding the set of real numbers less than $x^0(a)$, and in each case we choose a real number $s$ and a point $d$ such that $x^0(d) = s$:

(1) There is a number $s < x^0(a)$ such that the half-open interval $[s, x^0(a))$ is contained in $J$. Then we can take $d \in P \cap D$ with $x^0(d) = s$, because $s \in J$. We have $d\tau a$ since $a, d \in P \cap D \subset P \cap X$, which implies $d < a$ because $x^0(d) = s < x^0(a)$.

(2) There is a number $s < x^0(a)$ such that the half-open interval $[s, x^0(a))$ is disjoint from $J$. Since $D$ is a neighborhood of $a$, there are points $d \in D$ with $d < a$, arbitrarily close to $a$. By moving $s$ nearer to $x^0(a)$ if necessary, we can have $d \in D$, $d < a$, $[s, x^0(a))$ disjoint from $J$, and $x^0(d) = s$.

(3) There is an infinite monotonically increasing sequence $u_1, u_2, u_3, \ldots$ converging to $x^0(a)$, such that $u_1 \in J$, $u_2 \notin J$, $u_3 \in J$, $u_4 \notin J$, etc. In this case, let $s = u_1$, and take $d \in P \cap D$ with $x^0(d) = s$, which is possible since $s = u_1 \in J$. As in case (1), $d < a$.

The same three possibilities exist regarding the set of real numbers greater than $x^0(a)$, and we choose a real number $t$ and a point $e$ such that $x^0(e) = t$, subject to the conditions above with order reversed (call these reversed conditions on $t$ and $e$ (1'), (2'), and (3'), respectively. Note that $d, e \in D$ and that $d < a < e$ in all cases. Let $W$ be the diamond $TA_d \cap TB_e$, and note that $a \in W \subset W^- \subset D$, where $W^-$ denotes the closure of $W$ (to prove that $W^- \subset D$, one uses the fact that $D$ is a diamond and

# LIGHT SIGNALS AND THE STRUCTURE OF TIME   357

contains $d$ and $e$, the fact that $LA_d$ and $LA_e$ are the respective boundaries of $TA_d$ and $TA_e$, and Lemma 3.16).

We will need the following fact: if $c$ is any point in $P \cap D$ satisfying $s < x^0(c) < t$, then $c$ is in $W$. To prove this, note first that $a\tau c$ because $a, c \in P \cap D \subset P \cap X$. Hence either $a < c$ or $c < a$. Also, $x^0(c) \in J$ because $c \in P \cap D$. If (2) holds, we must have $x^0(d) \in [x^0(a), t)$, because $(s, x^0(a))$ is disjoint from $J$ in that case. Thus $c < a$ is impossible, and we have $d < a < c$. On the other hand, if (1) or (3) holds, then $d \in P \cap D \subset P \cap X$, so that $d\tau c$. In view of $x^0(d) = s < x^0(c)$, this shows that $d < c$. Similarly, by reversing directions, we have $x^0(c) \in (s, x^0(a)]$ in case (2'), whence $c < a < e$, and $c\tau e$ in cases (1') and (3'), whence $c < e$. Consequently, we have $d < c < e$ in all cases, so that $c \in TA_d \cap TB_e = W$.

In view of the three possibilities for $s$ and the three possibilities for $t$, there are a total of nine cases to consider. However, I will show that (3) is impossible, that (3') is impossible, and that (2) and (2') cannot both be true. The only cases that remain are (1)(1'), (1)(2'), and (2)(1'). I do this by showing that it is not possible to have $s < w_1 < w_2 < w_3 < t$ with $w_1 \notin J$, $w_2 \in J$, and $w_3 \notin J$ (in case (2)(2') we would be able to find such numbers by taking $w_2 = x^0(a)$ with $w_1$ and $w_3$ being two nearby numbers on either side, and in case (3) or case (3') we would take $w_1, w_2, w_3$ from the infinite monotonically increasing or decreasing sequence $u_2, u_3, u_4 \ldots$).

Suppose that we did have such numbers $w_1, w_2, w_3$. Let $F$ be the open set $W \cap \{b \in U: w_1 < x^0(b) < w_3\}$. Take $c \in P \cap (F^-)$, where $F^-$ is the closure of $F$. Then $c \in P \cap (W^-) \subset P \cap D$ and $s < w_1 \leq x^0(c) \leq w_3 < t$. But we have shown that any point $c \in P \cap D$ with $s < x^0(c) < t$ must be in $W$. Also, since $c \in P \cap D$, we have $x^0(c) \in J$, which shows that $x^0(c) \neq w_1$ and $x^0(c) \neq w_3$. Hence $w_1 < x^0(c) < w_3$, which shows that $c$ lies in $F$ itself. In other words, no point of $P$ can lie on the boundary of $F$. But then $F$ and its complement disconnect $P$. Furthermore, there is at least one point of $P$ which lies inside $F$: since $w_2 \in J$, there is a point $c \in P \cap D$ such that $x^0(c) = w_2$, and this point $c$ is in $W$, and therefore in $F$, because it satisfies $s < x^0(c) < t$. On the other hand, there is at least one point of $P$ which lies outside of $F$, because $F \subset D$, and $D$ was chosen to be small enough that there is a point outside. This contradicts the connectedness of $P$, and shows that $w_1, w_2, w_3$ cannot exist.

Now we define the finite interval $I$: $I = (s, t)$ in case (1)(1'), $I$

$=(s, x^0(a)]$ in case (1) (2'), and $I = [x^0(a), t)$ in case (2) (1'). Then $I \subset J$, so for every $u \in I$ there is a point $b \in P \cap D$ with $x^0(b) = u$. Only one such point $b$ can exist, because if $b' \in P \cap D \subset P \cap X$ were another, we would have $b \tau b'$ and $x^0(b) = x^0(b')$, a contradiction. Write $b$ in the form $(u, v_1, v_2, v_3)$ in terms of the coordinate system, and let $f: I \to R^3$ be the function that assigns to $u$ the triple $(v_1, v_2, v_3)$. Then $b = (u, f(u))$. Note that $b$ is in $W$, because it is in $P \cap D$ and satisfies $s < u = x^0(b) < t$. Also, note that every point of $P \cap W$ is of this form: if $c \in P \cap W$, then $x^0(c) \in J$ and $s = x^0(d) < x^0(c) < x^0(e) = t$, so that $x^0(c) \in I$ by the way $I$ was chosen, which shows that $c = (x^0(c), f(x^0(c)))$ by the uniqueness argument above. Only the continuity of $f$ remains to be shown.

To show that $f: I \to R^3$ is continuous, it is sufficient to consider the function $k: I \to P \cap W$ which maps $u$ to $(u, f(u))$. If $b \in P \cap W$ and $Y$ is a neighborhood of $b$, then we must find an interval $(v, w)$ containing $k^{-1}(b) = x^0(b)$, such that $k(u) \in Y$ whenever $u \in (v, w) \cap I$. Let $TA_g \cap TB_h$ be any diamond such that $b \in TA_g \cap TB_h \subset Y$. Let $G$ be any $B$-wedge in the smaller diamond $TA_g \cap TB_b$ (see Definition 3.15b and the discussion which follows it). Similarly, let $H$ be an $A$-wedge in $TA_b \cap TB_h$. Then $G \subset TA_g \cap TB_b \subset TA_g \cap TB_h \subset Y$, and $H \subset TA_b \cap TB_h \subset TA_g \cap TB_h \subset Y$. Let $v$ and $w$ be the respective cutoffs of $G$ and $H$, so that $G = \{c : v < x^0(c) \text{ and } c < b\}$ and $H = \{c : b < c \text{ and } x^0(c) < w\}$. Take $u \in (v, w) \cap I$. Since $k(u) \in P \cap W \subset P \cap D \subset P \cap X$ by the definition of the function $k$, we have $k(u) = b$ or $k(u) \tau b$. If $k(u) = b$, then $k(u) \in Y$ because $Y$ is a neighborhood of $b$. If $k(u) < b$, then $x^0(k(u)) = u > v$, so that $k(u) \in G \subset Y$. Finally if $k(u) > b$, then $x^0(k(u)) = u < w$, so that $k(u) \in H \subset Y$. Consequently $k(u) \in Y$ in all cases, as was to be shown.

*Proof that (ii) and (iii) jointly imply (i):* We must provide an open covering of $P$ by local timepaths. Take $a \in P$, let $D$ be a local diamond containing $a$, and let $U$ be the nice open set associated with $D$. By (ii), there is an open set $V$ with $a \in V \subset D \subset U$ such that any two points of $P \cap V$ can be connected by a segment of a timelike geodesic lying in $V \subset D$. By (iii), using the Gaussian coordinate system of $U$ restricted to $V$, there is an open set $W$ with $a \in W \subset V \subset D$ such that $P \cap W = \{(x, f(x)) : x \in I\}$ for some interval $I$ and continuous function $f$. Then $P \cap W$, being a continuous image of an interval, is connected. By Proposition 3.28 applied to the local diamond $D$ and the set $P \cap W \subset D$, we see that $P \cap W$ is a local timepath. Furthermore, $P \cap W$ contains $a$, and

$P \cap W$ is open in $P$ because $W$ is an open set. Doing this for every $a \in P$, we have the required open covering of $P$. ∎

Proposition 3.29 tells enough about timepaths so that the rest of our task is relatively easy. Since everything has been stated in local terms, our first step is to obtain some global information about timepaths.

We prove that timepaths are connected one-dimensional manifolds with boundary. Thus any timepath $P$ must be homeomorphic to the real line, the circle, the closed interval, or the half-open interval (if $M$ does not satisfy the second axiom of countability, then there could be three other possibilities: $P$ could be the long line in one direction, combined with the long line, the ordinary line, or the closed interval in the other direction).

COROLLARY 3.30. *Any timepath $P$ is a connected one-dimensional topological manifold with boundary.*

*Proof*: The set $P \cap W$ in the statement of 3.29(iii) is open in the relative topology of $P$, is homeomorphic to the open or half-open interval $I$ under the correspondence $(x, f(x)) \leftrightarrow x$, and contains the given point $a$. By varying $a$, we get an open covering of $P$ by sets that are each homeomorphic to an open or half-open interval in the real numbers. ∎

We have asserted above only that $P$ is a topological manifold, not that it is a differentiable manifold. In fact, the obvious coordinate charts which are given by 3.29(iii) (namely, projection of each point of $P \cap W$ to its time coordinate in $I$) lead to nondifferentiable coordinate transformations on $P$ when we pass to a new Gaussian coordinate system, if the function $f$ is not differentiable.

Next, we prove that timelike geodesic segments are timepaths. Several restrictions must be made to this statement, however. First, the relativistic torus discussed in Section I provides examples of timelike geodesics which are dense subsets of the manifold, and consequently not timepaths. The difficulty here is that the geodesic can pass through a given region an infinite number of times. This can be avoided by limiting our attention to a *finite* geodesic segment, that is, to the set of all points that lie on a geodesic between two specified endpoints. It simplifies matters without any significant reduction in generality to require that the endpoints always be included in the segment, making it a closed (in fact, compact) geodesic segment. The other possible difficulty is that geodesics can cross themselves. A segment on which this happened would also not be a timepath. This can be avoided by requiring that there be a *one-to-one* parameterization of the closed geodesic segment.

PROPOSITION 3.31. *A closed segment of a timelike geodesic is a timepath, provided that there is a* one-to-one *continuous function onto it from the closed unit interval* [0, 1].

*Proof*: Let $P$ be the geodesic segment and $f: [0, 1] \to P$ the function. We prove that $P$ is a timepath by showing that 3.29(ii) holds.

Suppose that we are given a point $a \in P$ and a neighborhood $V$ of $a$. Since $f$ is one-to-one and onto $P$, there is a unique inverse element $f^{-1}(a)$ in [0, 1]. Let $U$ be any nice open set with $a \in U \subset V$. Then $f^{-1}(P \cap U)$ is open by the continuity of $f$, so we may find an open interval $(v, w)$ satisfying $f^{-1}(a) \in (v, w) \subset f^{-1}(P \cap U)$. Note that if $b, c \in f((v, w))$, then $b$ and $c$ can be connected by a segment of a timelike geodesic within $U$: in that situation the closed interval $[f^{-1}(b), f^{-1}(c)]$ or $[f^{-1}(c), f^{-1}(b)]$ is contained in $(v, w) \subset f^{-1}(P \cap U)$, and its image under $f$ is then the required geodesic segment joining $b$ and $c$ within $U$.

The closed set $[0, 1] - (v, w)$ is compact and does not contain $f^{-1}(a)$, whence $f([0, 1] - (v, w))$ is compact and does not contain $a$. Thus we may let $Y$ be an open set which contains $a$ and is disjoint from $f([0, 1] - (v, w))$, and let $W$ be any nice open set satisfying $a \in W \subset Y$. Now if $b, c \in P \cap W \subset P \cap Y$, we have $b, c \in f((v, w))$ because $Y$ is disjoint from $f([0, 1] - (v, w))$. Therefore $b$ and $c$ can be connected by a segment of a timelike geodesic in $U$. But $b$ and $c$ can certainly be connected by some kind of a geodesic in the nice open set $W$, and the two geodesics must be the same by geodetic uniqueness in $U$, which shows that $W$ satisfies the requirement of 3.29(ii). ∎

We conclude by proving Theorems 3.9, 3.13, 3.11, and 3.12, which were claimed at the beginning of Section III. Theorem 3.10 has already been established by Proposition 3.29.

Regarding the proof of Theorem 3.9, it should perhaps be pointed out that 'boundary point' is being used here in the sense of manifold theory, *not* in the topological space meaning in which we have used it previously. Thus a point $c$ will be a boundary point if it has a neighborhood which is homeomorphic to the half-open interval with $c$ appearing at the closed end of the interval, and $c$ will be a nonboundary point if it has a neighborhood which is homeomorphic to the open interval. This depends only on internal properties of the one-dimensional manifold involved, not on properties of some larger topological space in which the manifold may be imbedded.

*Proof of Theorem* 3.9: $P$ is locally connected because it is a connected one-dimensional manifold with boundary by Corollary 3.30, and $P \cap U$ is open in $P$. Hence any component $S$ of $P \cap U$ is open in $P$, as well as being connected. Thus $S$ itself is a connected one-dimensional manifold with boundary. We have already given a list of all such manifolds: either $S$ is embeddable in the line or the long line as a finite or infinite interval of some kind, or else $S$ is homeomorphic to the circle. In either case the following statement is true: given distinct points $b$ and $c$ in $S$, there is a nonempty *open set* $T$ of $S$ which contains no boundary points of $S$ and has the property that $T \cup \{b\} \cup \{c\}$ is *compact* (if the manifold $S$ is linear, then $T$ consists of the set of all points of $S$ which lie strictly between $b$ and $c$, and if $S$ is circular, then $T$ consists of $S$ itself).

Let $I = \{x^0(c) : c \in S\}$, and let $g : S \to I$ be the restriction to $S$ of the $x^0$-coordinate function. Then $g$ is continuous and onto. Since $I$ is the continuous image of a connected topological space, it is connected and therefore an interval (open, half-open, or closed; finite or infinite). Given any $a \in S$, we know by 3.29(iii) that $g$ gives a homeomorphism of some neighborhood of $a$ in $S$ onto some neighborhood of $g(a)$ in $I$ (because $g$ has the continuous inverse $x \to (x, f(x))$ when restricted to a suitable neighborhood). From this local homeomorphism property, it follows that $g$ is an open map into $I$, and that $g$ carries nonboundary points to nonboundary points, where 'nonboundary point' is being used in the manifold theoretic sense. Finally, $g$ is one-to-one, because if $b$ and $c$ were distinct points of $S$ such that $g(b) = g(c)$, then we would be able to find a nonempty open set $T$ of $S$ which contains no boundary points, such that $T \cup \{b\} \cup \{c\}$ is compact. But then $g(T)$ would be a nonempty open set of $I$, containing no boundary points of $I$ and therefore open as a set of real numbers, such that $g(T)$ can be made compact by adding the single element $g(b) = g(c)$. In the real numbers, this is clearly impossible. Therefore $g : S \to I$ is a homeomorphism. Let $h : I \to S$ be its inverse. Then $h(x)$ has $x$ as its time coordinate for all $x \in I$. Letting $f(x)$ stand for the space coordinates, we have $h(x) = (x, f(x))$ for all $x \in I$, and this proves the theorem. ∎

*Proof of Theorem* 3.13: Suppose first that $c$ is $U$-locally temporally between $a$ and $b$. Then we have timepaths $P$, $R$ and $P \cup R$ satisfying the conditions of Definition 3.5. Since $a$, $b$, and $c$ all belong to $P \cup R$, any pair of them is $U$-locally time connected by the timepath $P \cup R$. We have $P \cup R = \{(x, f(x)): x \in I\}$ for some interval $I$ and continuous function $f$ because of Theorem 3.9. Let $g: P \cup R \to I$ be the restriction of the $x^0$-coordinate function. Then $g$ is a homeomorphism, because it has the continuous inverse $x \to (x, f(x))$. By one-to-oneness, we have $g(c) \in g(P) \cap g(R)$, $g(a) \in g(P) - g(R)$, and $g(b) \in g(R) - g(P)$. By continuity, $g(P)$ and $g(R)$ are connected subsets of $I$ and therefore intervals. This shows that $g(P) \cap g(R)$ lies between $g(P) - g(R)$ and $g(R) - g(P)$ on the real line. In particular, $x^0(c) = g(c)$ is between $x^0(a) = g(a)$ and $x^0(b) = g(b)$.

Conversely, suppose that each pair of $a$, $b$, and $c$ are $U$-locally time connected, and that the time coordinate of $c$ is between that of $a$ and $b$. Without loss of generality, we may take the order as being $x^0(a) < x^0(c) < x^0(b)$. By Definition 3.6 and Theorem 3.9, $a$ and $c$ belong to some timepath $\{(x, f_1(x)): x \in I_1\}$, and $b$ and $c$ belong to some timepath $\{(x, f_2(x)): x \in I_2\}$. If we let $P = \{(x, f_1(x)): x^0(a) \leqslant x \leqslant x^0(c)\}$ and $R = \{(x, f_2(x)): x^0(c) \leqslant x \leqslant c^0(b)\}$, we obtain a pair of possibly smaller timepaths $P$ and $R$, such that $P$ contains $a$ and $c$ but not $b$, and $R$ contains $b$ and $c$ but not $a$. Furthermore, these two timepaths can be 'pasted together' at $c$: for any point of $P \cup R$ other than $c$, the condition of 3.29(ii) is true for $P \cup R$, because it is true for each of the given timepaths. On the other hand, at $c$, one takes an open set $W_1$ which satisfies 3.29(ii) with respect to $P$, and an open set $W_2$ which satisfies 3.29(ii) with respect to $R$, and then lets $W$ be any nice open set such that $c \in W \subset W_1 \cap W_2$. Then if $d \in P \cap W$ and $e \in R \cap W$ we have $d < c < e$ (using notation with respect to the nice set $W$ and assuming that $d$ and $e$ are distinct from each other and from $c$), and if $d$ and $e$ belong to the same one of the sets $P \cap W$ and $R \cap W$, we have $d \tau e$ by 3.29(ii). In either case, 3.29(ii) is satisfied by the set $P \cup R$. Hence $P \cup R$ is a timepath, which shows that $c$ is $U$-locally temporally between $a$ and $b$. ∎

*Proof of Theorem* 3.11. Suppose that $a$, $b$, and $c$ are three points of $U$ such that each two of them are $U$-locally time connected. It then follows from the explicit representation for timepaths given in Theorem 3.9 that the time coordinates $x^0(a)$, $x^0(b)$, and $x^0(c)$ are distinct, since each pair of $a$, $b$, and $c$ lies on some timepath contained in $U$. Hence precisely one of the points $a$, $b$, $c$ has a time coordinate between that of the other two. By Theorem 3.13, this point, and only it, is $U$-locally temporally between the other two. Therefore $U$ is admissible. ■

*Proof of Theorem 3.12a:* Suppose that $a$ and $b$ are connected by a segment $P$ of a timelike geodesic lying in $U$. Let $u$ be any parameter for $P$, so that $P = \{f(u): u \in I\}$ for some interval $I$ and continuous function $f$. If we can show that $f$ is one-to-one, then it will follow by Proposition 3.31 that the (closed) segment of $P$ lying between $a$ and $b$ is a timepath, so that $a$ and $b$ are $U$-locally time connected.

If $f$ were not one-to-one, then we could find $u_1 < u_2$ such that $f(u_1) = f(u_2)$. Note that $f$ is one-to-one if we restrict it to a small enough interval surrounding $u_1$ (take the interval small enough that $f(u)$ must lie in some nice open set surrounding $f(u_1)$, and then apply geodetic uniqueness in that set). By the continuity of $f$, there must be a smallest number $w > u_1$ such that $f$ fails to be one-to-one on the closed interval $[u_1, w]$. Similarly, there is a largest number $v < w$ such that $f$ fails to be one-to-one on $[v, w]$. Then $f(v) = f(w)$, but $f$ is one-to-one on every proper subinterval of $[v, w]$.

Take $x$ and $y$ such that $v < x < y < w$. Then $f$ is one-to-one on each of the intervals $[v, x]$, $[v, y]$, $[x, y]$, $[x, w]$, and $[y, w]$. By Proposition 3.31, the image of each of them under $f$ is a timepath. Hence $f(x)$ is $U$-locally temporally between $f(v)$ and $f(y)$, and $f(y)$ is $U$-locally temporally between $f(x)$ and $f(w)$, by Definition 3.5. But then both $f(x)$ and $f(y)$ have the property that they are $U$-locally temporally between the other two of the three points $f(v) = f(w)$, $f(x)$, and $f(y)$, which contradicts the hypothesis that $U$ is admissible. ■

*Proof of Theorem 3.12b:* By Theorem 3.11, the nice open set $U$ is admissible, so the 'if' part has been proved by Theorem 3.12(a). Introduce notation relative to the nice open set $U$. To prove the 'only if' part, we must show that if $a$ and $b$ lie on a timepath $P$ contained in $U$, then $a \tau b$.

Write $P$ in the form $\{(x, f(x)): x \in I\}$, and assume without loss of generality that $x^0(a) < x^0(b)$. By 3.29(ii), there is an open covering of $P$ (in its relative topology) by open sets $P \cap W$, such that $d \tau e$ whenever $d$ and $e$ belong to the same open set in this covering. Using the connectedness of the interval $I$, we see that for any $c \in P$ satisfying $x^0(a) < x^0(c)$, there is a finite sequence $d_1 = a, d_2, d_3, \ldots, d_n = c$ of points of $P$, such that $x^0(a) = x^0(d_1) < x^0(d_2) < \ldots < x^0(d_n) = x^0(c)$, with adjacent terms in this sequence always belonging to a member of the open covering (because the set of all $x^0(c)$ which can be reached by a sequence of this form, together with all $x^0(c) \leqslant x^0(a)$, forms a set that is both open and closed in $I$). In terms of the time order relation $<$ in $U$, we then have $a = d_1 < d_2 < \ldots < d_n = c$. It now follows from Lemma 3.16 that we have $a\tau c$ for any $c \in P$ satisfying $x^0(a) < x^0(c)$. Consequently $a\tau b$. ∎

## IV. CONCLUDING REMARK

The ideas in this paper first had their origin in speculations about time travel (or the impossibility thereof). Naturally, if one wishes to consider paths through the space-time continuum that would carry an object into regions of its own past (perhaps under the influence of strange and as yet unknown forces), one would want to know the extent to which time order is a consequence of the basic geometry of the general relativistic continuum, independent of coordinate systems and of the usual causal principles and the usual behavior of the more common types of particles (i.e., those that we have observed to date).

The author wishes to express his gratitude to many writers of science fiction for suggesting this line of thought.

*Stanford University*

## BIBLIOGRAPHY

Adler, R., Bazin, M., and Schiffer, M., *Introduction to General Relativity*, McGraw-Hill, New York 1965.
Basri, S. A., *A Deductive Theory of Space and Time*, North-Holland Publ. Co., Amsterdam, 1966.
Carathéodory, C., 'Zur Axiomatik der speziellen Relativitätstheorie', *Sitzungsberichte der Preussischen Akademie der Wissenschaften, Sitzung der physikalisch-mathematischen Klasse,* 1924, pp. 12–27.
Carnap, R., *Introduction to Symbolic Logic and Its Applications*, Dover, New York, 1958.
Gödel, K., 'An Example of a New Type of Cosmological Solutions of Einstein's Field Equations of Gravitation', *Reviews of Modern Physics* **21** (1949), 447–450.
Helgason, S., *Differential Geometry and Symmetric Spaces*, Academic Press, New York, 1962.
Hocking, J. G. and Young, G. S., *Topology*, Addison-Wesley, Reading, Mass., 1961.
Kelly, J. L., *General Topology*, Van Nostrand, Princeton, N. J., 1955.
Reichenbach, H., *Axiomatik der relativistischen Raum-Zeit-Lehre*, Friedr. Vieweg and

Sohn, Braunschweig, 1924; English translation by Maria Reichenbach, *Axiomatization of the Theory of Relativity*, Univ. of California Press, Berkeley, 1969.

Robb, A. A., *A Theory of Time and Space*, Cambridge Univ. Press, Cambridge, 1914; also, *The Absolute Relations of Time and Space*, Cambridge Univ. Press, Cambridge, 1921; and *Geometry of Time and Space*, Cambridge Univ. Press, Cambridge, 1936.

Schnell, K., *Eine Topologie der Zeit in logistischer Darstellung*, Inaugural-Dissertation, Westfälischen Wilhelms-Universität zu Münster, 1938.

Walker, A. G., 'Axioms for Cosmology', in Leon Henkin, P. Suppes and A. Tarski (eds.), *The Axiomatic Method with Special Reference to Geometry and Physics*, North-Holland Publ. Co., Amsterdam, 1959, pp. 308-21.

Zeeman, E. C., 'Causality Implies the Lorentz Group', *Journal of Mathematical Physics* **5** (1964), 490–493.

Zeeman, E. C., 'The Topology of Minkowski Space', *Topology* **6** (1967), 161–170.

RICHARD H. HUDGIN

# COORDINATE-FREE RELATIVITY

## I. PRELIMINARY

The assumption that the universe can be described using a coordinate system is very strong. To reveal some of the structure in such an assumption, I present in this paper axioms for flat space-time which allow the coordinate system to be *derived*. In this way a structure more general than that of a coordinate system is suggested.

From five simple axioms we prove the validity of the usual coordinate description with Minkowski metric and the usual geodesic equation of motion.

## II. INTRODUCTION

The most basic question to address is what more fundamental structure will replace the coordinate system? In effect, the set of timelike geodesics is used. These paths fill the space and completely characterize the geometry; they are thus a sufficient substitute.

More precisely, let $P = \{i, j, k, ...\}$ be the set of timelike geodesics labeled by indices $i, j, k, ....$ (Each label is for a different geodesic – no repeats.) Each geodesic will have a proper time parameter $s$ which labels points along its path. Thus the ordered pair $(i, s)$ will represent a particular point in space-time labeled by proper time $s$ on geodesic $i$.

Thus far nothing relates one geodesic to another, and to serve this task, communication functions are introduced. One could imagine that signals (perhaps light pulses) join every pair of geodesics. Thus a signal leaving $j$ at $j$-time $s_j$ will hit $i$ at some $i$-time $s_i$. The communication function $\phi_{ij}$ is defined by $\phi_{ij}(s_j) = s_i$.

The properties of these functions are entirely defined by the five axioms where the word 'signal' never appears. Nevertheless this physical interpretation will be useful in following the development of the paper.

Milne (1948) used this communication function in his treatise on kinematic relativity, and Synge (1960) later investigated some properties of a related function in general relativity.

Milne was quick to point out that one big advantage of these communication functions was that a reflected communication, as shown in the diagram, would give primitive time and distance estimates.

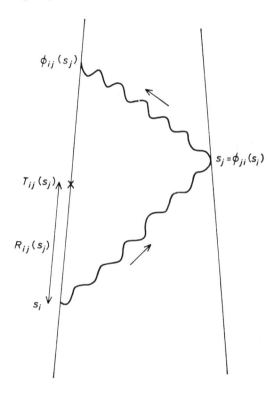

Two linear combinations of $\phi$ and $\phi^{-1}$ are defined:

$$T_{ij}(s) = \tfrac{1}{2}[\phi_{ij}(s) + \phi_{ij}^{-1}(s)]$$
$$R_{ij}(s) = \tfrac{1}{2}[\phi_{ij}(s) - \phi_{ij}^{-1}(s)].$$

Physically, $T_{ij}(s)$ is the time $i$ assigns to the instant $s$ at $j$. $R_{ij}(s)$ may likewise be interpreted as the distance from $j$ to $i$ at $j$-time $s$. If we are dealing with light signals in special relativity, then $T_{ij}(s)$ and $R_{ij}(s)$ are the time and distance measured in $i$'s rest frame.

## III. AXIOMS

The first axiom gives the communication functions their usual differentiability properties. One notes that velocity is discontinuous whenever $R_{ij}(s) = 0$.

*Axiom* 1: Communication

> For all $i, j \in P$ there is a continuous, monotonically increasing function $\phi_{ij}$ which is twice continuously differentiable for $R_{ij}(s) \neq 0$.

The second axiom says that a signal sent from $i$ to $k$ always arrives before or simultaneously with one sent from $i$ to $j$ to $k$. Among other things, this keeps a particle from communicating with its past.

*Axiom* 2: Causality

> For all $i, j, k \in P$ $\quad \phi_{ij} \circ \phi_{jk}(s) \geq \phi_{ik}(s)$.

Note: $f \circ g(s) = f(g(s))$.

The next axiom replaces the concept of inertial frame with a set of rest geodesics (or test particles) which fill the space.

*Axiom* 3: Rest Set

> A subset of $P$, denoted $P_o$, is called a rest set, such that
> (1) For any $i, j \in P_o$,
> $$T_{ij}(s) = s$$
> $$\frac{dR_{ij}}{ds}(s) = \dot{R}_{ij}(s) = 0.$$
> (2) For any $k \in P$ and any $k$-time $\hat{s}$ there is an $i \in P_o$ for which $R_{ik}(\hat{s}) = 0$.

Note: A dot (·) over a function will be standard notation for a derivative.

The fourth axiom, the symmetry axiom, is a very important one since it will yield with little effort time dilatation and other SR results. This axiom replaces Einstein's frame equivalence postulate and states that up to a simple translation of $j$-time $\phi_{ij}(s) = \phi_{ji}(s)$.

*Axiom* 4: Symmetry

> For any $i \in P_o$, $j \in P$ there is a constant $b$ such that $\phi_{ij}(s - b) = \phi_{ji}(s) + b$.

Note: To clarify the form of the axiom we mention that if a new $j$-time is defined by $s' = s + b$ then the new communication functions are

$$\phi'_{ij}(s) = \phi_{ij}(s - b)$$
$$\phi'_{ji}(s) = \phi_{ji}(s) + b.$$

The last axiom is like Euclid's line postulate. It will allow geometrical concepts such as lines, angles, and midpoints. One assumes that there exist arbitrarily slow test particles running between any two points of the rest set.

*Axiom* 5: Connectivity

For any $i, j \in P_o$ and any constant $b > 0$ there is some $k \in P$ and $k$-times $s_1$ and $s_2$ such that

$$R_{ik}(s_1) = R_{jk}(s_2) = 0,$$

and

$$\dot{R}_{ik}(s_1^+) < b.$$

One should mention that while the set $P_o$ seems to occupy a special position in the axioms and thereby may consitute a preferred frame, this is not at all the case. It is possible to reformulate the axioms so that all rest sets are obviously equivalent. However, this would be a stronger set of axioms, and in the interest of economy the above set is preferred.

## IV. PLAN OF ACTION

We begin by investigating the simpler rest set results and the functional relations when $R_{ij}(s) = 0$. This quickly escalates to a global condition on $T_{ij}(s)$, namely, $\ddot{T}_{ij}(s) \equiv 0$. With this condition, the functional equation $\phi(s) + \phi^{-1}(s) = 2bs$ for $b \geq 1$ is studied, and $R(s) = \frac{1}{2}[\phi(s) - \phi^{-1}(s)]$ is shown to be approaching SR form.

The rest of Section V investigates lines in a rest set and, using them, verifies the usual SR equations of motion.

Section VI immediately sets out to make a rest set into a Hilbert space. All the different concepts of a norm, an inner product, addition, etc., are derived from the geometry developed in Section V. Once linear motion for the test particles is demonstrated in the Hilbert space, it is a simple step to the usual metric and Lorentz transforms.

## V. KINEMATICS

*Coincidence*

If two points are coincident (zero distance apart), then the points are intuitively supposed to have the same properties. The first theorem proves this by showing that if $R_{ij}(\hat{s})=0$, then $i$ and $j$ at $j$-time $\hat{s}$ communicate in the same way with all other geodesics.

*Theorem 1*:

For any $i,j,k \in P$; $R_{ij}(\hat{s}) = 0 \Rightarrow R_{kj}(\hat{s}) = R_{ki} \circ \phi_{ij}(\hat{s})$
$$T_{kj}(\hat{s}) = T_{ki} \circ \phi_{ij}(\hat{s}).$$

*Proof*:
The causality axiom gives

$$\phi_{ki} \circ \phi_{ij}(s) \geq \phi_{kj}(\hat{s}) \geq \phi_{ki} \circ \phi_{ji}^{-1}(\hat{s}).$$

But

$$R_{ij}(\hat{s}) = 0 \Rightarrow \phi_{ij}(\hat{s}) = \phi_{ji}^{-1}(\hat{s}).$$

Thus the inequality bounds are tight and

$$\phi_{ki} \circ \phi_{ij}(\hat{s}) = \phi_{kj}(\hat{s}).$$

Since the same proof holds for $\phi_{jk}^{-1}$, one takes linear combinations to get $T_{kj}$ and $R_{kj}$ and is done.

A simple corollary shows that a particle is self-coincident.

*Corollary 1*:
For any $j \in P$ $\phi_{jj}(s) \equiv s$.

*Proof*:
Choose any time $\hat{s}$. From Axiom 3 there is some $i \in P_o$ for which $R_{ij}(\hat{s}) = 0$, i.e., for which $\phi_{ij}(\hat{s}) - \phi_{ji}^{-1}(\hat{s}) = 0$.

This implies that $\phi_{ji} \circ \phi_{ij}(\hat{s}) = \hat{s}$.

Theorem 1 says that (with $j = k$) $\phi_{ji} \circ \phi_{ij}(\hat{s}) = \phi_{jj}(\hat{s})$ and $\phi_{jj}(\hat{s}) = \hat{s}$ is proved.

*Triangle Inequalities*

This next theorem begins the rest set geometry. It shows that $P_o$ has a triangle inequality using the $R_{ij}$ distance function. Causality is again the key.

*Theorem 2:*
For any $i, j \in P_o$ and for any $k \in P$ and $k$-time $\hat{s}$

$$R_{ij} + R_{jk}(\hat{s}) \geqslant R_{ik}(\hat{s}) \geqslant |R_{ij} - R_{jk}(\hat{s})|.$$

(Proof omitted.)

*SR at a Point*

Here the power of the symmetry axiom begins to appear. The usual time dilatation relation

$$\frac{dT}{ds} = \left[1 - \left(\frac{dR}{dT}\right)^2\right]^{-1/2}$$

is derived for particle $j$ just before it arrives at (or just after it leaves) particle $i$.

*Theorem 3:*
For any $i \in P_o$, $j \in P$, $R_{ij}(s_o) = 0 \Rightarrow \dot{R}_{ij}^2(s_o^\pm) = \dot{T}_{ij}^2(s_o^\pm) - 1$ ($s_o^\pm$ means the limit as $s$ approaches $s_o$ from above or below).
*Proof:*
The symmetry axiom says that for some constant $b$

$$\phi_{ji}^{-1}(\phi_{ij}(s) - b) \equiv s + b.$$

In terms of $T_{ij}$ and $R_{ij}$ this becomes

$$(T_{ij} - R_{ij}) \circ (T_{ij}(s) + R_{ij}(s) - b) \equiv s + b.$$

Differentiating this last equation once and letting $s \to s_o^\pm$, we have the theorem.

We note without proof that differentiating twice yields the following result:

*Theorem 4:*
If $R_{ij}(s_o) = 0$, then $\ddot{T}(s_o^\pm)$ $[(\dot{T}(s_o^\pm) + \dot{R}(s_o^\pm))^3 + 1]/[(\dot{T}(s_o^\pm) + \dot{R}(s_o^\pm))^3 - 1] = \ddot{R}(s_o^\pm)$ (redundant $ij$ dropped).

This is not an SR result by itself since $\ddot{R}(s_o^\pm)$ and $\ddot{T}(s_o^\pm)$ both vanish in special relativity. However, the final result follows shortly in Theorem 7.

*Smoothness of $T_{ik}$*

Some simple facts about $T_{ik}$ and its derivatives are pointed out in this section. Trivially, Axiom 3 implies that for any $i, j, k \in P_o$ $T_{jk} \equiv T_{ik}$.

Using coincidence and the denseness of $P_o$ in $P$, allowing $k$ to be in $P$ instead of $P_o$ will give the same result, namely,

$$T_{jk}(s) \equiv T_{ik}(s) \quad \text{for any } i, j \in P_0.$$

An immediate ramification is

*Theorem 5*:

For any $i \in P_o, j \in P$ $T_{ij}(s)$ is twice continuously differentiable everywhere.

*Proof*:

The communication axiom suffices unless $R_{ij}(s) = 0$. Suppose $i$ and $j$ are coincident as $s = s_o$. Then choose any point $k \in P_o$ which is not coincident with $i$. $T_{kj}(s)$ and both derivatives are continuous in a neighborhood of $s_o$ by Axiom 1; thus the same holds for $T_{ij}$ since $T_{ij} \equiv T_{kj}$.

*Velocity Restriction*

Several results can now be assembled to limit the particle motion. The first theorem is a consequence of the triangle inequality and SR at a point.

*Theorem 6*:

For any $i, j \in P$ $\dot{R}_{ij}^2(s) \leq \dot{T}_{ij}^2(s) - 1$.

*Proof*:

For any $j$-time $s_o$ there is an $l \in P_o$ for which $R_{lj}(s_o) = 0$. From Theorem 2

$$R_{il} + R_{lj}(s) \geq R_{ij}(s) \geq R_{il} - R_{lj}(s).$$

Since equality holds for $s = s_o$, the inequality can be differentiated to get

$$\pm \dot{R}_{lj}(s_o^\pm) \geq \pm \dot{R}_{ij}(s_o^\pm) \geq \mp \dot{R}_{lj}(s_o^\pm).$$

With Theorem 3 and the continuity of $\dot{T}_{lj}$, one has

$$|\dot{R}_{ij}(s_o^\pm)| \leq (\dot{T}_{lj}^2(s_o) - 1)^{1/2}.$$

Since $T_{ij} \equiv T_{lj}$, we are done.

*T Restriction*

$\ddot{T}_{ij}(s)$ can now be shown to vanish by differentiating Theorem 6 and using continuity of the derivatives.

*Theorem 7*:

For any $i \in P_o, j \in P$, $\ddot{T}_{ij}(s) \equiv 0$.

*Proof*:

Let $k \in P_o$ be coincident with $j$ at $s_o$. If $\ddot{T}_{kj}(s_o) = 0$, then $\ddot{T}_{ij}(s_o) = 0$ too since $T_{ij} \equiv T_{kj}$. The Theorem 6 inequality has equality at $s = s_o$ when applied to $j$ and $k$ and can thus be differentiated to get

$$\pm \dot{R}(s_o^\pm) \ddot{R}(s_o^\pm) \leq \pm \dot{T}(s_o^\pm) \ddot{T}(s_o^\pm),$$

where the redundant $jk$ index has been dropped.

Since $\dot{R}(s_0^+) = -\dot{R}(s_0^-)$ from Theorem 3 and the continuity of $\dot{T}$, and since $\ddot{R}(s_0^+) = -\ddot{R}(s_0^-)$ from the continuity of $\ddot{T}$ and Theorem 4, the above inequality becomes

$$\dot{T}(s_o) \ddot{T}(s_o) = \dot{R}(s_o^+) \ddot{R}(s_o^+).$$

This equation plus the equation of Theorem 4 applied to $j$ and $k$ at $s = s_o$ gives a system of two homogeneous equations in $\ddot{T}$ and $\ddot{R}$. Either $\ddot{T} = \ddot{R} = 0$ or else the determinant of the coefficients vanishes. One can check that the determinant vanishes only for $\dot{R}(s_o) = 0$ which itself implies that $\ddot{T} = 0$.

*Motions Restrictions*

The symmetry axiom says that for $i \in P_o, j \in P$ $j$-time can be translated so that $\phi_{ij} = \phi_{ji}$. (Call them just $\phi$.) Then the previous theorem tells us that

$$T(s) = \tfrac{1}{2}[\phi(s) + \phi^{-1}(s)] = as + b$$

where $a$ and $b$ are constants and $a \geq 1$ (Theorem 6).

Thus the problem of calculating $\phi_{ij}(s)$ has been reduced to solving the functional equation

$$\phi(s) + \phi^{-1}(s) = 2as + 2b.$$

If there is a point $\hat{s}$ where $R(\hat{s}) = \tfrac{1}{2}[\phi(\hat{s}) - \phi^{-1}(\hat{s})]$ vanishes, then the Appendix shows that the solution is

$$R(s) = (a^2 - 1)^{1/2} |s - \hat{s}|.$$

If there is no point where $R(s) = 0$, then the solution is no longer unique but at least satisfies

$$|R(s) - (R^2(\hat{s}) + (s - \hat{s})^2 (a^2 - 1))^{1/2}| \leq NR(\hat{s})(a^2 - 1)^{1/4},$$

where $R(\hat{s}) = \inf_s R(s)$ and $N$ is a constant of order one.

Note: There is always a finite $\hat{s}$ where $R(s)$ takes its minimum. To see this let $k \in P_o$ be coincident with $j$ at $s_0$. Then $R_{kj}(s) \to \infty$ as $s \to \pm \infty$. The triangle inequality then has $R_{ij}(s) \to \infty$ as $s \to \pm \infty$. $R_{ij}(s)$ must therefore have its minimum at some finite value of $s$.

The important point is that the error term vanishes as $a \to 1$. It will be shown that in $P_o$ all particles running between two rest points follow the same path. Then using arbitrarily slow particles running along a fixed path (i.e., letting $a \to 1$), perfect SR motion will be demonstrated for all particles.

*Lines*

First it is shown that all particles moving through $P_o$ trace out lines, which are defined to be ordered sets of collinear points.

*Definition*:

Let $\beta(s)$ be a $P_o$-valued function of $s$. $\beta(s)$ is a line if for some constant $b$

$$R_{\beta(s_1)\beta(s_2)} = b|s_1 - s_2| \quad \text{for all} \quad s_1, s_2.$$

Note: Any linear reparametrization of a line is also a line.

*Theorem 8*:

If $\beta(s)$ is the path traced in $P_o$ by a particle $k \in P$ (i.e., if $R_{\beta(s)k}(s) = 0$), then $\beta(s)$ is a line.

*Proof*:

Choose any $s_1$ and $s_2$. Since $k$ is coincident with $\beta(s_j)$ at $s_j$,

$$R_{\beta(s_1)\beta(s_2)} = R_{\beta(s_1)k}(s_2) = (a^2 - 1)^{1/2} |s_1 - s_2|,$$

where $a = T_{\beta(s)k}$, a constant independent of $s$.

*Line Uniqueness*

The next step toward special relativity is to show that two distinct rest points determine a unique line.

*Theorem 9*:

Given two distinct points $i, j \in P_o$, there is a unique line $\beta(s)$ passing through both points in the sense that if $\hat{\beta}(s)$ is another such line there are constants $a, b$ for which

$$\hat{\beta}(s) = \beta(as + b).$$

*Proof*:

The existence of a line follows from the connectivity axiom and Theorem 8. It will be assumed that all lines have been linearly reparametrized so that $\beta(0) = i$ and $\beta(1) = j$. Then one must prove that $\hat\beta(s) = \beta(s)$.

Let $k = \beta(s_o)$. Then $k$ is a point collinear with $i$ and $j$. (Assume $R_{ij} + R_{jk} = R_{ik}(= s_o)$ for concreteness.) Let $m_n$ be a sequence of test particles in $P$ connecting $i$ to $j$ tracing out reparametrized lines $\beta_n(s)$ where $\dot{T}_{im_n} \to 1$ as $n \to \infty$. The connectivity axiom allows this.

We know that

(1) $\quad R_{k\beta_n(0)} = R_{ik}$
(2) $\quad R_{k\beta_n(1)} = R_{jk} = R_{ik} - R_{ij}$
(3) $\quad R_{k\beta_n(s)}/R_{ij}|s| \to 1$ as $s \to \pm\infty$.

(Condition 3 comes from $R_{i\beta_n(s)} + R_{ik} \geqslant R_{k\beta_n(s)} \geqslant R_{i\beta_n(s)} - R_{ik}$.)

If one had perfect SR motion, these three conditions would fix the motion, and one could conclude $R_{k\beta_n(R_{ik})} = 0$. As it is, SR motion is approached as $n \to \infty$.

Thus

$$\lim_{n \to \infty} R_{k\beta_n(s_o)} = 0 \quad \text{or} \quad \lim_{n \to \infty} \beta_n(s_o) = k.$$

If $\hat\beta(s)$ is another reparametrized line through $i$ and $j$, then $\beta_n(s_o)$ converges to both $\hat\beta(s_o)$ and $\beta(s_o)$ which means that $\hat\beta(s_o) = \beta(s_o)$.

For later use in constructing a coordinate system we state the following corollary and omit its proof.

*Corollary 9*:

For any $i, j \in P_o$ there is a unique $k \in P_o$ for which $R_{ik} = R_{jk} = \frac{1}{2}R_{ij}$. (Unique midpoint.)

*Perfect SR Motion*

Since all moving particles trace out lines, and since two distinct points determine a unique line, one concludes that any particle line is retraced by arbitrarily slow particles. This is used to prove:

*Theorem 10*:

For any $i \in P_o, j \in P$ $R_{ij}(s)$ has the form

$$R_{ij}^2(s) = R_o^2 + a^2(s - s_o)^2,$$

where $a^2 = \dot{T}_{ij}^2 - 1$.

*Proof*:

Let $m_n$ be a sequence of particles retracing $j$'s path arbitrarily slowly, i.e., $a_n = (\dot{T}_{im_n}^2 - 1)^{1/2} \to 0$ as $n \to \infty$. For convenience let $s = 0$ at the point of the path closest to $i$, and call the closest distance $R_o$.

Then

$$\lim_{n \to \infty} |R_{im_n}(s) - (R_o^2 + a_n^2 s^2)^{1/2}| = 0$$

uniformly in $s$.

But

$$R_{ij}(s) - (R_o^2 + a^2 s^2)^{1/2} = R_{im_n}\left(\frac{as}{a_n}\right) - (R_o^2 + a^2 s^2)^{1/2}$$

from our reparametrization condition in Theorem 9.

Since the right side goes uniformly to zero as $n \to \infty$, the left side is zero.

## VI. COORDINATE SYSTEM

The easiest way to construct a coordinate system is to make a Hilbert space out of $P_o$. Then the orthonormal basis will yield the usual Cartesian coordinate patch.

*Operations*

First choose any point $\mathfrak{O} \in P_o$ as an origin. Then define the following operations:

*Definition*:

For any $x \in P_o$ let $\beta(s)$ be the unique line with $\beta(0) = \mathfrak{O}$ and $\beta(1) = x$. Define $ax = \beta(a)$.

*Definition*:

For any $x, y \in P_o$ let $x + y$ be the unique midpoint of $2x$ and $2y$.

*Definition*:

For any $x, y \in P_o$ let

$$(x, y) = \tfrac{1}{2}(R_{\mathfrak{O}x}^2 + R_{\mathfrak{O}y}^2 - R_{xy}^2).$$

Now the various properties of a linear space with a positive definite inner product must be demonstrated. For ease a norm is defined that is clearly positive definite.

*Definition*:
For any $x \in P_o$ $|x| = (x, x)^{1/2} = R_{\mathfrak{D}x}$.
Some facts are obvious, e.g.,

$$(x, y) = (y, x) \qquad x + y = y + x$$
$$ax + bx = (a + b) x \qquad \mathfrak{D} = 0x$$
$$|ax| = |a| \, |x| \qquad \mathfrak{D} + x = x$$
$$x + y = 0 \Rightarrow x = (-1) y.$$

What must be shown are (1) linearity of (...)
(2) associativity
(3) distributivity.

Actually (2) and (3) follow easily from (1) and the positive definiteness. For (2) one shows that

$$|[(x + y) + z] - [x + (y + z)]|^2 = 0$$

by expanding the inner product. For (3) one shows

$$|a(x + y) - (ax + ay)|^2 = 0.$$

Item (1) has two parts:

(1a) $(x + y, z) = (x, z) + (y, z)$
(1b) $(ax, z) = a(x, z)$.

Both results will follow simply from the next theorem.

*Theorem 11*:
For any $x \in P_o, j \in P$ and for $\beta_j(s) \in P_o$ coincident with $j$ at $s$

$$\frac{d^2}{ds^2}(x, \beta_j(s)) = 0.$$

*Proof*:
By definition $(x, \beta_j(s)) = \frac{1}{2}(R_{\mathfrak{D}x}^2 + R_{\mathfrak{D}j}^2(s) - R_{xj}^2(s))$.
But $(d^2/ds^2) R_{yj}^2(s) = 2(\dot{T}_{\mathfrak{D}j}^2 - 1)$ for any $y \in P_o$. Substitute. Finally linearity can be shown.

*Theorem 12*:
The inner product is linear.

*Proof*:
Theorem 11 proved that $(x, \beta(s))$ is a linear function for any line $\beta(s)$. Let $\beta(0) = \mathfrak{O}$ and $\beta(1) = y$. Then

$$(x, ay) = (x, \beta(a)) = a(x, \beta(1)) = a(x, y).$$

Now let $\beta(s)$ be a line joining $2x$ and $2y$. Again $(z, \beta(s))$ is a linear function which implies that

$$\tfrac{1}{2}[(z, 2x) + (z, 2y)] = (z, x + y),$$

since $x + y$ is the midpoint of the $2x$ to $2y$ line segment. Using $(z, 2w) = 2(z, w)$, we substitute and are done.

## Hilbert Space

Now $P_0$ has been shown to be a pre-Hilbert space (everything except completeness). If the space is finite dimensional, we are done. If the space is infinite dimensional, then one can use a general theorem:[1]

> A pre-Hilbert space is isomorphic and isometric to a dense subset of a Hilbert space.

Physically, a pre-Hilbert and a Hilbert space are indistinguishable, and one may as well assume completeness. This prevents dealing with convergent sequences all the time.

Now taking an orthonormal coordinate system for the space $P_o$, one has a usual Cartesian type patch where moving particles follow straight lines (Theorem 11). Also $dR^2 = |d\vec{x}|^2$ in this patch since

$$|\vec{x} - \vec{y}|^2 = R_{xy}^2.$$

If a time coordinate $x^o(k, s) = T_{\mathfrak{O}k}(s)$ is added, then the usual relation to proper time will appear for every moving particle $k$. Namely

$$\eta_{\mu\nu} \frac{dx^\mu}{ds} \frac{dx^\nu}{ds} = 1$$

where $n_{\mu\nu}$ is the Minkowski metric with signature $1, -1, -1, \ldots$.

This result comes immediately from substituting

$$\dot{T}_{ik}(\hat{s}) = \frac{dx^o}{ds}(\hat{s})$$

$$\dot{R}_{ik}^2(\hat{s}) = \left|\frac{d\vec{x}}{ds}\right|^2(\hat{s})$$

for $i \in P_o$ and $R_{ik}(\hat{s}) = 0$ into

$$\dot{R}_{ik}^2(\hat{s}) = \dot{T}_{ik}^2(\hat{s}) - 1.$$

Thus it has been shown that in $P_o$ there is a space-time coordinate system in which all particles follow straight lines with the correct metric relating motion to proper time.

Now if the usual set of flat space test particles is denoted by $F$, the question is how $P$ fits into $F$. What has been proved is that $P \subseteq F$. However, $P$ is not necessarily all of $F$. To guarantee this a stronger connectivity axiom, such as the following, is necessary.

*Axiom 5'*:

For any $i, j \in P_o$ and for any constant $b > 0$, there are a test particle $k \in P$ and a $j$-time $\hat{s} > 0$ such that

$$R_{ki}(0) = R_{kj}(\hat{s}) = 0$$

and

$$R_{ki}(0^+) = b.$$

The simpler Axiom 5 was used instead since it alone was required to verify coordinatization and SR motion.

## VII. CONCLUSIONS

The five axioms used have been shown to imply that the rest set can be coordinatized with the Minkowski metric and that the geodesics in $P$ are the usual timelike straight lines. The axioms surprisingly do not involve equations of motion for the geodesics nor do they involve properties of coordinate patches. They are five qualitative statements about the geodesics which completely characterize flat space time. In this respect, this axiomatic treatment differs dramatically from the usual presentation in terms of neighborhoods and curvature tensors.

It should be emphasized that the set $P$ of geodesics or test particles is not meant to represent paths of physical masses. Rather, it is planned that $P$ will replace the coordinate system in that all interactions will be referred to the $P$ net of geodesics instead of $x, y, z, t$.

In this way, one will have a more general framework for space-time than the usual coordinate system. Perhaps then one can begin to address questions about why we have the space-time we do or in what manner space-time may break down in the microscopic limit. To attack these questions one should have a tool available more general than the usual Riemannian manifold. The hope is that the geodesic description offered above will provide such a tool.[2]

## APPENDIX

In this section are collected the purely mathematical results concerning the functional equation

$$\phi(s) + \phi^{-1}(s) = 2as + b \quad a \geq 1,$$

where $\phi(s)$ satisfies Axiom 1, $\phi(s) \geq s$ (Causality), and $|\dot{R}(s)| \leq (a^2 - 1)^{1/2}$ (Theorem 3). This last restriction on $R(s)$ is implied by the first two, but since it has been shown by other means, we just state it without a redundant proof.

First, the simpler case where $R(s)$ passes through zero is treated.

*Theorem A1:*
If $R(\hat{s}) = 0$, then $R(s) = (a^2 - 1)^{1/2}|s - \hat{s}|$.
*Proof:*
First make $\hat{s} = 0$ by translating the argument.

$$\bar{\phi}(s) = \phi(s + \hat{s}) - \hat{s}.$$

Then

$$\bar{\phi}(s) + \bar{\phi}^{-1}(s) = 2as \quad \bar{R}(s) = R(s + \hat{s}).$$

Dropping the ¯ now, the equation is rewritten as

$$\phi \circ \phi(s) = 2a\phi(s) - s.$$

Differentiating, one finds

$$\phi' \circ \phi(s) = 2a - 1/\phi'(s).$$

This is a recursion relation with the property that for $a - (a^2 - 1)^{1/2} < \phi'(s) \leq a + (a^2 - 1)^{1/2}$

$$\lim_{n \to \infty} \phi' \circ \phi^n(s) = a + (a^2 - 1)^{1/2}.$$

If we let $n \to \infty$ and keep $\phi^n(s)$ fixed at $s_o > 0$, then $s \to 0^+$ (or another $R(s) = 0$ point), $\phi'(s) \to a + (a^2 - 1)^{1/2}$ from below since $\phi'(0^+) = a + (a^2 - 1)^{1/2}$ and $\phi'(s)$ is always $\leq a + (a^2 - 1)^{1/2}$.

Thus $\phi'(s_o)$ must be $a + (a^2 - 1)^{1/2}$ and

$$\phi(s) = [a + (a^2 - 1)^{1/2}]s \quad \text{for} \quad s \geq 0.$$

Similarly,

$$\phi(s) = [a - (a^2 - 1)^{1/2}]s \quad \text{for} \quad s \leq 0.$$

Together the results imply $R(s) = (a^2 - 1)^{1/2}|s|$.

Finally the case where the $\inf_s R(s) = R(\hat{s})$ is studied.

*Theorem A2*:

If $\inf R(s) = R(\hat{s}) > 0$, then

$$|R(s) - (R^2(\hat{s}) + (a^2 - 1)(s - \hat{s})^2)| \leq NR(\hat{s})(a^2 - 1)^{1/4},$$

where $N$ is a constant of order unity.

*Proof*:

First let $\hat{s} = 0$ by translating the argument as in Theorem 1. The recursion relation $\phi \circ \phi(s) = 2a\phi(s) - s$ gives $\phi(s)$ on the interval $[\phi(s_o), \phi \circ \phi(s_o)]$ in terms of $\phi(s)$ on the interval $[s_o, \phi(s_o)]$ (or vice versa). Thus if $\phi(s)$ is specified for $[O, \phi(O)]$, $\phi(s)$ is known everywhere. Here the $R(s) \neq 0$ condition insures that the intervals fill the entire real line.

Now the coupled recursion relations

$$\phi_{n+1} = 2a\phi_n - s_n$$
$$s_{n+1} = \phi_n$$

have as a general solution

$$\phi_n = A\alpha^n + B/\alpha^n$$
$$s_n = A\alpha^{n-1} + B/\alpha^{n-1},$$

where $\alpha = a + (a^2 - 1)^{1/2}$ and $A, B$ are constants.

With some algebra one can show that

$$\phi_n = as_n + \sqrt{C + (a^2 - 1)s_n^2}.$$

$C$ is a constant for each sequence, but still depends upon which sequence is involved, One can write

$$R(s) = [C(s) + R_o^2 + (a^2 - 1) s^2]^{1/2},$$

where

$$C \circ \phi(s) = C(s) \quad \text{and} \quad R_0 = R(0).$$

Now for $s \in [\phi^{-1}(O), \phi(O)] \equiv I$, the $R$ restriction gives

$$|R(s) - R_o| \leqslant (a^2 - 1)^{1/2} \circ 2R_o.$$

This plus $R(s) \geqslant R_o$ gives for $s \in I$ bounds on $C(s)$ in terms of $a$ and $R_o$. The result is that for $a$ in a neighborhood of 1

$$|C(s)| \leqslant N^2 (a^2 - 1)^{1/2} R_o^2$$

for some constant $N$. ($C(s)$ takes all its values in $I$.)

The restriction on $C(s)$ gives the final result using $|\sqrt{x+y} - \sqrt{x}| \leqslant \sqrt{|y|}$.

*State University of New York*

### ·NOTES

[1] Any textbook on functional analysis will suffice. See for instance Yosida (1968, p. 56).
[2] The more general manifold using a geodesic net instead of a coordinate system has now been developed complete with curvature tensor and Bianchi identities. The work is shortly to be published.

### BIBLIOGRAPHY

Milne, E. A., *Kinematic Relativity*, Clarendon Press, Oxford, 1948.
Synge, J. L. *Relativity: The General Theory*, North-Holland Publ. Co., Amsterdam, 1960.
Yosida, K., *Functional Analysis*, Springer-Verlag, New York, 1968.

PATRICK SUPPES

# SOME OPEN PROBLEMS IN THE PHILOSOPHY OF SPACE AND TIME

ABSTRACT. This article is concerned to formulate some open problems in the philosophy of space and time that require methods characteristic of mathematical traditions in the foundations of geometry for their solution. In formulating the problems an effort has been made to fuse the separate traditions of the foundations of physics on the one hand and the foundations of geometry on the other. The first part of the paper deals with two classical problems in the geometry of space, that of giving operationalism an exact foundation in the case of the measurement of spatial relations, and that of providing an adequate theory of approximation and error in a geometrical setting. The second part is concerned with physical space and space-time and deals mainly with topics concerning the axiomatic theory of bodies, the operational foundations of special relativity and the conceptual foundations of elementary physics.

Philosophical analysis and speculation about the concepts of space and time are as old as philosophy itself. Concurrent with the astounding technical development of Greek mathematical and observational astronomy, a polished and carefully articulated theory of space and time was set forth early in the Hellenistic period by Aristotle, especially in the *Physics*. Aristotle's *Physics*, Euclid's *Elements* and Ptolemy's *Almagest* form a triad that elaborate the philosophical, mathematical and physical foundations of space and time in ancient philosophy. Although Ptolemy was Aristotelian in his philosophical attitudes, a clear divergence between Aristotle on the one hand and Euclid and Ptolemy on the other is obvious. These two quite distinct traditions, one Aristotelian and philosophical, and the other mathematical and Euclidean or Ptolemaic, continued in the succeeding millenium and a half leading up to the outburst of modern science in the seventeenth century.

The separate life of the two traditions did not stop even there. Descartes' *Principles of Philosophy*, for example, is much closer in spirit to Aristotle's *Physics* than to Euclid's *Elements* or Ptolemy's *Almagest*. In spite of the fact that in other domains Descartes was a creator of new mathematical concepts, in his *Principles* there is no genuine mathematical organization or development of ideas. From a philosophical standpoint, it is evident that the closeness of argument characteristic of Aristotle's *Physics* is not matched, and there is a general degradation of intellectual standard.

The continuation of the Euclidean-Ptolemaic tradition is quite otherwise. Newton's *Principles*, first published in 1687, is very much in the spirit of Ptolemy's *Almagest* and satisfies a standard of intellectual rigor and clarity that would have been acceptable in Alexandria in Ptolemy's time. In Newton's *Principles* there is no sharp separation of mathematics and physics, or of mathematics and astronomy. To a large extent this fusion of mathematical and astronomical investigations was continued a hundred years later in Laplace's *Celestial Mechanics*.

The separation of mathematical investigations on the one hand, and physical or astronomical investigations on the other, did not really occur until the nineteenth century. By the latter half of that century there were very few examples of individuals making original contributions both to mathematics and to physics. Separation of the intellectual traditions was nearly complete by the beginning of this century. I perhaps need to be more explicit in defining this separation. Certainly physicists continued to use mathematics, and to use it with great power and sophistication, but original contributions to the foundations of mathematics and original contributions to the conceptual foundations of physics were not made by the same people. Of course a small number of individuals like von Neumann and Hermann Weyl made significant contributions to both domains, but still generalization is, I think, a sound one, and midway through the last half of the twentieth century it is more valid than earlier.

This scientific separation has given rise to a separation within philosophy, so that to a large extent the philosophical foundations of mathematics, including the foundations of geometry, are now an almost totally separate subject from the philosophy of space and time. In this article I would like to describe some open problems in the philosophy of space and time that require the methods characteristic of mathematical traditions in the foundations of geometry for their solution, and thereby to encourage within philosophy a fusion of the two traditions.

I have organized the analysis of problems under two main headings. In the first section I am concerned with the geometry of space and deliberately deal with classical questions that do not take into account the theory of relativity. In the second section I turn to physical space and space-time, including such classical problems as the formulation of an adequate theory of bodies.

## I. GEOMETRY OF SPACE

Because of extensive development, the foundations of geometry are a natural proving ground for more general philosophical concepts in the philosophy of science, but surprisingly little has been done to use this proving ground. I restrict myself to two classes of problems. The first class deals with the attempt to give operationalism a sharp foundation in the case of the measurement of spatial relations or, more generally, in terms of the geometry of space. The second class deals with combining measurement and error to yield some systematic theory of approximation. The idea of such approximations is familiar both in physics and in psychology. What is not familiar in either discipline is the development of geometrical foundations of such approximations.

### A. *Operational Foundations*

Compared with the notion of constructivity in the foundations of mathematics, there have been few attempts to give a sharp formulation of the concept of an operational definition, or of operationalism, in the philosophy of science. The foundations of geometry provide an excellent place to give such a formulation. In the first place, perhaps the oldest issues concerning constructivity in mathematics are to be found in the foundations of geometry. Certainly the three classical problems of Greek elementary geometry – squaring a circle, duplicating a cube and trisecting an angle – are examples of constructive problems. The beautiful thing about these problems is that we can approach the foundations of geometry in a qualitative way, but with the objective of providing a precise solution to the problems. Such a solution, of course, is negative for elementary operations. Problems of a comparable sort have not been formulated in the foundations of physics, but I do not explore that aspect of the problem in this section. In the present context, I want to concentrate only on the purely geometrical aspects of constructivity.

Reflection on the three classical problems or, at a more mundane level, examination of the propositions in the early books of Euclid's *Elements* suggests that existential statements are always backed up by a highly constructive sequence of operations. From a mathematical standpoint, especially from an algebraic one, the natural idea then is to formulate the qualitative foundations of geometry in terms of operations rather than in terms of relations and existential statements about these relations. From a general philosophical standpoint the problem can be expressed as follows: *Characterize operations that can be performed on spatial points so that from the known properties of the operations, the usual properties of space can be derived.*

A first thought might be that these operational problems are already solved by the standard formulations of axioms for vector spaces, but this is not the case for two reasons. First, the vector spaces themselves are not the same in structure as the ordinary Euclidean spaces, because of the distinguished point of origin. Second, the operations in the vector spaces do not correspond to the operations needed, for example, to solve the problems formulated in Book 1 of Euclid's *Elements*, and it is clear that from a geometrical standpoint operations that are closer to the Euclidean constructions are available with many alternative possibilities open.

Put another way, an operationally satisfactory formulation of the constructive part of Euclidean geometry should be a theory in standard formulation, that is, a theory that is formulated within first-order logic with identity and that is also quantifier-free. The reason for the quantifier-free requirement should be apparent from what has already been said. An existential statement in constructive parts of geometry is misleading, because a specific and definite constructive method of finding the point existentially postulated is known. A conceptual discrepancy exists between the axioms and the methods of construction when a general existential statement rather than a specific sequence of constructive operations is postulated. A reason for insisting on such a viewpoint in geometry is that it is possible to get a thorough understanding of the operational situation in a way that is not at present possible in physics. We can realistically hope to give a theory with standard formalization that fully characterizes constructive Euclidean geometry and that does so in an elementary way. It is not yet clear that we understand how to do this in any thoroughgoing fashion for substantial parts of physics, although this is a topic on which I shall have more to say later.

In an earlier paper (Moler and Suppes, 1968), a constructive formulation of geometry in the sense just defined was given. This formulation depends on two primitive operations: one the operation of finding the point of intersection of two line segments; and the other, the operation of laying off one line segment on another. Both of these operations are discussed in some detail in Hilbert's *Foundations of Geometry*, but our task was to give an explicit axiomatic formulation in terms of just these two operations and in quantifier-free form. The axioms turn out to be complicated, and a simpler and more elegant quantifier-free formulation in terms of other primitive operations is needed. For example, let $I$ be the intersection operation and $S$ the laying-off operation so that $I(xyuv)$ is the point of intersection of the line determined by $x$ and $y$ with the line determined by $u$ and $v$, and $S(xyuv)$ is the point as distant from $u$ in the direction of $v$ as $y$ is from $x$. Then betweenness is defined by:

$B(xyz)$ iff [if $x \neq z$ then $S(xyxz) = y = S(zyzx)$] & $[x = z \to x = y]$,

collinearity is defined by:

$L(xyz)$ iff $S(xyxz) = y$ or $S(zyzx) = y$ or $x = z$,

and noncollinearity of four points is defined by:

$NL(xyuv)$ iff not $(L(xyu)$ or $L(yuv)$ or $L(xuv)$ or $L(xyv))$.

Euclid's axiom, the most complicated of the 18 axioms of the system, then has the following formulation:

if $NL(xyuv)$ & $B(x, I(x, S(xyuv), y, u), S(xyuv))$ & $S(y, S(xyuv), x, u) \neq u$ then $L(x, y, I(xyuv))$,

which is far from transparent in its content, although we know an axiom of approximately this sort is necessary.

One conjecture is that it is a mistake to take points as the primitive objects. The difficulty with the intersection and laying-off operations formulated in terms of points is that these are quaternary operations, and the properties of quaternary operations as opposed to binary operations are inevitably somewhat complex. In subsequent thinking about the problem, I have looked at axioms based upon the primitive objects' being directed line segments with an operation of addition for such segments. An additional unary operation is that of taking the inverse of a directed line segment. Thus, for example, a line segment plus its inverse yields simply the point of origin of the first line segment. The natural axioms here on addition and the inverse operation are close to those for an additive group as in the case of vector addition, but in the present instance they do not actually satisfy all the axioms. For example, the addition of a directed line segment and its inverse yields not the identity of the group, but the particular point of origin of the first segment. (In fact, we do not even get a Brandt groupoid, because the left-cancellation axiom is not satisfied.) In such a geometry it is also natural to add an operation of a qualitative comparison of length, easily represented by a binary ordering relation. Additional constructive operations, like that of one directed line segment being perpendicular to another, are also easily added. However, I am not satisfied with the full set of axioms I have put together – they are too complicated and again too awkward as in the case of the earlier work.

I am persuaded that with additional effort and insight natural and quantifier-free axioms on simple geometric operations can provide an adequate formulation of constructive Euclidean geometry.

B. *Geometry of Approximations and Errors*

The theory of error in astronomical observations, and more generally in any sort of numerical observations, dates from the work of Simpson, Lagrange and Laplace in the eighteenth century. Their efforts were aimed at problems that arise especially in astronomical observations. Dating from the latter part of the nineteenth century there is also a tradition in psychology concerned with the phenomenon that it is easy to judge, for example, tone $A$ to be just as loud as tone $B$, tone $B$ to be just as loud as tone $C$, but tone $A$ to be strictly louder than tone $C$. This phenomenon of just noticeable differences and the related phenomenon of the nontransitivity of judgments of indifference has received considerable attention, and there is much that is common to the formal theory as applicable to both physics and psychology. However, the extension of the formal theory to spatial concepts, and thereby to geometry, has as yet been inadequately developed, in spite of the considerable conceptual interest in understanding what it is like to have directly a qualitative geometrical theory of error or approximation. About the only qualitative part of the theory that is thoroughly understood is the theory of order. Even then, until the relatively recent discussion by Luce (1956) the problem of formulating the theory of order was not properly considered in explicit fashion. Luce's axioms for semiorders were modified and simplified in Scott and Suppes (1958). The theory is developed for a binary relation in one dimension. In the following definition, I call a binary structure an ordered pair $\mathfrak{A} = \langle A, R \rangle$

such that $A$ is a nonempty set and $R$ is a binary relation on $A$. The definition of semiorders is then easily given in elementary form.

DEFINITION 1. *A binary structure $\mathfrak{A} = \langle A, R \rangle$ is a semiorder if and only if the following axioms are satisfied for every $x$, $y$, $z$ and $w$ in $A$:*

1. *Not $xRx$.*
2. *If $xRy$ and $yRz$ then either $xRw$ or $wRz$.*
3. *If $xRy$ and $zRw$ then either $xRw$ or $zRy$.*

The following representation theorem for such semiorders can then be proved.

THEOREM 1. *If $\mathfrak{A} = \langle A, R \rangle$ is a finite semiorder, that is, $A$ is a finite set, then there is a real-valued function $\varphi$ such that for every $x$ and $y$ in $A$,*

$$xRy \quad \text{iff} \quad \varphi(x) > \varphi(y) + 1.$$

The closely related binary relation $I$ of indistinguishability has been thoroughly investigated by Roberts (1970). (We define $I$ as follows in terms of $R$: $xIy$ iff not $xRy$ and not $yRx$.) The surprising thing Roberts shows is that indistinguishability, unlike semiorders, is not axiomatizable in an elementary fashion by a finite set of open sentences. It is of course clear that a semiorder is not a natural relation in geometry because a direction on the line is assumed, and clearly the binary relation of indistinguishability by itself does not have very much geometrical content, although its topological properties have been developed in a surprising way by Zeeman (1962).

The next natural thing is to ask for order on the line. Classical axioms for betweenness on the line may be stated in terms of a ternary structure that is a nonempty set $A$ and a ternary relation $B$, interpreted as betweenness. A variant of axioms that may be found in the literature is given in the following definition:

DEFINITION 2. *A ternary structure $A = \langle A, B \rangle$ is a one-dimensional betweenness structure if and only if the following five axioms are satisfied for every $x$, $y$, $z$ and $w$ in $A$:*

1. *If $B(xyx)$ then $x = y$.*
2. *If $B(xyz)$ then $B(zyx)$.*
3. *If $B(xyz)$ and $B(ywz)$ then $B(xyw)$.*

4. *If $B(xyz)$ and $B(yzw)$ and $y \neq z$ then $B(xyw)$.*
5. $B(xyz)$ *or* $B(yzx)$ *or* $B(zxy)$.

On the basis of these axioms, it is straightforward to prove the following theorem.

THEOREM 2. *Let $\mathfrak{A} = \langle A, B \rangle$ be a one-dimensional betweenness structure and let $A$ be a finite set. Then there is a real-valued function $\varphi$ such that for all $x$, $y$ and $z$ in $A$*

$$B(xyz) \quad \text{iff} \quad [\varphi(x) \leq \varphi(y) \leq \varphi(z) \quad \text{or} \quad \varphi(z) \leq \varphi(y) \leq \varphi(x)].$$

To express the idea of approximation, we can use the notion of ε-betweenness, following the developments in Roberts (1972). The intuitive idea is that the relation of betweenness holds to within a small physical or perceptual error. Formally this is caught in the following condition, which replaces the equivalence of the preceding theorem.

(1) $\quad B(xyz)$ iff $|\varphi(x) - \varphi(y)| + |\varphi(y) - \varphi(z)| < |\varphi(x) - \varphi(z)| + \varepsilon$.

For the formulation of Roberts' axioms we need the additional notion of an indistinguishability relation as discussed above, defined in terms of betweenness: $xIy$ iff $B(xyx)$. Of a number of different formulations of indifference graphs given in Roberts (1970), perhaps the simplest one is this. A binary structure $\mathfrak{A} = \langle A, I \rangle$ is an indifference graph iff any subgraph, that is, any subset of $A$, call it $A_1$, is connected, that is any two points in $A_1$ are related by some power of the relation $I$; more precisely, for any $x$ and $y$ in $A_1$, there is an $n$ such that $xI^n y$, and any such connected subgraph has at most two extreme points that are not equivalent. (Two points are said to be equivalent if they stand in the relation $I$ to exactly the same points in a graph, and an element $e$ of $A$ is an extreme point if whenever $x$ and $y$ are in $A$, and both stand in relation $I$ to $e$, but are not equivalent to $e$, then $x$ stands in relation $I$ to $y$, and moreover, there is another element in $A$ that stands in relation $I$ to $x$ and $y$ but not to $e$.)

The axioms for ε-betweenness are then embodied in the following definition.

DEFINITION 3. *A ternary structure $\mathfrak{A} = \langle A, B \rangle$ is a one-dimensional ε-betweenness structure iff the following axioms are satisfied for every $x$, $y$, $z$, $u$ and $v$ in $A$:*

1. $(A, I)$ is an indifference graph.
2. If $B(xyz)$ then $B(zyx)$.
3. If $B(xyz)$ and $B(ywz)$ and not ($yIz$ and $wIz$) then $B(xyw)$.
4. If $B(xyz)$ and $B(yzw)$ and not $yIz$ then $B(xyw)$.
5. If $B(wyz)$ and $B(yxz)$ then $xIy$ or ($zIx$ and $zIy$).
6. If $xIy$ then $B(xyz)$.
7. $B(xyz)$ or $B(xzy)$ or $B(yxz)$.

On the basis of this definition Roberts proves the following theorem:

THEOREM 3. *Let $\mathfrak{A} = \langle A, B \rangle$ be a one-dimensional $\varepsilon$-betweenness structure, let $A$ be a finite set, and let $\varepsilon > 0$ be given. Then there is a real-valued function $\varphi$ on $A$ satisfying* (1) *above*.

Unfortunately, as is evident from the above axioms, even the theory of $\varepsilon$-betweenness is relatively complicated. The axioms in terms of $\varepsilon$-betweenness and what we can call $\varepsilon$-equidistance, corresponding to the two primitive relations used by Tarski (1959), seem to lead to an extremely complicated set of axioms in order to characterize the 'approximation version' of the Euclidean plane. The problem is open of finding a reasonable set of axioms for the Euclidean plane in terms of $\varepsilon$-approximations to standard geometric relations or operations.

## II. PHYSICAL SPACE AND SPACE-TIME

In this section I discuss open problems connected with the following topics: the theory of bodies, the operational foundations of special relativity and the conceptual foundations of elementary physics.

### A. *Theory of Bodies*

One program of research investigated from a number of perspectives over many years is that of replacing the classical notion of point or line as primitive concepts in geometry and constructing three-dimensional geometry from the concept of a solid object or body. Fairly extensive efforts in this direction were made, for example, by Whitehead (1919, 1920), who regarded his efforts as a significant application of his method of extensive abstraction.

A brief, but classical, article on this subject is Tarski's 'Approach to the Foundations of the Geometry of Solids', which takes only Lesniewski's relation of part and the geometrical concept of sphere as primitive. A translation of this work from the twenties may be found in Tarski (1956, pp. 24–29).

The classical tendency has been to impose increasingly strong axioms on bodies in order to obtain ordinary three-dimensional Euclidean space. In Tarski's axiomatization, for example, axioms in terms of the primitive concepts of part and sphere actually play a minor role, for in terms of these concepts he defines the concept of point and the ordinary geometric relations between points.

Of particular philosophical interest is a more restricted theory of bodies. A useful beginning in this direction is provided by Noll (1966). I shall not follow through all of Noll's work, because he extends his axioms to obtain a foundation of mechanics and introduces thereby spatial concepts in an interesting indirect way in terms of representing the force exerted on a body at a given instance by a vector, that is, an element of an ordinary vector space. The initial elementary axioms are close to the ideas of Lesniewski, but almost certainly the theory has been constructed independent of Lesniewski. Noll begins with the relation *part of*. There seem to be good philosophical reasons for substantially changing some of Noll's approach, but the spirit of what I give below draws directly on his work. Although I begin with modified versions of Lesniewski's and Noll's axioms, I add other axioms and concepts that are not at all in the spirit of their developments. What is given here is incomplete and thus perhaps suggestive of some interesting open problems.

Let $\pi$ be the relation of *part*; in other words, in the intended interpretation $A\pi B$ iff $A$ is a part of $B$. If $B\pi A$ and $C\pi A$, then $A$ is an *envelope* of $\{B, C\}$. Moreover, $A$ is *the least envelope* of $\{B, C\}$ iff $A$ is an envelope of $\{B, C\}$ and for any $D$ that is an envelope of $\{B, C\}$, $A\pi D$.

Some additional definitions are useful. Their intuitive content is obvious. $A$ is a *common part* of $\{B, C\}$ iff $A\pi B$ and $A\pi C$. Bodies $A$ and $B$ are *separate* iff they have no common part. Body $A$ is *a least part* of $B$ iff $A\pi B$ and there is no body $C$ such that $C\pi A$ and $C \neq A$. (The clause that $C \neq A$ is required because $\pi$ is taken to be reflexive and thus every body is a part of itself.) Body $A$ is *the greatest common part* of $\{B, C\}$ iff $A$ is a common part of $\{B, C\}$ and for every body $D$ if $D$ is a common part of $\{B, C\}$, $D\pi A$.

We also define partial operations of join and meet. If $A$ is the least envelope of $\{B, C\}$, then $B \cup C = A$, and we say that $A$ is the *join* of $B$ and $C$. If $A$ is the greatest common part of $\{B, C\}$, then $B \cap C = A$, and we say that $A$ is the *meet* of $B$ and $C$. The operations are partial, because separated bodies do not have joins and meets.

Finally, let $A_1, ..., A_n$ be parts of $B$, let $A_1 \cup ... \cup A_n$ exist, and let $A_1 \cup ... \cup A_n = B$, then we say that $\{A_1, ..., A_n\}$ is a *finite dissection* of $B$.

My incomplete set of axioms for bodies is embodied in two definitions. The first six axioms of Definition 4 are a weakened version of Lesniewski's axioms for mereology as formulated in Grzegorczyk (1955), although I use some of the rather natural terminology introduced by Noll (1966). The axioms are weaker than Lesniewski's in that products, sums and differences are not necessarily defined for any two bodies. Stronger conditions are imposed by my axioms for products, sums and differences to exist. These conditions, which seem physically natural, are similar to ones imposed by Noll. For example, for the product or greatest common part of two bodies to exist they must, according to the axioms given here, have a common part. On the other hand, the axioms diverge from Noll's in not postulating the body that is exterior to a given body. The existence of this possibly unlimited exterior seems dubious, and for many intuitive examples, it is not a natural physical object. For instance, the body that is the exterior of the earth or sun is not conceptually well defined in celestial mechanics. The import of the remaining axioms is discussed below.

DEFINITION 4. *A binary structure* $\chi = \langle X, \pi \rangle$ *is a structure of bodies if and only if the following axioms are satiesfied for every $A$, $B$, $C$ and $D$ in $X$*:

1. $A\pi A$.
2. *If $A\pi B$ and $B\pi A$ then $A = B$.*
3. *If $A\pi B$ and $B\pi C$ then $A\pi C$.*
4. *If $A$ and $B$ have a common part, then they have a greatest common part.*
5. *If $A$ and $B$ have an envelope, then they have a least envelope.*
6. *If $A$ is a part of $B$ and $A \neq B$, then there is a body $C$ in $X$ such that $B$ is the least envelope of $\{A, C\}$.*
7. *Every body has a least part.*
8. *Every body has a finite dissection of least parts.*

It should be apparent that it is easy to formulate all but the last of these eight axioms as first-order axioms. For example, Axiom 5 would read:

$(\exists C) (A\pi C \ \& \ B\pi C) \rightarrow (\exists D) (A\pi D \ \& \ B\pi D \ \& \ (\forall E) (A\pi E \ \& \ B\pi E \rightarrow D\pi E))$.

Axioms 7 and 8 are much stronger and restrictive in character than the first six axioms. They may be regarded as general axioms of abstract atomism. Thus, Axiom 7 might be interpreted as saying that every body contains at least one atom, and Axiom 8 that every body is made up of a finite number of atoms.

There are a number of different ways to extend the axioms of Definition 4, and by heavy-handed methods, we can reach ordinary Euclidean geometry fairly rapidly. We simply have to postulate enough bodies and atoms. We would not of course expect to get the full Euclidean space, because of the finite dissection property, but we would want to be able to imbed in three-dimensional space, and to get a representation of this imbedding unique up to the standard group of rigid motions.

There is no doubt that this program can be carried through. The techniques for the one-dimensional case of measurement exploited in many different directions in Krantz *et al.* (1971) provide more than adequate tools, but as yet I do not see how to pursue it in a simple and elegant fashion. At the same time I am beginning to see a philosophically interesting aspect of this program if it can be satisfactorily carried through. Properly carried out, it should provide a new way of looking at the nature of space.

For many technical reasons that were clear already in Greek geometry, it is much easier to start with points and to deal with the abstractions that follow not only from consideration of points, but also from consideration of points filling space. It is extremely hard to escape from this way of looking at things. The approaches to geometry that begin with a concept of body or solid, as, for example, those of Whitehead, Tarski, Lesniewski, Grzegorczyk or Noll, end up with a richness of structure that is essentially exactly equivalent to Euclidean three-dimensional space. On the other hand, this not an idle fact; it must be recognized that we have to come to terms with Euclidean geometry in some form. A theory that does not is obviously too weak to be of serious conceptual interest.

To begin with and to put it baldly, I propose looking at the intuitive concept of space as just a set of possible worlds. Of course, it is a rather special set of possible worlds. It is the set of all possible relative positions of bodies. But insisting on this viewpoint seems to me to clarify a number of problems. Certainly, it strikes down the container theory of space which, in spite of criticisms that go back to Aristotle, continues to be a perennially popular view of space. This viewpoint also gives a deeper analysis of relational theories of space. The difficulty with relational theories is that it is too easy to cast them in terms of actual relations. Rather, we need to think of the set of all possible relations between bodies, and this characterizes space. Where we get in trouble epistemologically is in beginning with points rather than with bodies. It is somewhat like the problem of constructing a sample space in probability theory. We understand the construction of the sample space best when we start with the method of generating the possible sequence of events and use this method of generation to describe the possible experimental outcomes, the set of which constitute the sample space.

In constructing space as a set of possible relative positions, it is not the concept of point as such that creates difficulties. Rather it is the classical concept of there being so many points. The points ordinarily postulated as existing in space have no more reality under the view advocated here than do the possible sequences in a large number of flips of a coin. The various sequences represent nicely possible experimental outcomes, but in themselves they have no concrete existence. Only one of them will come to represent the actual sequence, and I say the same is true of points. It is not possible here to develop this view thoroughly, but I do think that beginning with the kind of theory of bodies discussed above it is feasible to develop a theory of space from the theory of bodies and to get the concept of space itself out as a construction derived from the set of possible relative positions of bodies.

Moving from positions to trajectories we may obtain a characterization of space-time as the set of all possible trajectories of bodies, and this is probably more fundamental than the separate concept of space.

B. *Special Relativity*

On several past occasions I have stressed the significance of Robb's axiomatization (1936) of space-time in the sense of special relativity. His axiomatization is important, because of its completeness and the simplicity of its single primitive – the binary relation of *after* holding between space-time events. Robb's important work has been repeatedly ignored by philosophers, but I am happy to say that the long article by Domotor in this volume includes a detailed discussion of Robb's work. The article by Latzer also provides an axiomatic treatment different from, but very close to that of Robb.

As I have remarked in earlier discussions of Robb's axiomatization, the complexity of the axioms stands in marked contrast to the simplicity of his single primitive concept. The point I want to emphasize is the desirability of quantifier-free axioms of the sort discussed above for Euclidean geometry. It is almost paradoxical that no such axiomatizations have yet been given for special relativity. Given the enormous literature on operationalism, its relations to Mach and Einstein, and the extensive discussions of physicists like Bridgman, without knowing the literature one would anticipate that a number of different rigorous treatments of an operational approach to special relativity could be found.

One thing is evident. The kind of primitive operations I discussed earlier for Euclidean geometry do not seem intuitively appropriate for operations in a space-time manifold – I mean the operations of finding the intersection of two line segments and of laying off one line segment on another:

From the results in Suppes (1959) we should be able to establish a sufficient axiomatic base by considering just segments of inertial paths, because of the invariance of the relativistic measure of such segments. Moreover, as also shown in that article, we use in a natural way parallelogram constructions to get at the relativistic invariance of other segments that are not segments of inertial paths. The explicit proof in Suppes (1959) of the invariance of such inertial path segments being an adequate basis for deriving the Lorentz transformations requires the use of various elementary geometrical operations, like that of finding a midpoint that could be used in an operational, quantifier-free geometry of special relativity. However, I have been unable to find a transparent way to build up an adequate axiomatic construction from this approach.

The approach begun by Walker several years ago (1948, 1959) may possibly lead to more satisfactory results. Walker takes a richer set of primitives than Robb's, but one sthat are related. In addition to events he also has particles, an ordering relation of beforeness on events, and most importantly, a one-one signal-mapping from one particle onto another.

With this apparatus, he gives one of the few formal definitions of observables to be found anywhere in the literature of special relativity; namely, an observable is a mapping from the distinguished particle called the observer on to the observer, that is, from that particle on to itself, resulting from a chain of signal-mappings and inverse signal-mappings. My central reservation about Walker's approach is that the signal-mappings are in fact complex functions that do all the work at once that should be done by a painstaking buildup of more elementary operations. At least that is my perspective on the intuitively correct approach. Another remark is that several of his axioms are very powerful; for example, his notion of a particle's being dense makes each particle ordinally equivalent to the continuum of real numbers. All the same, Walker's work, which conceptually derives from the earlier intuitive ideas of Milne, is a clear conceptual alternative to Robb's and marks a clear advance over the level of rigor and explicitness found in most of the literature.

As the discussion in Latzer's paper in this volume shows, the mathematical problems of finding an adequate qualitative axiomatic basis for the general theory of relativity are complex and formidable. But this is certainly not the case for the special theory of relativity, and it is surprising that so few axiomatic results of a definite nature have as yet been achieved. The absence of such explicit work indicates how poorly we understand in any deep conceptual way the ideas of operationalism that have been current for almost a hundred years. I shall say more about special relativity in the next section on elementary physics.

## C. *Elementary Physics*

Talk about some parts of physics being elementary is fairly frequent, and presumably there is an effort on the part of textbook writers to restrict themselves to that part of physics that is elementary. Actually, the situation is not clear. While a modern secondary-school textbook will probably contain a chapter on quantum mechanics, its discussion is purely qualitative and no actual numerical exercises are worked out.

It is my conviction, reinforced by a number of conversations with Seymour Papert, that the concept of elementary physics can be made an intellectually respectable one, with a precise formulation of what its range of subject matter is. I should make it clear at once that I do not think there is any unique approach to elementary physics; several different ways of formulating the domain are possible. I do think a kind of representation result can be given prominence, and that I want to describe. However, I want to approach that representation theorem somewhat indirectly and begin with a characterization that is natural in the context of the great emphasis on first-order logic in the philosophy of mathematics and science.

One natural approach would be to say that a part of physics is elementary if it can be expressed as a theory with standard formalization in first-order logic. Several of the problems discussed earlier in this article have that character, and it is certainly a framework familiar enough in the philosophy of science. Although organizing much of geometry in the first-order framework is easy, it is hard to point to significant examples of physics that have been axiomatized with this restriction. To some extent, this may be due to a lack of sustained effort, and I have the conviction that much real physics can be put within a first-order framework.

Another approach that is closely related but that can get us more quickly into a formulation of several parts of physics, and that is probably at the present time considerably more practical as an actual way of marking off in some systematic fashion elementary parts of physics, is to restrict ourselves to an elementary algebraic approach, in particular, to restrict our field of numbers to an ordered Euclidean field. (An ordered field in the sense of modern algebra is Euclidean if whenever a positive element $a$ is in the field then there is an element $b$ such that $b^2 = a$, i.e., we can take square roots.) We get all the vector space apparatus we need by considering vector spaces over such Euclidean fields, and we then introduce elementary laws of physics by means of special functions which take values either in the field or in a three-dimensional vector space over the field. Simple formulations of the conservation laws of momentum, for example, can easily be made within such a framework.

A second example may be found in the foundations of special relativity as discussed above. It is clear that an elementary geometric foundation can be given for special relativity that has as its representation theorem isomorphism to a four-dimensional vector space over a Euclidean field. The proof in Suppes (1959) that invariance of relativistic distance along inertial paths is sufficient to derive the Lorentz transformations can be carried through over such a Euclidean field. Such a field can of course be denumerable, and consequently, the results are also interesting from the standpoint of the large philosophical literature on the problems of a metric or a measure in relativity. The intuitive reason that the proof can be carried through with just the apparatus of a Euclidean field available is that all the assumptions needed are macroscopic in character, and the algebraic methods of argument, although complicated in spots, are elementary, for example, familiar facts needed in the argument about affine spaces holding for affine spaces over Euclidean fields and not just over the field of the real numbers. From a pedagogical standpoint, this means that we should be able to teach the central mathematics of special relativity to students who have a good background in linear algebra, but who do not necessarily have any knowledge of the differential and integral calculus. However, I shall not push this point further here.

A third example is the algebra of physical quantities. By physical quantities I mean things such as lengths, times and masses; for instance 5 meters, 10 seconds and 15 grams are all examples of physical quantities. A detailed study of the algebra of such quantities is to be found in Chapter 10 of Krantz et al. (1971). Restricting ourselves only to square roots, for elementary purposes, we can easily give elementary axioms for physical quantities over an ordered Euclidean field. In these axioms, which are modifications of those given in Krantz et al. (1971), the set $A$ is the set of physical quantities, in which fall the different 'dimensions' of physical quantities that ordinarily occur in physics. We also include in the primitive notions the set $A^+$ for the positive physical quantities, a binary operation $*$ of multiplication of physical quantities, a unary operation $^{-1}$ for finding inverses and a unary operation $^{1/2}$ for finding square roots. Also, in stating the axioms the elements 0 and 1 of the given field are referred to. For more elaborate applications we will want to extend ourselves beyond a Euclidean field, but for elementary applications, this apparatus is sufficient. Consequently, I refer to the structures characterized in the axioms as elementary structures of physical quantities.

DEFINITION 5. *A structure* $\mathfrak{A} = \langle A, A^+, *, ^{-1}, ^{1/2} \rangle$ *is an elementary structure of physical quantities* (*relative to an ordered Euclidean field $\mathscr{E}$*) *iff, for all* $x, y, z$ *in* $A$:

1. $x*y = y*x.$
2. $x*(y*z) = (x*y)*z.$

3. If $\alpha \in \mathscr{E}$ then $\alpha \in A$.
4. If $\alpha \in \mathscr{E}$ and $\alpha \in A^+$ then $\alpha \in \mathscr{E}^+$.
5. $0*x = 0$.
6. $1*x = 1$.
7. If $x \neq 0$ then exactly one of $x$ and $(-1)*x$ is in $A^+$.
8. If $x, y$ are in $A^+$ then $x*y$ is in $A^+$.
9. If $x \neq 0$, $x*x^{-1} = 1$.
10. $x^{1/2} * x^{1/2} = x$.

We may introduce the physical concept of dimension for such structures in the following way. If $x \neq 0$, the dimension of $x$ is defined as:

$$[x] = \{\alpha * x \mid \alpha \in \mathscr{E}\}.$$

In other words, the dimension of $x$ is just the set of physical quantities obtainable from $x$ by multiplying $x$ by a number, i.e., an element of the field $\mathscr{E}$. Of course, if we do not want to escalate the type of objects considered in elementary physics, we can introduce an equivalence relation instead of the set $[x]$. Physical quantities $x$ and $y$ have the same dimension, e.g., length, time, mass, force, etc., if there is a number $\alpha$ such that $\alpha * x = y$.

It is shown in Krantz et al. (1971) that an arbitrary structure of physical quantities can be represented as a multiplicative vector space over the rationals, or more exactly, a set of dimensions of such a structure is a multiplicative vector space over the rationals. Given this apparatus, we can then go on to elementary dimensional analysis, and more importantly, develop the elementary theory of the laws of similitude and exchange developed in Krantz et al. (1971).

I emphasize of course that I have given only a few samples of elementary physics in this brief discussion. It is, I think, worth finding out just exactly how much can be done within such a framework. One possibility, however, needs to be mentioned for enlarging the framework. If we want to make an exact connection with first-order logic on the one hand, and the usual background of real numbers on the other, it is natural to extend ourselves from Euclidean fields to real closed fields. Such fields are Euclidean, but they also have the property that every polynomial of an odd degree with coefficients in the field also has a zero in the field. A fundamental result of Tarski's decision procedure for elementary algebra and geometry is that any first-order sentence that holds for the field of real numbers also holds for real closed fields. By this extension, which takes us somewhat deeper into algebraic methods, we can get an exact correspondence between the two senses of elementary physics introduced at the beginning of this discussion.

*Stanford University, Stanford*

### BIBLIOGRAPHY

Grzegorczyk, A., 'The Systems of Lesniewski in Relation to Contemporary Logical Research', *Studia Logica* **3** (1955), 77–95.

Krantz, D. H., Luce, R. D., Suppes, P., and Tversky, A., *Foundations of Measurement*, Vol. 1, Academic Press, New York, 1971.

Luce, R. D., 'Semiorders and a Theory of Utility Discrimination', *Econometrica* **24** (1956), 178–191.

Moler, N. and Suppes, P., 'Quantifier-Free Axioms for Constructive Plane Geometry', *Composito Mathematica* **20** (1968), 143–152.

Noll, W., 'The Foundations of Mechanics', in C. Truesdell and G. Grioli (coordinators), *Non-linear Continuum Theories*, Edizioni Cremonese, Rome, 1966, pp. 159–200.

Robb, A. A., *Geometry of Space and Time*, Cambridge University Press, Cambridge, 1936.

Roberts, F. S., 'On Nontransitive Indifference', *Journal of Mathematical Psychology* **7** (1970), 243–258.

Roberts, F. S., 'Tolerance Geometries', *Notre Dame Journal of Formal Logic*, 1972, in press.

Scott, D. and Suppes, P., 'Foundational Aspects of Theories of Measurement', *Journal of Symbolic Logic* **23** (1958), 113–128.

Suppes, P., 'Axioms for Relativistic Kinematics With or Without Parity', in L. Henkin, P. Suppes and A. Tarski (eds.), *The Axiomatic Method*, North-Holland, Amsterdam, 1959, pp. 291–307.

Tarski, A., *Logic, Semantics, Metamathematics*, Oxford University Press, Oxford, 1956.

Tarski, A., 'What Is Elementary Geometry?', in L. Henkin, P. Suppes and A. Tarski (eds.), *The Axiomatic Method*, North-Holland, Amsterdam, 1959, pp. 16–29.

Walker, A. G., 'Foundations of Relativity', *Proceedings of the Royal Society of Edinburgh* **62** (1948), 319–335.

Walker, A. G., 'Axioms for Cosmology', in L. Henkin, P. Suppes and A. Tarski (eds.), *The Axiomatic Method*, North-Holland, Amsterdam, 1959, pp. 308–321.

Whitehead, A. N., *An Inquiry Concerning the Principles of Natural Knowledge*, Cambridge University Press, Cambridge, 1919.

Whitehead, A. N., *The Concept of Nature*, Cambridge University Press, Cambridge, 1920.

Zeeman, E. C., 'The Topology of the Brain and Visual Perception', in M. K. Fort (ed.), *The Topology of Three Manifolds*, Prentice-Hall, Englewood Cliffs, N. J., 1962, pp. 240–256.

ERNEST W. ADAMS

# THE NAIVE CONCEPTION OF THE TOPOLOGY OF THE SURFACE OF A BODY*

### I. MOTIVATION

The aim of this paper is to sketch aspects of the analysis of some 'naive' topological concepts which enter into everyday descriptions of features of the surfaces of physical bodies. Examples of such descriptions and of the reasoning involving them are to say that: (a) Feature A (say a reddish region on the surface of a cube) *touches* but does not completely *cover* feature B (say an edge of the cube), from which it can be inferred that feature B touches the edge or *boundary* of feature A. (b) Several features meet in a *common point*, from which it can be inferred that any two of them touch. (c) The surface of a transparent object has *places* or *points* on it where there are no features (at any rate, no visually discernible ones). It is evident that not only are such concepts as 'point' and 'boundary' *used* in everyday descriptions, but they also are assumed to conform to certain general principles which underlie *inferences*. These general principles can be regarded as elements of an implicit 'naive topological theory' of the surface of a body, which I hope to clarify in the present analysis.

A partial motivation for undertaking an analysis of the naive conception of the topology of a surface is that it is arguable that this conception is presupposed in typical *applications* of elementary geometry and even elementary physics. This cannot be demonstrated without a detailed examination of such applications, which cannot be undertaken here. One point worth making, however, is that though the concepts of *point* and *neighborhood* are usually taken as primitive or undefined in geometrical and topological theory, it is far from evident what 'empirical meanings' they have in application. Hence the analysis of *applied* geometry requires the clarification of these concepts, and, if the application is to features on surfaces (like diagrams drawn with ruler and compass), this is a matter of our conception of points on those surfaces. It will be seen in fact that the 'naive' conception of a point or a neighborhood (more exactly, the *interior* of a region) is by no means 'simple' or 'primitive'.

## II. THE ANALOGY TO THE NAIVE CONCEPTION OF A SET

Though I have called the everyday use of topological concepts in describing features of surfaces 'naive' (because it is not grounded in mathematical topology), this use is actually quite sophisticated. It is clear that we do not conceive of points *as* concrete physical *things* or features of them, since we are prepared to recognize the possibility that there should be points or places on surfaces which are feature*less*. It is more plausible that point and place are normally conceived as *abstract* 'entities' which are not physical things, but which stand in particular *relations* to concrete observables. There is in fact a suggestive partial analogy between the naive conception of points and places on surfaces and their relations to concrete things, and the intuitive conception of a *set* and its relation to the individuals which may or may not be members of it.

The fundamental relation between individuals and sets is the membership relation, and the plausible analogue to this is the relationship of an individual thing being *at* a place. Sets are basically referred to by use of *set descriptions* of the form 'the set whose members are ...' which specify necessary and sufficient conditions for membership in them, and it is arguable that particular places are referred to in such a way as to make it possible to determine when concrete things are at them. The basic existence postulate of naive set theory – the Axiom of Abstraction – simply assures the *existence* of sets answering to basic modes of set description, and there are analogous, though less easily formulable, postulates of 'intuitive place theory' which assure the existence of places corresponding to fundamental modes of place description. Perhaps the most essential of these 'implicit place-existence postulates' which are presupposed by our common modes of place reference is that for any feature on a surface at a time, there *is* a place where it is or was at the time. Finally, as identity of sets is defined in terms of their members, we can expect identity of place to be similarly analyzable in terms of the concrete things which are or could be at those places.

The set-place analogy is in fact a more useful heuristic guide towards the analysis of the concept of a *place* on a surface than it is to the concept of a *point* on a surface. The reason is that *place*, like *set*, is a fundamental category of abstract entity, not borrowing its 'criteria' of existence and identity from any more general categories of which places can be regarded as particular species. Points on the other hand can be regarded as places of particular kinds (the 'smallest' ones), having the same criteria of identity as places, and the special problem of analysis for the concept of a point is that of specifying the defining characteristics which distinguish points from other kinds of places. As we shall here want to skip over consideration of places in general as fast as possible, in order to discuss points in particular, we shall not make much further direct use of the set-place analogy. One thing which the analogy does suggest, however, which applies to the 'postulates' implicit in ordinary descriptions and reasoning concerning places or points on surfaces, is that these can be expected to have a logical status similar to the axioms of Abstraction and Extensionality of intuitive set theory. Such axioms – that there exists a place where any concrete feature of a surface is – are neither empirical laws nor derivative logical truths capable of being justified by appeal to more fundamental laws of logic. What modes of evaluation are appropriate to laws of this type is a very difficult question which will not be entered into here. It will be sufficient simply to *observe* that ordinary descriptions, say that a particular feature now on the surface is at the place where another feature had previously been (say the other feature was a chalk mark which was subsequently erased), implicitly assume that there *are* places answering to the forms of description employed.

The concept of *place* differs importantly from that of a set in two ways. One is that while a thing can be at a place at one time but not at another, naive set theory does not admit the possibility that a thing can be a member of a set at one time but not at another. The possibility of change of place leads to the problem of clarifying the 'criteria of place reidentification', for determining the truth of propositions of the form 'the place where feature A was at time $t$ is the same as the place where feature B is now'. This is obviously very close to the problem of defining motion and rest relative to a particular surface (for objects on the surface). Though it seems plausible that rest can be defined relative to a surface (in contrast to the possibility of defining rest relative to 'space'), nonetheless the problem presents considerable difficulties which we shall simply leave aside here. Idealizing, we shall assume that objects once on a surface do not change their positions on it.

The second way in which the analogy of place to set breaks down is in the fact that there can be two distinct places, neither of which has discernible concrete features at them (there can be two different places in 'empty space'). The explanation for this lies in the fact that not only do we conceive of places in terms of where concrete things *are* or *were* at some previous time, but also in terms of where they *could be*. Thus, we may describe a surface as being featureless at some places where features (e.g., chalk marks on a blackboard) could be made. In fact, it will be seen that the criteria of identity for places are to be specified not in terms of the concrete things which actually *are* (or have been at some previous time) at the places in question, but rather in terms of the concrete things which *could be produced* at those places. In this respect, places and points are to be likened more to the *properties* of which particular sets are the extensions than they are to the sets themselves (for example, two colors may be distinct, even though no concrete thing has either of those colors). In any case, it is suggested that the naive conception of a point or place has an essential *modal* aspect which merits examination, since it is arguable that the same type of modality is involved in the Euclidean formulation of elementary geometry.

### III. GEOMETRICAL POSSIBILITY

The most natural interpretation of the Euclidean formulations of the postulates of geometry involves modality. Thus, Heath (1956, pp. 195, 196) translates Postulate 1 as: "Let the following be postulated: to draw a straight line from any point to any point." He explains this as meaning that it is postulated that *it is possible to draw* a straight line from any point to any poin . Modern formulations appear to avoid modalities by postulating instead something like 'there exists a straight line between any two points'. Propositions about what can be produced are replaced by ones about what exists. There is a sense in which either formulation is correct, because it follows from the fact that if a concrete thing with such and such properties *can be produced* there *exists* a corresponding abstract entity with analogous properties. That is, modern geometry can be regarded as the theory of the abstract objects (points, places, and so on) which are where concrete things might be produced, whereas Euclidean geometry is about what can be produced. The drawback to the modern formulation *is* its complete abstractness, which takes no explicit cognizance of the *relation* between the abstract entities with which it deals and the concrete things to which they apply. On the other hand, the modern formulation at least avoids modalities and so no problems of modal logic arise in analyzing the deductive relations among its propositions. To analyze the Euclidean formulation of geometry (as well as our naive conception of a space of points on the surface of a body), we must therefore pay special attention to what may be called the notion of *geometrical possibility* involved in the statement of many of its propositions.

It will be hazarded that the possibility involved in the 'constructibility quantifier' "it is possible to produce (draw) an object such that..." is best regarded as a *primitive concept* of Euclid's formulation of geometry. Clearly, neither logical nor technical possibility is meant (though both are interestingly related to geometrical possibility), and this quantifier is not defined explicitly in the Euclidean corpus. On the other hand, examination of the arguments (proofs) into which the quantifier enters shows that it conforms to quite definite logical laws. These laws cannot be discussed here, but it will simply be claimed that they are such that the expression 'it is possible to produce an object such that...' can be *paraphrased* as saying 'there exists a member of the *class of constructible objects* such that...', where the class of (geometrically) constructible objects now becomes an explicit primitive of geometrical theory. Note that if the foregoing claim is correct, Euclidean geometry cannot be criticized for lack of rigor simply on the grounds of its modal formulation, whatever other faults it may have in this respect, since the logical laws to which the constructibility quantifier conforms are quite clear.

The foregoing observations bear on the analysis of the naive conception of the topology of a surface, since the most natural way to analyze its implicit postulates, like postulates of identity for places described in different ways, is in terms of the features which it is possible to produce on those surfaces. In fact, we shall treat the appropriate constructibility quantifiers as being 'primitive for present purposes', though deeper analysis would require inquiry into them as well. It will, however, be assumed, as in Euclid's geometry that constructibility propositions can be paraphrased into ones about the existence of objects with particular properties in a class of features which can be constructed (constructible features) on a surface. Thus, we shall shortly formulate a theory somewhat on a par with Euclidean geometry, which is based on the unanalyzed notion of a class of features that it is possible to construct on a surface. One thing to be noted about this is that it idealizes by excluding the possibility of change in the same way that we have left aside the possibility of an actual thing's changing its place on a surface: we will not allow that what can be drawn on a surface at one time differs from what can be drawn at another – i.e., that the entire topology of the surface changes over time.

## IV. PLACES, POINTS, DIVISIBILITY AND COMMON POINTS

As noted, the natural way to analyze the intuitive idea of a point on a surface would be first to analyze that of a place, and then to analyze *point* as a species of the genus *place*. There are difficulties in how to clarify the concept of a place, however, which makes this course impracticable. This much can be said about places. First, implicit in our talk about them is the 'postulate' that for any feature of a surface there exists a place which is *coextensive* with it in the sense that any feature which could be drawn on the surface would *touch* that place if and only if it touched the feature defining the place. The 'touching' or 'contact' relation between features is assumed to be observationally determinable. Second, for any two places, they are the same if and only if it would not be possible to produce a feature touching one of them but not the other. The foregoing existence and identity postulates are analogues to the abstraction and identity postulates for set theory.

The difficulty which now arises is whether or not the place-of-feature postulate above exhausts the place-existence assumptions implicit in our conception of places on surfaces, and in particular, whether it follows from this postulate that there exist places which satisfy the requirements for being points. What are those requirements? Roughly that the place be *indivisible*. A sense in which a place, such as that occupied by a small dot on a sheet of paper, is divisible is that it may be possible to produce two features each of which touches the place but which do not touch each other (if the place is that occupied by the dot, the features both touch the dot but do not touch each other – which is what shows that the place occupied by the dot is not a geometrical point). Let us call this requirement (that it not be possible to produce features touching the place but not touching each other) the *indivisibility requirement*. The question is: what assures the existence of places which are indivisible in this sense?

If features are taken to be things like marks, then it is clear that no matter how small they are, the places they occupy cannot satisfy the indivisibility requirement, hence they cannot be single points. If we allow that 'features' include things like *edges*, and intersections of edges as well, then it is more plausible that these should 'individuate' single points (at least provided we idealize by assuming that edges are perfectly sharp). However, we do not want to presuppose a concept of an edge, at least in the sense of a *boundary*, if we want to later subject this topological concept to analysis.

Instead of attempting to deduce the existence of places satisfying the indivisibility requirement from the basic place-of-feature existence postulate, it is more plausible to look for further, independent, place-existence postulates from which the existence of indivisible places could be inferred. A plausible candidate for such a postulate is one that entails the existence of a unique place which is *common* to a number of features that partly overlap one another, roughly in the way regions in a Venn diagram do. In fact, if we imagine that the features are something like the slices of a pie that meet at the center, it is plausible that the place common to these features should be indivisible, and hence be a point.

The problem that arises in attempting to formulate a postulate assuring the existence of a place which is common to a number of partially overlapping features is that of specifying the conditions that the features must satisfy in order that there *exist* some place common to all of them. It is not enough to require that any one feature of the set of features touch every other one, as Figure 1 illustrates, since this shows features A, B and C all touching, but not meeting in any common point. A further

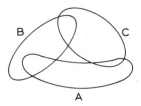

Fig. 1.

requirement which must also be satisfied is that it should not be possible to 'cover' any one of the features with smaller ones in such a way that none of the covering features touches all of the original features. Figure 2 illustrates how feature A in Figure 1 can be covered with smaller features (oval-shaped regions) D and E, where neither D nor E touches all three

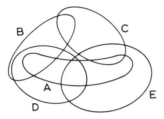

Fig. 2.

of the original features A, B and C. A 'covering' like the two features D and E in Figure 2 which shows that the original features do not meet in a common point, will be said to *separate* the original features, and it is plausible to postulate that a (finite) set of features all meet in a common point if and only if it is not possible to construct a covering of any one of them which separates the set.

The separation 'test' for whether several features have a point in common is itself quite complicated, and it is desirable to digress briefly to discuss some of its properties.

## V. THEORY OF THE SEPARATION TEST

The theory, which will be described somewhat informally, takes as primitive a class $\mathscr{C}$ of *constructible features* (of a surface) and a binary relation, $C$, over this domain, which is assumed to be symmetric and reflexive in $\mathscr{C}$. $xCy$ means that feature $x$ is *in contact with* (touches) feature $y$. We want also to consider a finite set of constructible features *covering* another feature, and in so doing we will be treating the set as extending over the *union* of the regions covered by the individual features belonging to it. Accordingly, we define:

DEFINITION 1. A *U-system* is a finite (possibly empty) set of constructible features; a *U-atom* is a singleton set of constructible features; two U-systems are in *U-contact* if some member of one is in contact with some member of the other; one U-system *U-covers* another if and only if there is no U-system in U-contact with the second which is not in U-contact with the first.

Trivially, U-contact is symmetric and reflexive in the class of nonempty U-systems, and U-covering is transitive and reflexive in the class of all U-systems.

We want also to introduce a kind of system which acts as an *intersection* of the features of which it is composed, but for this purpose we need to apply the separation test in order to state the conditions under which such a class of features has an *empty* intersection. Intuitively, a class of U-systems (or constructible features) has an empty intersection if it is possible to separate them by constructing a U-covering of one of them, no member (U-atom) of which is in U-contact with all of the original U-systems.

DEFINITION 2. Let $X$ be a set of U-systems. Then $X$ is *separable* if and only if there is a U-system $x$ in $X$ and a U-system $y$ such that $y$ U-covers $x$, but no U-atom of $y$ is in U-contact with all U-systems in $X$.

Certain properties which we would expect to hold for the concept of separability follow just from the definition. For instance, it is trivial that if any set of U-systems is not separable, then any two of its members must be in U-contact. Also, the following *compactness* property holds: A set of U-systems is separable if and only if some finite subset of it is separable. Certain other properties of separability do not follow from the definition alone, but rather depend on the following *separation postulate*:

*Separation postulate*

>Let $X$ be a separable set of U-systems, and let $x$ be a U-system in $X$. Then there exists a U-system $y$ which U-covers $x$, and no U-atom of which is in U-contact with all members of $X$.

The separation postulate is as a matter of fact derivable from a much more 'elementary' axiom to the following effect: If $x$, $y$ and $z$ are U-atoms and $w$ is a U-system such that $x$ and $y$ are not in U-contact and $z$ and $w$ are not in U-contact, then there exists a U-system $t$ which: (a) U-covers $z$, (b) is not in U-contact with $w$, and (c) no U-atom of $t$ is in U-contact with both $x$ and $y$. The reduction of the separation postulate to the more elementary axiom involves a series of very tedious inductions, so this will be left aside here. In what follows, the separation postulate will be assumed. We are now ready to introduce 'Intersection systems' or 'I-systems'.

DEFINITION 3. An *I-system* is a nonempty (possibly infinite) class of U-systems; an *I-atom* is a singleton set of U-systems; two I-systems are in *I-contact* if and only if their union is not separable; one I-system *I-covers* another I-system if and only if every I-system in I-contact with the second system is in I-contact with the first.

Once again, expected theorems are easy to derive. I-contact is reflexive and symmetric in the class of inseparable I-systems, and I-covering is reflexive and transitive in the class of all I-systems. Also I-contact and I-covering reduce to U-contact and U-covering in the case of I-atoms (which are singleton sets of U-systems). The fact that theorems like the above are provable does not yet show that I-systems and I-contact act like true pointset intersections; in fact, it is difficult to formulate the question of whether they do in the absence of the concept of a *point*. Some light is shed on the question, however, by considering a special class of 'pointlike' I-systems, which we shall call *I-points*. In essence, I-points are sets of I-systems which satisfy the indivisibility requirement discussed in Section IV:

DEFINITION 4. An *indivisible* I-system is an I-system which is not separable, and which is such that any two I-systems in I-contact with it are in I-contact with each other; an *I-point* is a maximal set of indivisible I-systems, any two of which are in I-contact with one another; an I-point is an I-point *of* a constructible feature $x$ if and only if any member of it is in I-contact with $\{\{x\}\}$; it is an I-point *of* a U-system $x$ if and only if any member of it is in I-contact with $\{x\}$; it is an I-point *of* an I-system $X$ if and only if any member of it is in I-contact with $X$.

Note that the foregoing definition first defines an indivisible system and then defines an I-point as the class of all indivisible I-systems which, in a sense, 'individuate' the same point. This allows that different indivisible I-systems may individuate the same point, and that it is the set of them, rather than any one of them, which is taken to be the I-point in question.

Though we shall omit the proofs, it is possible to use a transfinite construction to prove not only that indivisible systems and nonempty I-points exist, but in fact they are 'everywhere' on a surface. Moreover, it can be shown that U-systems and I-systems act like unions and intersections of I-points.

THEOREM 1

(1.1) Every I-point is an I-point of some constructible feature and for every constructible feature there is an I-point which is an I-point of it.

(1.2) The I-points of a U-system are the union of the I-points of the constructible features belonging to it, and the I-points of an I-system are the I-points of the intersection of the I-point-sets of the U-systems belonging to it.

(1.3) Two constructible features, U-systems, or I-systems are in contact, U-contact, or I-contact if and only if there is an I-point common to both. An I-system is separable if and only if there is no I-point which is an I-point of it.

(1.4) Given two U-systems or two I-systems, the first U-covers or I-covers the second if and only if every I-point of the second is an I-point of the first.

Theorem 1 is far from showing that I-points *are* points, though it suggests that for certain purposes it may be safe to describe features of surfaces and their relations in terms of points, since at least we know (assuming the separation postulate) that there must, logically, exist 'systems' which act *like* points. We now want to consider briefly how far we can go towards demonstrating not only that I-points act like points, but that they *individuate* points in the sense that for each I-point there is an abstract point where it is, and conversely.

## VI. ABSTRACT POINTS AND THEIR RELATION TO I-POINTS

Let us begin by postulating about abstract points of a surface only what appears to be presupposed by naive usage of 'point talk' in our descriptions of surface features. Here are four abstract point postulates, of which some may be questionable as idealizations, but which will be accepted for present purposes:

*Existence postulate*

>Every point of a surface is a point of some constructible feature of it, and for every constructible feature of it, there is a point of the surface which is a point of the constructible feature.

*Identity postulate*

>Two points of a surface are identical if and only if there is no constructible feature which one is a point of, but not the other.

*Contact postulate*

>Two constructible features are in contact if and only if there is a point which is a point of both of them.

*Covering postulate*

>One U-system U-covers another if and only if every point of the second is a point of the first, where the points of a U-system are the union of the points of the constructible features belonging to it.

If we now define the points of an I-system as the intersection of the pointsets of the U-systems belonging to it, we can ask in a precise way: Is it the case that every I-point individuates one and only one point in the sense that there is exactly one point which is a point of it (and, conversely, every point is individuated by a unique I-point)? That the answer to this question is *No* follows easily by considering certain *topological models* which satisfy simultaneously the 'empirical' assumptions about contact and separability given in the last section, and the abstract point postulates listed in this section. Take any noncompact topological space $\langle X, \Omega \rangle$, let $X$ be the class of abstract points, let constructible features be all nonempty *closed* sets in this space, and define contact between such 'features' as nonempty intersection. It is easily verified that the contact and separability postulates, as well as the abstract point postulates, are satisfied. On the other hand, the I-systems and I-points which are 'generated' in such a model do not individuate points in the set $X$: roughly, they individuate points in a *compactification* of the topology.

The foregoing does not settle the question whether 'it is part of our conception' of the points of a surface that I-points of the surface should individuate abstract points of it, since it could be that we have failed to include all of the assumptions relating abstract points and concrete features among the four abstract point postulates listed above. It is clear that certain of the topological models described above would not be accepted as representing 'possible systems of constructible features'. For example, the discrete topology is one such model, which would in effect represent any single point of the surface as (the location of) a constructible feature, though this runs strongly against our conceiving constructible features as being, in a sense, 'two dimensional'. Without running in detail through possible further abstract point assumptions which might be added to our present list of postulates, we shall conclude this section by citing one highly speculative one which would indeed entail that points were individuated by I-points.

A *compactum* is by definition a topological space which is compact and metrizable. A *basis* of a compactum (or any topological space) is any set of nonempty open sets in the space, such that any open set of the space is a union of sets in the basis.

*Compactum basis postulate*

> The interiors of the pointsets of the constructible features of a surface form the basis of a compactum over the set of points of the surface.

It is easily demonstrable that the compactum basis postulate, together with the other abstract point postulates, not only entails the separation postulate, but also entails that every point of a surface is individuated by an I-point of the surface, and conversely, every I-point of the surface individuates a unique abstract point. Whether the compactum basis postulate itself is in some sense implicit in our ordinary ways of describing features and the places and points where they are is a question not easily answered. The most obvious difficulty is that the postulate is stated in terms of the notion of the interior of a constructible feature, which is a topological concept which we have not yet scrutinized. Accordingly, we now turn to the analysis of the 'topological' concepts in terms of which we describe features of surfaces.

## VII. CONNECTEDNESS AS THE BASIC CONCEPT OF NAIVE TOPOLOGY

Even assuming that abstract points are individuated by I-points, we still lack an explicit analysis of the concept of a neighborhood or of an interior point, which would then permit other topological concepts to be defined in terms of these basic notions in the way which is usual in mathematical topology. Without attempting to argue for it here, we will in fact suggest that the topological concept whose empirical meaning is more immediately given is that of a feature of a surface being *connected*, and that in consequence the proper order of analysis should begin with intuitive connectedness, and then try to show how other topological concepts are reduced to that.

Intuitively, a feature is *connected* if it is not composed of spatially separate parts. It is clear in fact that something like 'apparent connectedness' is fundamental to our individuating a feature of a surface as *one* thing. Thus, to know *what* mark is being referred to when someone points and says 'that mark', we must know what counts as part of the mark and what does not, and this is ascertained by 'tracing' through the parts of the mark which appear to be connected to that part of it actually pointed at. Similarly, it can be argued that physical connectedness is essential to the most fundamental ways of individuating physical bodies. Not only is connectedness fundamental to individuation of concrete individuals, it is fundamental to ascertaining important relations between them such as contact and separation. For instance, observing two apparently connected (continuous) chalk marks (one red and one green) to cross, one would affirm that they touched. If, on the other hand, microscopic examination of the marks were to show that they were composed of visually discrete bits of colored chalk, this very discovery would cause one to reconsider his original affirmation that the marks touched one another. In other words, the 'topological' concept of connectedness is not merely another concept requiring analysis in application to features on surfaces, it is really fundamental to the conception of a thing or feature itself, and to properties and relations among features.

The kind of connectedness essential to individuation of particulars is considerably more complex than its topological analogue. It probably suffices that connectedness (say of a chalk mark) be judged from the point of view of a 'typical observer', so it does not matter that what appears connected to such an observer might appear disconnected when viewed up close or under a microscope. This 'observer's situation dependence' introduces complications which we shall ignore here, however, and we shall idealize by assuming that the connectedness of one or a group of features is something definable independent of the observer's point of view (which indeed influences his judgments as to what the features even *are*). We now want to consider how the idealized conception of connectedness can be used as the basis for analyzing other naive topological concepts applying to features of surfaces.

### VIII. A SURFACE TOPOLOGY

Let us begin by giving an appropriate definition of connectedness, applicable in the first instance to constructible features, U-systems and I-systems, but derivatively applicable to arbitrary sets of I-points.

DEFINITION 5. A constructible feature, U-system or I-system $x$ is *connected* if and only if there do not exist I-separate I-systems $Y$ and $Z$, each of which is in contact with $x$, such that every I-system in contact with $x$ is in I-contact with either $Y$ or $Z$. A set of I-points $P$ is *connected* if and only if for any two I-points $\alpha$ and $\beta$ in $P$, there is a connected I-system $X$, all I-points of which are in $P$, containing $\alpha$ and $\beta$.

Following our suggestion in the previous section that concrete features must be connected, we postulate:

*Connectedness postulate*

    All constructible features of a surface are connected.

# NAIVE CONCEPTION OF TOPOLOGY OF BODY SURFACE

It is also convenient to assume that we are working with surfaces of limited extent, hence we can stipulate:

*Covering postulate*

> There exists a U-system which U-covers all constructible features of the surface.

Given the definition of connectedness and the two postulates just stated, various desirable 'properties' of connectedness follow. They also follow from the fact that the presently defined concept of connectedness closely (but not exactly) coincides with the standard topological definition of connectedness in a topology to be constructed shortly, but it is worth citing some of the properties here. U-systems prove to be connected in the above sense if and only if they are not 'partitionable' into nonempty, U-separate U-systems. For instance, a U-system consisting of just two constructible features is connected if and only if those features are in contact. I-systems are connected if and only if it is not possible to construct two U-separate *U-systems*, each of which is in contact with the I-system, and whose union covers the I-system. Hence, in a sense, if I-systems are not connected, then it is possible to 'cover' their separate 'parts' with U-systems which are separate. The *union* of two connected U-systems is itself connected. And so on.

Now we want to inquire how the concept of connectedness just defined can be used to define a 'natural' topology over the space of I-points of a surface. The idea is to take as the basic neighborhoods in the topology not the sets of points *interior* to constructible features, but rather the connected sets of points *exterior* to a constructible feature, U-system or I-system. For example, a 'natural' open set is the set of points *not* on the circular mark in Figure 3, which are connectible to the dot marked 'a' by a connected I-system which is separate from the circular mark:

The set of points thus described would ordinarily be called the interior of the circle outlined by the circular line, but the analysis of this description actually leads back to our more basic characterization. In order to define these open sets in a precise way, we need the notion of an *external component* of a given I-system.

DEFINITION 6. A *component* of an I-point set is a maximal connected subset of the set; an *external component* of an I-system is a component of the set of I-points which are not I-points of the I-system.

External components satisfy the following requirements which are necessary and sufficient that a set of pointsets constitute a basis for a topology: (a) for every I-point, there is an external component containing it, and (b) for any two external components containing an I-point, there is an external component which is a subset of their intersection, containing the I-point. Also, it is interesting to note that the sets of I-points which are *complements* of external components are in fact the I-point sets of I-systems. Accordingly we define:

DEFINITION 7. The *external component topology* of a surface is the system $\langle P, \Omega \rangle$, where $P$ is the set of I-points of the surface, and $\Omega$ is the set of all external components of I-systems of the surface.

Some of the properties of the external component topology are listed in the next theorem, which is stated without proof:

THEOREM 2.
- (2.1) A set of I-points is closed in the external component topology if and only if it is the set of I-points of some I-system.
- (2.2) The external component topology is Hausdorf and compact.
- (2.3) A set of I-points which is connected is connected in the external component topology; the set of I-points of an I-system is connected if and only if it is connected in the external component topology.
- (2.4) The set of I-point sets $A$ such that $A$ is the set of all I-points internal to the set of I-points of some U-system is a basis of the external component topology.

(2.5) The external component topology is metrizable if and only if there is a countable set of U-systems such that for every U-system $x$ and every internal I-point $\alpha$ of $x$, there is a member of the countable set which is U-covered by $x$, and which contains $\alpha$ as an internal point.

As in the case of the construction of I-points, the fact that we are able to prove that the external component topology has certain of the properties which we would expect of our 'naive' conception of the topology of a surface does not prove that it is *the* naive topology of a surface. In order to consider whether this is indeed the 'true' naive topology, we must consider whether definitions of such immediately 'meaningful' concepts as those of a *boundary* which are formulated in terms of this topology agree with our intuitions.

## IX. BOUNDARY CONCEPTS AND DIMENSION

We begin by noting a difficulty which arises immediately when we try to compare the definition of a boundary (of a pointset) defined in the usual topological way with the intuitive concept of a boundary. This is that intuitively we are apt to speak of the boundaries of a surface itself, whereas if the 'space' consists of all points of a surface, then that whole space cannot have a 'boundary' in the topological sense. Thus, it is plausible that the intuitive concept of a boundary simply differs from the topological one, at least with respect to its application to the possible boundaries of an entire 'space' or surface. Leaving the problem of characterizing the boundaries of an entire surface aside, however, we will now show that the concept of a boundary definable in the standard way in terms of the external component topology actually agrees fairly closely with intuition.

We begin by defining a *boundary cover* for an I-point set.

DEFINITION 8. Let $A$ be an I-point set. $p(A)$ is the set of boundary points of $A$ in the external component topology. An I-point set $B$ is a *boundary cover* of $A$ if and only if $p(A) \subseteq B$. A U-system $x$ is a *U-boundary cover* of $A$ if and only if the I-points of $x$ form a boundary cover of $A$.

We shall also speak of boundaries and boundary covers of constructible features, U-systems and I-systems, which are defined as boundaries and boundary covers of the sets of I-points of those systems. Here are some theorems.

THEOREM 3

(3.1) Let $X$ and $Y$ be I-systems. Then $Y$ is a boundary cover of $X$ (i.e., the I-points of $Y$ are a boundary cover of $X$) if and only if there does not exist a connected U-system $z$ in contact with but not covered by $X$, which is not in contact with $Y$.

(3.2) If $X$ and $Y$ are I-systems, then $X$ contains a boundary point of $Y$ if and only if every U-boundary cover of $Y$ is in contact with $X$.

(3.3) I-systems $X$ and $Y$ have a common boundary point if and only if every U-boundary cover of $X$ is in contact with every U-boundary cover of $Y$.

(3.4) A U-system $x$ contains a common boundary point of I-systems $X$ and $Y$ if and only if for any U-boundary covers $x$ and $y$ of $X$ and $Y$, respectively, the I-system $\{x, y, z\}$ is inseparable.

(3.5) I-systems $X$ and $Y$ have at most $n$ common boundary points if and only if for any U-systems $z_1, ..., z_{n+1}$, all of which contain common boundary points of $X$ and $Y$, at least two of these U-systems are in U-contact.

(3.6) An I-system $X$ is internal to an I-system $Y$ (i.e., all I-points of $X$ are internal points of $Y$) if and only if $Y$ I-covers $X$ and there exists a U-boundary cover of $Y$ which is not in contact with $X$.

(3.7) I-system $X$ is internal to I-system $Y$ if and only if there exists a U-system $z$ such that $X$ is internal to $z$ and $z$ is internal to $Y$.

Theorem (3.1) says that one feature completely covers the boundary of another if it is not possible to 'draw a line' from a point 'inside' the second feature to 'outside' it without having the line intersect the boundary cover, which is at least a necessary condition which we would expect the 'boundary covering feature' to satisfy. Theorems (3.2)–(3.7) in effect 'translate' theoretical boundary relations into relations definable in terms of easily constructible and observable U-boundary covers (which could be marks drawn over the outlines of features in such a way as to completely cover their boundaries). Theorem (3.5) is particularly significant because the property of boundaries intersecting in only a finite number of points is involved in the definition of a topological space's being two dimensional. To show that features have only two common boundary points it is necessary and sufficient to show that there cannot be three separate features, each of which contains a common boundary point of the original features, where the 'test' of whether a feature contains a common boundary point of two other features is given in Theorem (3.4).

To show that the entire 'space' of points on a surface is two dimensional, it is sufficient to show that there is a 'basis set of U-systems' (whose interiors are nonempty, and such that any nonempty external component of an I-system contains such a U-system) such that any two of them have boundaries intersecting in only a finite number of points. It is at least intuitively plausible that in fact such a basis set could be taken to be constructed from 'arbitrarily small' features approximating circles, the boundaries of which would intersect in at most two points. Thus, we can account for the two dimensionality of a surface in terms of the observational tests for features' having only a finite number of common boundary points given in (3.4) and (3.5), which assume that the external component topology is essentially 'the' naive topology of a surface.

The foregoing is far from showing that 'space' is three dimensional. We simply lack a characterization of 'space' as distinct from the 'spaces' of points on surfaces, and it is plausible to suppose that this idea can only be arrived at through a consideration of specifically *geometrical* as opposed to *topological* concepts.

## X. TOWARDS GEOMETRY PROPER – SOME UNFINISHED BUSINESS

The fact that spatial measurement generally involves *superposition* suggests that the way to extend the foregoing analysis of topological concepts applicable to surfaces to include geometrical ones is to take into account the possibility of superposing one surface on another. Given two or more distinct surfaces, each with its own 'naive topology', it is easy (given certain 'compatibility assumptions') to characterize the composite *superposition space* which is formed temporarily when these surfaces are fitted together in certain ways, of which the individual spaces are subspaces. The idea of a possible superposition of a number of surfaces, which can be taken as a new primitive in an axiomatic formulation of the theory, is in fact applicable to possible ways of fitting together deformable surfaces, such as sheets of rubber. In order to arrive at *geometry* it is necessary to single out a class or classes of 'relatively rigid surfaces' which are distinguished by the fact that such surfaces can be fitted to other surfaces of the class in a *minimum* number of ways, and which in a sense act like rigid bodies. The fundamental problem of analysis, which is presently unsolved, is that of giving an appropriate precization of vague idea of 'minimum superposability' which characterizes rigidity, and then showing that this definition, perhaps augmented by suitable existence postulates concerning possible superpositions has the consequences we would expect. Among these consequences are that rigid surfaces are not (relatively) compressible or deformable, and that two such surfaces form a rigid composite surface when they are 'rigidly attached'.

Once in possession of a definition of relative rigidity, we should be in a position to examine a number of geometrical ideas which are related to it. Perhaps the most important of these is that of a point *off* the surface of a rigid body (i.e., a body with a rigid surface) in the *space* of that body. The most essential feature of reference to such points (e.g., a point one foot perpendicularly above some point *on* the surface) is that these implicitly describe the rigid extensions of the surface which would 'reach' the point in question, which would have to be made in order to verify that any particular thing was *at* that point.

Also, given a definition of relative rigidity, it should be possible to characterize *congruence* in terms of possible superpositions (though congruence is not the same as possible exact superposability), and investigate its properties.

Finally, one would hope to be able to connect the present analysis with plane geometrical *theory*, first defining the notion of a *flat* surface, and then defining that of a *plane* in terms of flat extensions of such surface. Such an approach might be expected to throw light not only on the logical status of such postulates of plane geometry as being able to draw a straight line connecting any two points, but possibly also to make sense out of the much criticized 'definitions' of Euclid's geometry such as 'A straight line is a line which lies evenly with the points on itself' (Heath, 1956, p. 155). Further difficulties, well known to classical geometers, arise in giving an acceptable definition of flatness from which it is possible to derive all of the intuitively recognized properties of planes (e.g., that they should be two dimensional). As in the case of the analysis of rigidity, the fact that these problems remain unsolved means that the actual 'tying up' of our analysis of surfaces and their topologies with standard geometrical theory remains unfinished business. In justification of discussing these here, I hope that the foregoing remarks at least make it plausible that the foundations of classical geometry *can* be analyzed within a framework in which superposition of surfaces of physical bodies and their features plays a fundamental rôle.

A final word should be said by way of justifying the undertaking of what proves to be a very difficult analysis of our 'naive' topological and geometrical concepts, when that analysis proves to be based on a conception of a physical body (as having a definite surface with features which can be directly superimposed on features of other surfaces) which modern science shows to be at least oversimple. A positive result of such an analysis would be to make clear the interdependence of geometrical concepts and concepts of 'the structure of matter'. Once this is made clear, it becomes possible to consider in detail how the proposed analysis might be modified to accommodate the fact that material bodies do not have well-defined surfaces and the idealization of assuming that they do (which arguably underlies our confident use of naive geometrical concepts in describing bodies) becomes worse and worse the smaller the bodies in question are.

*University of California, Berkeley*

### NOTE

* The present paper summarizes in very condensed form results of research carried out over several years. This research has been supported in part by grants from the National Science Foundation (1965–66) and a Guggenheim Fellowship (1967–68). I have benefited

from comments of many colleagues, and in particular from extensive discussions with Professor David Shwayder. Lack of space precludes detailed discussion of the relation between the present work and related work of others, and in particular with Shwayder's (as yet unpublished) work on the analysis of the concept of a body.

## BIBLIOGRAPHY

Heath, T. L., *The Thirteen Books of Euclid's Elements*, Vol. 1, (2nd ed.), Dover, New York, 1956.

# SYNTHESE LIBRARY

Monographs on Epistemology, Logic, Methodology,
Philosophy of Science, Sociology of Science and of Knowledge, and on the
Mathematical Methods of Social and Behavioral Sciences

*Editors:*

DONALD DAVIDSON (The Rockefeller University and Princeton University)
JAAKKO HINTIKKA (Academy of Finland and Stanford University)
GABRIËL NUCHELMANS (University of Leyden)
WESLEY C. SALMON (Indiana University)

ROLAND FRAÏSSÉ, *Course of Mathematical Logic.* Volume I: *Relation and Logical Formula.* 1973, XVI + 186 pp.

I. NIINILUOTO and R. TUOMELA, *Theoretical Concepts and Hypothetico-Inductive Inference.* 1973, X + 259 pp.

RADU J. BOGDAN and ILLKA NIINILUOTO (eds.), *Logic, Language, and Probability.* 1973, X + 323 pp.

GLENN PEARCE and PATRICK MAYNARD (eds.), *Conceptual Change.* XII + 282 pp.

M. BUNGE (ed.), *Exact Philosophy – Problems, Tools, and Goals.* 1973, X + 214 pp.

ROBERT S. COHEN and MARX W. WARTOFSKY (eds.), *Boston Studies in the Philosophy of Science.* Volume IX: *A. A. Zinov'ev: Foundations of the Logical Theory of Scientific Knowledge (Complex Logic).* Revised and Enlarged English Edition with an Appendix by G. A. Smirnov, E. A. Sidorenka, A. M. Fedina, and L. A. Bobrova. 1973, XXII + 301 pp. (Also in paperback.)

K. J. J. HINTIKKA, J. M. E. MORAVCSIK, and P. SUPPES (eds.), *Approaches to Natural Language. Proceedings of the 1970 Stanford Workshop on Grammar and Semantics.* 1973, VIII + 526 pp. (Also in paperback.)

WILLARD C. HUMPHREYS, JR. (ed.), *Norwood Russell Hanson: Constellations and Conjectures.* 1973, X + 282 pp.

MARIO BUNGE, *Method, Model and Matter.* 1973, VII + 196 pp.

MARIO BUNGE, *Philosophy of Physics.* 1973, IX + 248 pp.

LADISLAV TONDL, *Boston Studies in the Philosophy of Science.* Volume X: *Scientific Procedures.* 1973, XIII + 268 pp. (Also in paperback.)

SÖREN STENLUND, *Combinators, λ-Terms and Proof Theory.* 1972, 184 pp.

DONALD DAVIDSON and GILBERT HARMAN (eds.), *Semantics of Natural Language.* 1972, X + 769 pp. (Also in paperback.)

MARTIN STRAUSS, *Modern Physics and Its Philosophy. Selected Papers in the Logic, History, and Philosophy of Science.* 1972, X + 297 pp.

‡STEPHEN TOULMIN and HARRY WOOLF (eds.), *Norwood Russell Hanson: What I Do Not Believe, and Other Essays.* 1971, XII + 390 pp.

‡ROBERT S. COHEN and MARX W. WARTOFSKY (eds.), *Boston Studies in the Philosophy of Science.* Volume VIII: *PSA 1970. In Memory of Rudolf Carnap* (ed. by Roger C. Buck and Robert S. Cohen). 1971, LXVI + 615 pp. (Also in paperback.)

‡YEHOSUA BAR-HILLEL (ed.), *Pragmatics of Natural Languages.* 1971, VII + 231 pp.
‡ROBERT S. COHEN and MARX W. WARTOFSKY (eds.), *Boston Studies in the Philosophy of Science.* Volume VII: *Milič Čapek: Bergson and Modern Physics.* 1971, XV + 414 pp.
‡CARL R. KORDIG, *The Justification of Scientific Change.* 1971, XIV + 119 pp.
‡JOSEPH D. SNEED, *The Logical Structure of Mathematical Physics.* 1971, XV + 311 pp.
‡JEAN-LOUIS KRIVINE, *Introduction to Axiomatic Set Theory.* 1971, VII + 98 pp.
‡RISTO HILPINEN (ed.), *Deontic Logic: Introductory and Systematic Readings.* 1971, VII + 182 pp.
‡EVERT W. BETH, *Aspects of Modern Logic.* 1970, XI + 176 pp.
‡PAUL WEINGARTNER and GERHARD ZECHA (eds.), *Induction, Physics, and Ethics, Proceedings and Discussions of the 1968 Salzburg Colloquium in the Philosophy of Science.* 1970, X + 382 pp.
‡ROLF A. EBERLE, *Nominalistic Systems.* 1970, IX + 217 pp.
‡JAAKKO HINTIKKA and PATRICK SUPPES, *Information and Inference.* 1970, X + 336 pp.
‡KAREL LAMBERT, *Philosophical Problems in Logic. Some Recent Developments.* 1970, VII + 176 pp.
‡P. V. TAVANEC (ed.), *Problems of the Logic of Scientific Knowledge.* 1969, XII + 429 pp.
ROBERT S. COHEN and RAYMOND J. SEEGER (eds.), *Boston Studies in the Philosophy of Science.* Volume VI: *Ernst Mach: Physicist and Philosopher.* 1970, VIII + 295 pp.
‡MARSHALL SWAIN (ed.), *Induction, Acceptance, and Rational Belief.* 1970, VII + 232 pp.
‡NICHOLAS RESCHER *et al.* (eds.), *Essays in Honor of Carl G. Hempel. A Tribute on the Occasion of his Sixty-Fifth Birthday.* 1969, VII + 272 pp.
‡PATRICK SUPPES, *Studies in the Methodology and Foundations of Science. Selected Papers from 1911 to 1969.* 1969, XII + 473 pp.
‡JAAKKO HINTIKKA, *Models for Modalities. Selected Essays.* 1969, IX + 220 pp.
‡D. DAVIDSON and J. HINTIKKA (eds.), *Words and Objections: Essays on the Work of W. V. Quine.* 1969, VIII + 366 pp.
‡J. W. DAVIS, D. J. HOCKNEY and W. K. WILSON (eds.), *Philosophical Logic.* 1969, VIII + 277 pp.
‡ROBERT S. COHEN and MARX W. WARTOFSKY (eds.), *Boston Studies in the Philosophy of Science.* Volume V: *Proceedings of the Boston Colloquium for the Philosophy of Science 1966/1968.* VIII + 482 pp.
‡ROBERT S. COHEN and MARX W. WARTOFSKY (eds.), *Boston Studies in the Philosophy of Science.* Volume IV: *Proceedings of the Boston Colloquium for the Philosophy of Science 1966/1968.* VIII + 537 pp.
‡NICHOLAS RESCHER, *Topics in Philosophical Logic.* 1968, XIV + 347 pp.
‡GÜNTHER PATZIG, *Aristotle's Theory of the Syllogism. A Logical-Philological Study of Book A of the Prior Analytics.* 1968, XVII + 215 pp.
‡C. D. BROAD, *Induction, Probability, and Causation. Selected Papers.* 1968, XI + 296 pp.
‡ROBERT S. COHEN and MARX W. WARTOFSKY (eds.), *Boston Studies in the Philosophy of Science.* Volume III: *Proceedings of the Boston Colloquium for the Philosophy of Science 1964/1966.* 1967, XLIX + 489 pp.
‡GUIDO KÜNG, *Ontology and the Logistic Analysis of Language. An Enquiry into the Contemporary Views on Universals.* 1967, XI + 210 pp.
*EVERT W. BETH and JEAN PIAGET, *Mathematical Epistemology and Psychology.* 1966, XXII + 326 pp.
*EVERT W. BETH, *Mathematical Thought. An Introduction to the Philosophy of Mathematics.* 1965, XII + 208 pp.
‡PAUL LORENZEN, *Formal Logic.* 1965, VIII + 123 pp.

‡Georges Gurvitch, *The Spectrum of Social Time*. 1964, XXVI + 152 pp.
‡A. A. Zinov'ev, *Philosophical Problems of Many-Valued Logic*. 1963, XIV + 155 pp.
‡Marx W. Wartofsky (ed.), *Boston Studies in the Philosophy of Science*. Volume I: *Proceedings of the Boston Colloquium for the Philosophy of Science 1961/1962*. 1963, VIII + 212 pp.
‡B. H. Kazemier and D. Vuysje (eds.), *Logic and Language. Studies dedicated to Professor Rudolf Carnap on the Occasion of his Seventieth Birthday*. 1962, VI + 256 pp.
*Evert W. Beth, *Formal Methods. An Introduction to Symbolic Logic and to the Study of Effective Operations in Arithmetic and Logic*. 1962, XIV + 170 pp.
*Hans Freudenthal (ed.), *The Concept and the Role of the Model in Mathematics and Natural and Social Sciences. Proceedings of a Colloquium held at Utrecht, The Netherlands, January 1960*. 1961, VI + 194 pp.
‡P. L. Guiraud, *Problèmes et méthodes de la statistique linguistique*. 1960, VI + 146 pp.
*J. M. Bochenski, *A Precis of Mathematical Logic*. 1959, X + 100 pp.

ST. CHARLES BORROMEO SEMINARY
QC173.59.S65 S96
Space, time and geometry./ Suppes, Patrick,

3 7227 00005731 1

RYAN MEMORIAL LIBRARY
ST. CHARLES SEMINARY
OVERBROOK, PHILA., PA. 19151